SUPERCONDUCTIVITY
IN d- and f-BAND METALS

Academic Press Rapid Manuscript Reproduction

SUPERCONDUCTIVITY
IN *d-* and *f-*BAND METALS

Edited by
HARRY SUHL
M. BRIAN MAPLE

Institute for Pure and Applied
Physical Sciences
University of California, San Diego
La Jolla, California

ACADEMIC PRESS

A Subsidiary of Harcourt Brace Jovanovich, Publishers

New York   London   Toronto   Sydney   San Francisco   1980

ACADEMIC PRESS, INC.
111 Fifth Avenue, New York, New York 10003

United Kingdom Edition published by
ACADEMIC PRESS, INC. (LONDON) LTD.
24/28 Oval Road, London NW1 7DX

Library of Congress Cataloging in Publication Data

Conference on Superconductivity in d− and f−Band Metals,
    University of California, San Diego, 1979.
    Superconductivity in d− and f−band metals.

    1. Superconductivity−Congresses. 2. Free electron
theory of metals−Congresses. 3. Energy−band theory of
solids−Congresses. I. Suhl, Harry. II. Maple, M. Brian, Date.
III. Title. IV. Title: d- and f−band
metals.
QC610.9.C63   1979   537.6'23   80−12907
ISBN 0−12−676150−7

Contents

List of Contributors

Numbers in parentheses indicate the pages on which the authors' contributions begin.

H. ADRIAN (369), Physikalisches Institut der Universität, Erlangen-Nürnberg, D 8520 Erlangen, Federal Republic of Germany.

PHILIP B. ALLEN (215, 291), Department of Physics, State University of New York, Stony Brook, New York 11794.

A.J. ARKO (87, 121), Argonne National Laboratory, Argonne, Illinois 60439.

GERALD B. ARNOLD (153, 159), Department of Physics, University of Notre Dame, Notre Dame, Indiana 46556.

S. ARYAINEJAD (479), Department of Physics, Indiana University, Bloomington, Indiana 47401.

M. ASHKIN (25), Westinghouse R&D Center, Pittsburgh, Pennsylvania 15235.

C.A. BALSEIRO (105), Department of Physics, University of California, Berkeley, California 94720.

C.D. BARNET (509), Northern Illinois University, DeKalb, Illinois 60115.

S. BERKO (281), Department of Physics, Brandeis University, Waltham, Massachusetts 02254.

J. BIEGER (369), Physikalisches Institut der Universität, Erlangen-Nürnberg, D 8520 Erlangen, Federal Republic of Germany.

J. BOSTOCK (153, 165, 259), Department of Physics, Massachusetts Institute of Technology, Cambridge, Massachusetts 02139.

LARRY L. BOYER (455), Naval Research Laboratory, Washington, D.C. 20375.

A.I. BRAGINSKI (25), Westinghouse R&D Center, Pittsburgh, Pennsylvania 15235.

B.S. BROWN (363), Argonne National Laboratory, Argonne, Illinois 60439.

W.H. BUTLER (215, 443), Metals and Ceramics Division, Oak Ridge National Laboratory, Oak Ridge, Tennessee 37830.

RANDALL CATON (355), Department of Physics, Clarkson Technical College, Potsdam, New York 13676.

R. CHEVREL (485), Laboratoire de Chimie Minérale B, Université de Rennes, Avenue du Général Leclerc, F 35000 Rennes, France.

D.K. CHRISTEN (433), Solid State Division, Oak Ridge National Laboratory, Oak Ridge, Tennessee 37830.

C.W. CHU (191), Department of Physics and Energy Laboratory, University of Houston, Houston, Texas 77004.

MARVIN L. COHEN (13), Department of Physics, University of California, Berkeley, California 94720 and Materials and Molecular Research Division, Lawrence Berkeley Laboratory, Berkeley, California 94720.

D.E. COX (335), Brookhaven National Laboratory, Upton, New York 11973.

G.W. CRABTREE (87, 113, 121), Argonne National Laboratory, Argonne, Illinois 60439.

W.P. CRUMMET (143), Solid State Division, Oak Ridge National Laboratory, Oak Ridge, Tennessee 37830.

J. DeBROUX (259), Department of Physics, Massachusetts Institute of Technology, Cambridge, Massachusetts 02139.

M. DECROUX (485), Départment de Physique de la Matière Condensée, Université de Genève, 24 Quai E. Ansermet, CH 1211 Geneva, Switzerland.

P. DESCOUTS (273), Department of Condensed Matter, Geneva University, Geneva, Switzerland.

DAVID DEW-HUGHES (355), Metallurgy and Materials Science Division, Department of Energy and Environment, Brookhaven National Laboratory, Upton, New York 11973.

V. DIADIUK (259), Department of Physics, Massachusetts Institute of Technology, Cambridge, Massachusetts 02139.

B.D. DUNLAP (495), Argonne National Laboratory, Argonne, Illinois 60439.

D.H. DYE (113), Argonne National Laboratory, Argonne, Illinois 60439.

R.C. DYNES (409), Bell Laboratories, Murray Hill, New Jersey 07974.

CHARLES P. ENZ (181), Départment de Physique Théorique, University of Geneva, Geneva, Switzerland.

L.M. FALICOV (105), Department of Physics, University of California, Berkeley, California 94720.

W.S. FARMER (281), Department of Physics, Brandeis University, Waltham, Massachusetts 02254.

J.L. FELDMAN (207), Naval Research Laboratory, Washington, D.C. 20375.

Ø. FISCHER (485), Départment de Physique de la Matière Condensée, Université de Genève, 24 Quai E. Ansermet, CH 1211 Geneva, Switzerland.

E.S. FISHER (143), Materials Science Division, Argonne National Laboratory, Argonne, Illinois 60439.

Z. FISK (87, 121, 427), Institute for Pure and Applied Physical Sciences, University of California, San Diego, La Jolla, California 92093.

R. FLÜKIGER (265), Department of Condensed Matter, Geneva University, Geneva, Switzerland.

S. FONER (265), Francis Bitter National Magnet Laboratory, Massachusetts Institute of Technology, Cambridge, Massachusetts 02139.

F.Y. FRADIN (495, 509), Argonne National Laboratory, Argonne, Illinois 60439.

ARTHUR J. FREEMAN (99, 521), Department of Physics and Astronomy, Northwestern University, Evanston, Illinois 60201 and Argonne National Laboratory, Argonne, Illinois 60439.

C.B. FRIEDBERG (473), Texas Instruments, Houston, Texas 77001.

J.R. GAVALER (25), Westinghouse R&D Center, Pittsburgh, Pennsylvania 15235.

T.H. GEBALLE (1), Department of Applied Physics and Center for Materials Research, Stanford University, Stanford, California 94305 and Bell Laboratories, Murray Hill, New Jersey 07974.

P. GEORGOPOULOS (363), Argonne National Laboratory, Argonne, Illinois 60439.

A.K. GHOSH (305), Brookhaven National Laboratory, Upton, New York 11973.

A.L. GIORGI (223), Los Alamos Scientific Laboratory, Los Alamos, New Mexico 87545.

ANDREAS GRÜTTNER (515), Institut für Kristallographie, Universität Freiburg i. Brsg., Freiburg, Federal Republic of Germany.

D.U. GUBSER (207), Naval Research Laboratory, Washington, D.C. 20375.

MICHAEL GURVITCH (317), Brookhaven National Laboratory, Upton, New York 11973.

E.L. HAASE (369), Institut für Angewandte Kernphysik der GFK, D 7500 Karlsruhe, Federal Republic of Germany.

R.R. HAKE (479), Department of Physics, Indiana University, Bloomington, Indiana 47401.

R. HAMM (57), Bell Laboratories, Murray Hill, New Jersey 07974.

WERNER HANKE (201), Max-Planck-Institut für Festkörperforschung, D 7000 Stuttgart, Federal Republic of Germany.

B.N. HARMON (173), Ames Laboratory-USDOE and Department of Physics, Iowa State University, Ames, Iowa 50011.

K.M. HO (173), Ames Laboratory-USDOE and Department of Physics, Iowa State University, Ames, Iowa 50011.

JOHN G. HUBER (71), Department of Physics, Tufts University, Medford, Massachusetts 02155.

Y. IMRY (207), Department of Physics and Astronomy, Tel Aviv University, Ramat Aviv, Israel.

G. ISCHENKO (369), Physikalisches Institut der Universität, Erlangen-Nürnberg, D 8520 Erlangen, Federal Republic of Germany.

T. JARLBORG (521), Department of Physics and Astronomy, Northwestern University, Evanston, Illinois 60201.

D.C. JOHNSTON (427), Corporate Research Laboratories, Exxon Research and Engineering Co., Linden, New Jersey 07036.

ROBERT JONES (355), Metallurgy and Materials Science Division, Department of Energy and Environment, Brookhaven National Laboratory, Upton, New York 11973.

R.T. KAMPWIRTH (363), Argonne National Laboratory, Argonne, Illinois 60439.

D.P. KARIM (113), Argonne National Laboratory, Argonne, Illinois 60439.

M.G. KARKUT (479), Department of Physics, Indiana University, Bloomington, Indiana 47401.

H.R. KERCHNER (433), Solid State Division, Oak Ridge National Laboratory, Oak Ridge, Tennessee 37830.

J.B. KETTERSON (113), Northwestern University, Evanston, Illinois 60201.

C.W. KIMBALL (495, 509), Northern Illinois University, DeKalb, Illinois 60115.

BARRY M. KLEIN (455), Naval Research Laboratory, Washington, D.C. 20375.

G.S. KNAPP (363), Argonne National Laboratory, Argonne, Illinois 60439.

W.C. KOEHLER (427), Solid State Division, Oak Ridge National Laboratory, Oak Ridge, Tennessee 37830.

DALE D. KOELLING (99), Argonne National Laboratory, Argonne, Illinois 60439.

G.H. LANDER (143), Materials Science Division, Argonne National Laboratory, Argonne, Illinois 60439.

M. LEHMANN (369), Physikalisches Institut der Universität, Erlangen-Nürnberg, D 8520 Erlangen, Federal Republic of Germany.

M. LEVINSON (165), Superconducting Materials Group, Massachusetts Institute of Technology, Cambridge, Massachusetts 02139.

M.L.A. MACVICAR (153, 165, 259), Department of Physics, Massachusetts Institute of Technology, Cambridge, Massachusetts 02139.

A.A. MANUEL (273), Department of Condensed Matter, Geneva University, Geneva, Switzerland.

M.B. MAPLE (427), Institute for Pure and Applied Physical Sciences, University of California, San Diego, La Jolla, California 92093.

B.T. MATTHIAS (23), Institute for Pure and Applied Physical Sciences, University of California, San Diego, La Jolla, California 92093 and Bell Laboratories, Murray Hill, New Jersey 07974.

E.J. McNIFF, JR. (265), Francis Bitter National Magnet Laboratory, Massachusetts Institute of Technology, Cambridge, Massachusetts 02139.

S. MOEHLECKE (335), Brookhaven National Laboratory, Upton, New York 11973 and Unicamp, Campinas, Brazil.

D.E. MONCTON (381), Bell Laboratories, Murray Hill, New Jersey 07974.

H.A. MOOK (427), Solid State Division, Oak Ridge National Laboratory, Oak Ridge, Tennessee 37830.

F.M. MUELLER (121), Physics Laboratory and Research Institute for Materials, University of Nijmegen, Toernooiveld, Nijmegen, The Netherlands.

P. MÜLLER (369), Physikalisches Institut der Universität, Erlangen-Nürnberg, D 8520 Erlangen, Federal Republic of Germany.

W. MÜLLER (65), Commission of the European Communities, Joint Research Center, European Institute for Transuranium Elements, Karlsruhe, Federal Republic of Germany.

ALEJANDRO MURAMATSU (201), Max-Planck-Institut für Festkörperforschung, D 7000 Stuttgart, Federal Republic of Germany.

H.W. MYRON (121), Physics Laboratory and Research Institute for Materials, University of Nijmegen, Toernooiveld, Nijmegen, The Netherlands.

K.P. NERZ (501), Institut für Festkörperforschung, Kernforschungsanlage Jülich, Jülich, Federal Republic of Germany.

L.R. NEWKIRK (335), Los Alamos Scientific Laboratory, Los Alamos, New Mexico 87545.

R.M. NICKLOW (143), Solid State Division, Oak Ridge National Laboratory, Oak Ridge, Tennessee 37830.

W. ODONI (403), Laboratorium für Festkörperphysik, ETH-Hönggerberg, Zürich, Switzerland.

H.R. OTT (403), Laboratorium für Festkörperphysik, ETH-Hönggerberg, Zürich, Switzerland.

C.S. PANDE (349), Brookhaven National Laboratory, Upton, New York 11973.

DIMITRIOS A. PAPACONSTANTOPOULOS (455), Naval Research Laboratory, Washington, D.C. 20375.

M. PETER (273), Department of Condensed Matter, Geneva University, Geneva, Switzerland.

WARREN E. PICKETT (77, 99), Department of Physics and Astronomy, Northwestern University, Evanston, Illinois 60201.

F.J. PINSKI (215), Department of Physics, State University of New York, Stony Brook, New York 11794.

F. POBELL (501), Institut für Festkörperforschung, Kernforschungsanlage Jülich, Jülich, Federal Republic of Germany.

U. POPPE (501), Institute für Festkörperforschung, Kernforschungsanlage Jülich, Jülich, Federal Republic of Germany.

M. POTEL (485), Laboratoire de Chimie Minérale B, Université de Rennes, Avenue du Général Leclerc, F 35000 Rennes, France.

A.F. REX (473), Physics Department, University of Virginia, Charlottesville, Virginia 22901.

HERMANN RIETSCHEL (465), Institut für Angewandte Kernphysik, Kernforschunszentrum Karlsruhe, Karlsruhe, Federal Republic of Germany.

J.M. ROWELL (57, 409), Bell Laboratories, Murray Hill, New Jersey 07974.

J. RUVALDS (473), Physics Department, University of Virginia, Charlottesville, Virginia 22901.

R. SACHOT (273), Department of Condensed Matter, Geneva University, Geneva, Switzerland.

S. SAMOILOV (273), Department of Condensed Matter, Geneva University, Geneva, Switzerland.

A.T. SANTHANAM (25), Westinghouse R&D Center, Pittsburgh, Pennsylvania 15235.

GERMAN SCHELL (465), Institut für Angewandte Kernphysik, Kernforschungszentrum Karlsruhe, Karlsruhe, Federal Republic of Germany.

P.H. SCHMIDT (57, 409), Bell Laboratories, Murray Hill, New Jersey 07974.

W.K. SCHUBERT (159), Ames Laboratory-USDOE and Department of Physics, Iowa State University, Ames, Iowa 50011.

B. SEEBER (485), Départment de Physique de la Matière Condensée, Université de Genève, 24 Quai E. Ansermet, CH 1211 Geneva, Switzerland.

S.T. SEKULA (433), Solid State Division, Oak Ridge National Laboratory, Oak Ridge, Tennessee 37830.

M. SERGENT (485), Laboratoire de Chimie Minérale B, Université de Rennes, Avenue du Général Leclerc, F 35000 Rennes, France.

G.K. SHENOY (495), Argonne National Laboratory, Argonne, Illinois 60439.

G. SHIRANE (381), Brookhaven National Laboratory, Upton, New York 11973.

F. SINCLAIR (281), Department of Physics, Brandeis University, Waltham, Massachusetts 02254.

S.K. SINHA (495), Argonne National Laboratory, Argonne, Illinois 60439.

H.G. SMITH (143), Solid State Division, Oak Ridge National Laboratory, Oak Ridge, Tennessee 37830.

JAMES L. SMITH (37, 65), Los Alamos Scientific Laboratory, Los Alamos, New Mexico 87545.

J.C. SPIRLET (65), Commission of the European Communities, Joint Research Center, European Institute for Transuranium Elements, Karlsruhe, Federal Republic of Germany.

BYRON STAFFORD (509), Northern Illinois University, DeKalb, Illinois 60115.

J.-L. STAUDENMANN (247), Ames Laboratory-USDOE and Department of Physics, Iowa State University, Ames, Iowa 50011.

G.R. STEWART (65), Los Alamos Scientific Laboratory, Los Alamos, New Mexico 87545.

B. STRITZKER (49), Institut für Festkörperforschung, Kernforschungsanlage Jülich, Jülich, Federal Republic of Germany.

MYRON STRONGIN (305), Brookhaven National Laboratory, Upton, New York 11973.

A.R. SWEEDLER (335), Brookhaven National Laboratory, Upton, New York 11973 and Department of Physics, California State University, Fullerton, California.

R. DEAN TAYLOR (419), Los Alamos Scientific Laboratory, Los Alamos, New Mexico 87545.

L.R. TESTARDI (247), Bell Laboratories, Murray Hill, New Jersey 07974.

W. THOMLINSON (381), Brookhaven National Laboratory, Upton, New York 11973.

P. THOREL (433), Solid State Division, Oak Ridge National Laboratory, Oak Ridge, Tennessee 37830.

C.C. TSUEI (233), IBM Thomas J. Watson Research Center, Yorktown Heights, New York 10598.

F.A. VALENCIA (335), Los Alamos Scientific Laboratory, Los Alamos, New Mexico 87545.

J.M. VANDENBERG (57), Bell Laboratories, Murray Hill, New Jersey 07974.

A.T. VAN KESSEL (121), Physics Laboratory and Research Institute for Materials, University of Nijmegen, Toernooiveld, Nijmegen, The Netherlands.

C.M. VARMA (391), Bell Laboratories, Murray Hill, New Jersey 07974.

N. WAKABAYASHI (143), Solid State Division, Oak Ridge National Laboratory, Oak Ridge, Tennessee 37830.

WERNER WEBER (131), Kernforschungszentrum Karlsruhe, Institut für Angewandte Kernphysik I, D 7500 Karlsruhe, Postfach 3640, Federal Republic of Germany.

M. WEGER (501), Hebrew University, Jerusalem, Israel.

DAVID O. WELCH (355), Metallurgy and Materials Science Division, Department of Energy and Environment, Brookhaven National Laboratory, Upton, New York 11973.

J.O. WILLIS (419), Los Alamos Scientific Laboratory, Los Alamos, New Mexico 87545.

HERMANN WINTER (465), Institut für Angewandte Kernphysik, Kernforschungszentrum Karlsruhe, Karlsruhe, Federal Republic of Germany.

E.L. WOLF (153, 159), Ames Laboratory-USDOE and Department of Physics, Iowa State University, Ames, Iowa 50011.

S.A. WOLF (207), Naval Research Laboratory, Washington, D.C. 20375.

M. WONG (165), Department of Physics, Massachusetts Institute of Technology, Cambridge, Massachusetts 02139.

H. WÜHL (501), Department of Physics, University of Stuttgart, 7000 Stuttgart, Federal Republic of Germany.

KLAUS YVON (515), Laboratoire de Cristallographie aux Rayons X, University of Geneva, Geneva, Switzerland.

JOHN ZASADZINSKI (153, 159), Ames Laboratory-USDOE and Department of Physics, Iowa State University, Ames, Iowa 50011.

Preface

This volume constitutes the Proceedings of the Conference on Superconductivity in d- and f-Band Metals that was held at the University of California, San Diego, June 21–23, 1979. The conference was the third in a series of conferences dealing with superconductivity in a particularly intriguing class of materials — the so-called d- and f-band metals. The previous two conferences in this series were held at the University of Rochester in 1971 and 1976 under the chairmanship of D. H. Douglass.

The emphasis of these conferences has been on the establishment of systematics among d- and f-band metals, with a view towards developing a basic theory with predictive capability for these complex materials, as well as on the discovery of materials with superior superconductive characteristics or with totally novel features. The remarkable physical properties of the d- and f-band metals clearly challenge our fundamental understanding of several general aspects of superconductivity. Subjects covered at the conference included: new superconductors; electronic structure; phonon effects; other mechanisms; A-15 systematics; disorder and transport; magnetism and superconductivity; parameters characterizing superconductivity; alloys; and ternary compounds.

The conference organizing committee consisted of the following members: D. H. Douglass, Z. Fisk, T. H. Geballe, J. K. Hulm, M. B. Maple, F. M. Mueller, H. Suhl, M. Swerdlow, and G. W. Webb. The program committee, chaired by J. K. Hulm, included P. B. Allen, M. L. Cohen, D. H. Douglass, R. C. Dynes, T. H. Geballe, A. L. Giorgi, M. B. Maple, B. T. Matthias, S. K. Sinha, and M. H. Strongin.

No one worked harder than Nancy McLaughlin in planning and coordinating the conference from the outset and ensuring that everything ran smoothly. Julie Bay, Susan Cowen, Ben Ricks, and Lee Stephens provided technical assistance, and Barbara Stewart and Marliss Gregerson handled the financial matters associated with the conference. We are especially indebted to Annetta Whiteman who skillfully edited the manuscripts for format, a long and arduous task, and masterfully assembled the proceedings. The conference programs were provided by J. K. Hulm with the assistance of Madeliene Manning. A number of graduate students generously helped with many other aspects of the conference: Chris Hamaker, Steve Lambert, Greg Meisner, Carlo Segre, Patty Tsai, Mike Wire, and Larry Woolf. Finally, we thank the University of California, San Diego for the use of its facilities, and the UCSD conference office for their advice and assistance.

We are grateful for the financial support of the United States Department of Energy, the National Science Foundation, and the Office of Naval Research without which this valuable conference would not have been possible.

d- AND f-BAND SUPERCONDUCTIVITY -
SOME EXPERIMENTAL ASPECTS[1]

T. H. Geballe

Department of Applied Physics
and
Center for Materials Research
Stanford University
Stanford, California
and
Bell Laboratories
Murray Hill, New Jersey

INTRODUCTION

Ours is a subject with roots that go back to the begin-
nings of superconductivity. In the first edition of
Shoenberg's *Superconductivity* (1938), following Kurti and
Simon (1935), the known superconductors were grouped according to the Periodic Table into transition metals and nontransition metals. Along the way to our present state, some
art, some fiction, and quite a bit of science has been produced. The Periodic Table has proved to be a most useful
vehicle for predicting new superconductors as evidenced in
the work of Matthias, Hulm and others (Roberts, 1976).
Douglass in initiating these conferences (1972, 1976) recognized that a branch of superconductivity exists called
"d- and f-band superconductivity" which has an identity of
its own. Nature provides a rich variety of superconductors
in which the d- and f-bands play an important role.

[1]*Work at Stanford University is supported in part by the
Air Force Office of Scientific Research Contract No. F49620-
78-C-0009.*

An attempt to identify the major themes of d- and f-band superconductivity is given in Fig. 1. The discussion given below follows from the outline.

OCCURRENCE AND VARIATIONS OF T_c

New superconducting systems are continuously being found in d- and f-band compounds, although it is not so easy to do so as it once was. As a result there has been a slow rise in

ATTRIBUTES WHICH DEFINE d- AND f-BAND SUPERCONDUCTING RESEARCH

1. Occurrence - Wide variety of solid solutions and intermediate phases; ternary compounds which admit chemical substitution; search for new systems and high T_c's.

2. Response of T_c to parameters under experimental control;
    - Physical variables such as high pressure;
    - Chemical substitution
    - Structural changes;
    - Competition and coexistence with other kinds of long range order, magnetic, sdw and cdw ferroelectricity (?)

3. Mechanisms (next talk);
    - connection with normal state microscopic parameters, band structure, phonon dispersion

4. Intrinsic superconducting properties;
    - large gap
    - high H_c, J_c

5. Experimental Idiosyncrasies: Special methods of synthesis due to;
    - difficulty in handling reactive refractory metals;
    -many competing phases; metastability; [necessity of using quench methods and highly specific annealing treatments].

FIGURE 1. *Attributes which define d- and f-band superconducting research.*

the highest known transition temperature (T_c) as a function of
time. It is reasonable to assume that modest increases in T_c
will continue. The most likely route is via further stabilization of one of the presently known high-T_c metastable A15
phases, or some derivative. However, there is always the re-
mote possibility that some completely unexpected new superconductor will provide much higher T_c's. CuS must have been
a complete surprise when Meissner found it to be superconducting 50 years ago and I suspect if CuCl were to be shown to be
superconducting today it would evoke even more surprise. The
large diamagnetic anomalies which have been observed above
liquid nitrogen temperatures as transient signals in CuCl
under pressure (Brandt *et al.*, 1978; Chu *et al.*, 1978) have
not yet been proven to be associated with superconductivity.
A summary of the status of CuCl research will be given by
Chu later in this conference.

The vast number of solid solutions and intermediate
phases formed by the transition metal elements show systematic superconducting behavior that Matthias first pointed
out could be predicted simply by averaging over the Periodic
Table and finding ratio of valence electrons to atoms.
The behavior is very well summarized by Hulm and Blaugher
(1972) in their paper at the first d- anf f-band conference. Little can be added except for the work going on
in the 5f series which is reviewed by Smith in the next
session.

The simple dependence of T_c upon the electron to atom
ratio is applicable to many structures when other variables
such as degree of order either do not affect T_c, or can be
maintained constant. However, there can be a strong dependence of T_c upon structure, and frequently is, when the structure gives rise to flat bands, soft phonons and nesting Fermi
surfaces that affect the microscopic superconducting parameters. One only has to note that amorphous and heavily disordered Nb_3X (X = Ge, Al, Sn) have transitions near 3K
whereas the corresponding compositions with the A15 structure
have transitions above 18K. A further illustration of the
effect of structure, or rather lack of structure, is gained
by comparing the Collver-Hammond (1973, 1977) curve for T_c's
found for the amorphous 4d and 5d series with their crystalline counterparts. The former give a T_c vs. composition
curve that maximizes at or near the half-filled d-band, in
contrast to the minimum found in the corresponding crystalline series. The T_c minimum and the adjacent maxima at e/a -
4.8 and 6.7 correlate well with the density of states at the
Fermi-level in the crystalline series as determined experimentally and can be understood from the theoretical model of

Varma and Dynes (1976). Empirical considerations similar to
those used in Varma and Dynes (1976) applied to the amorphous
4d and 5d series (i.e., the constancy of $<I^2>/M<\omega^2>$ in a
given class) would again lead one to expect T_c to correlate
with the density of states. The maximum would now be expected to be where the spherical Fermi surface has its maximum
area - near the half-filled band.

In the case of ternary compounds which have more than two
sites and elements per unit cell there is no good way that
I am aware of for predicting superconductivity. Empirical
generalizations will be forthcoming as new ternary systems
are found and, as Fischer shows in this conference, they can
already be made within given systems. Within the large unit
cells that exist there is room for ionic and covalent, as
well as metallic bonding and a wide range of chemical substitution can frequently be accomplished. The contact interaction of the conduction electron wave function at a given
site varies when, by the substitution, localized ionic or
covalent orbitals are caused to mix with the conduction band.
The valence electron to atom ratio then becomes an arbitrarily defined variable. The mixing of localized magnetic
states gives rise to interplay between superconductivity and
magnetism in the Chevrel and RE-boron phases about which much
fascinating information will be presented in the following
sessions.

There is also the possibility of keeping the essential
bonds fixed, while changing volume. TaS_2 can be intercalated
with the series of aliphatic amines containing from 1 to 18
carbon atoms (Gamble, *et al.*, 1971). At first, for the
shorter chains which lie with their axes parallel to the layers, the number of bonds per unit area between the nitrogen
of the amine and the metallic layers surrounding it decreases
as the length of the aliphatic chain increases; T_c gradually
decreases with chain length. On the other hand the larger
chains stand perpendicular to the layers, the number density
of nitrogen-layer bonds is constant; T_c remains constant
while volume per bond (layer spacing) increases with increasing chain length. More recently metallocences (Gamble and
Thompson, 1978) with and without uncompensated spins have
been intercalated between the layers. The former follow an
expected Curie law with little effect upon T_c. The insensi-
tivity of T_c to the presence of dense layers of spins suggests that the chalcogenide layers may not be Josephson-
coupled, leaving open the question of how the superconducting,
and the eventual magnetic ordering required by the third law
of thermodynamics, take place.

MECHANISMS AND MICROSCOPIC PARAMETERS

The wide range of homogeneity in d- and f-band systems
offers the opportunity for relating superconductivity to the
normal state microscopic parameters. Heat capacity measurements are particularly convenient when new and/or difficult-to-
prepare superconductors are studied. They can be used to
establish whether or not observed electrical and magnetic
transitions are representative of the bulk of the sample as
well as for extracting averaged normal state and superconducting microscopic parameters. We shall hear of one such investigation in the next session (Stewart *et al.*, 1979).

Superconducting tunnel junctions as discovered by Giaever
(1960) can have I-V characteristics that directly measure the
energy gap, and that can be analyzed, as shown by Rowell and
McMillan (1969) to give the spectral function of $\alpha^2 f(\omega)$, and
the Coulomb pseudopotential μ^*. Unfortunately the great re-
activity of the d- and f-band materials makes it very difficult to fabricate junctions that can be analyzed satisfac-
torily. Shen (1972) presented the first satisfactory tunnel
junction data for Nb_3Sn at the first d- and f-band conference.
Progress has been made since then using the coevaporation
techniques developed by Hammond (1975, 1978). Some recent
data are shown in Fig. 2. The 50-odd Nb_3Sn junctions studied
as a function of composition by Moore, *et al.* (to be published), are seen to become strongly coupled only at and near
the ordered stoichiometric structure. This is a microscopic
manifestation that special features of the A15 structure
which enhance the superconductivity are a property of the
ordered stoichiometric structure. Application of the analysis of Wolf and Arnold (to be published), which allows for
degradation of the A15 structure at the interface via the
proximity effect, gives reasonable values for μ^*. Also in
Fig. 2 are some data for Nb_3Al junctions being studied by Kwo
et al. (1979). Well-behaved junctions for off-stoichiometry
compositions have been obtained which suggest that Nb_3Al is
even more strongly coupled than Nb_3Sn.

The difficulty of fabricating junctions of compounds such
as Nb_3Sn has prevented their being considered in the present
stage of development of superconducting circuitry where reproducibility of the individual elements is of paramount importance. Nevertheless, the larger gaps and potentially
faster switching characteristics of A15 junctions should be
of increasing interest as more is learned about fabrication.
When the rather heroic methods used by Shen to prepare the
first junction are compared with the now routine methods used

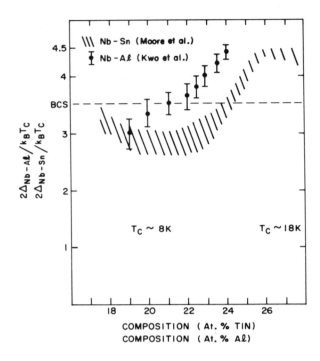

FIGURE 2. *Compositional dependence of energy gaps for*
A15 phases of Nb-Sn (Moore et al., 1979) and Nb-Al (Kwo
et al., 1979).

for the preparation of those shown in Fig. 2, it is evident
that substantial progress is being made.

INTRINSIC PROPERTIES

    The intrinsic superconducting properties of type II tran-
sition metal superconductors become of intense interest following the discovery of the existence of high-field, high-
current superconductivity by Kunzler and coworkers (1961).
The possibility of finding improved properties that can
directly be put to practical use offers support for this
field of research. Again the large homogeneous compositional
ranges available make systematic studies and quantitative
comparison with the theory of type II superconductivity
possible. Recent work of Orlando *et al.* (to be published)
for Nb_3Sn shows that it is necessary to bring strong-coupling
and the electron phonon enhancement in on an equal footing

with the spin-orbit scattering before the effects of Pauli
paramagnetism at high fields can be understood on a quantitative basis.

The maximum known H_{c2} as a function of time has increased
drastically since 1960. It is perhaps no coincidence that the
highest known critical fields and critical field slopes are
found for ternary compounds, for it is in these compounds with
large unit cells which contain electrically inactive sites
that unexpected behavior in high magnetic fields is found.
This is true for the Chevrel phases (Fischer, 1978), the RE
borides (Fertig *et al.*, 1977) and the intercalated layered
metal dichalcogenides (Prober *et al.*, 1975). The field is
able to penetrate perpendicular to the c-axis of the
dichalcogenides with so little cost in energy that enormous
critical field slopes

$$\frac{d_{H_{c2}}}{dT} > 2 \times 10^5$$

are found. This enormous slope has motivated theoretical
studies of critical fields of the layered compounds which
show how novel the superconducting properties of such open
structures can be (Klemm *et al.*, 1974). Possibly related,
but still not understood are the negative curvatures found in
the low field portion of the H_{c2}-T curves (Woollam, 1974). In
the Chevrel and rare earth boride phases it is not so straightforward from the structure alone to see why the critical
fields can be so high; however, the intercluster linkage must
play an important role.

It is clear that we have a long way to go before magnets
and other kinds of devices will be limited by an intrinsic
inability to remain superconducting in high magnetic fields.
The limitation is more with the mechanical properties of the
superconductors, the intrinsic properties of the various kinds
of pinning centers and the interaction of the flux-line-
lattice with distributed pinning centers. While the fabrication of composites seeking to optimize simultaneously all
these properties is more nearly at the center of gravity of
the Applied Superconducting Conference, there is plenty of
work still to be done of interest to this conference, in
developing a quantitative understanding of each of the parameters involved. Quantitative understanding can eventually
lead to much higher performance superconductors.

EXPERIMENTAL IDIOSYNCRASIES

Finally of great interest to this conference is the relationship of high temperature superconductivity with instabilities and metastable phases. It need only be noted that the three highest T_c superconductors, Nb_3Ge, Nb_3Ga, and Nb_3Al, are all metastable. Only over a narrow temperature range at the peritectic does their equilibrium composition even approach (but not include) stoichiometry (see for example Jorda *et al.*, 1978). It follows that quench techniques must be employed in order to obtain stoichiometry.

No convincing experimental data exist, to my knowledge, that show Nb_3Al, Nb_3Ge, Nb_3Ga, or Nb_3Si has been made homogeneous and stoichiometric in the ordered A15 structure, in either film or bulk form. It remains as a challenge to do so in order to be able to compare them with Nb_3Sn, for example. Epitaxial vapor phase growth by Dayem *et al.* (1978) is a potentially powerful method. The intent in this method is to prepare the substrate with a surface favoring the free energy of formation of the desired phase with respect to undesired more stable phases. The work on Nb_3Ge (Dayem *et al.*, 1978) showed that growth of the desired A15 phase and composition is favored when it takes place on prepared Nb_3Ir films of the proper lattice constant.

The favorable role of O_2 during growth can also be such as to encourage epitaxial growth as shown by Gavaler *et al.* (1979). Feldman *et al.* (1979) have found that the presence of O_2 during growth of Nb_3Si films favors the formation of the A15 structure much as it does for Nb_3Ge. The results obtained so far, shown in Fig. 3, are the first study of Nb_3Si in the A15 phase where systematic variations can be observed and reasonably interpreted.

Resistance measurements can be valuable for correlating T_c with the microscopic parameters of transport theory. However, in cases like Nb_3Al, Nb_3Ge, and Nb_3Si, which heat capacity measurements show are not homogeneous, both the superconductivity and the transport can be dominated by (not necessarily the same) minority phases or compositions. A more reliable set of parameters is the combination of lattice constant, and composition. Evidence is shown by Cox *et al.* (1979) at this conference for Nb_3Ge, and earlier for Nb_3Sn by Burbank *et al.* (1977) that for a given composition the lattice constant can be a single valued, experimentally determined function of disorder. There still can be ambiguity in comparing different films in that different concentrations of

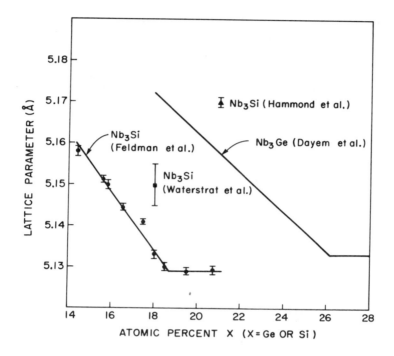

FIGURE 3. Lattice constant vs. composition for Nb_3Ge films (Dayem et al., 1978) compared with corresponding data for Nb_3Si (Hammond et al., 1972; Waterstrat et al., 1979; and Feldman et al., 1979).

vacancies, defects (antisite and displacement) and impurities can give the same lattice constant. Lattice constant vs. composition curves nevertheless form the most simple relia-ble basis for characterizing films. This is illustrated in the comparison of Nb_3Si and Nb_3Ge, Fig. 3, where the composi-tion determined by electron microprobe analysis gives the ratio of Nb to X. The Nb_3Si line extrapolates to $a_o = 5.08$ Å, which is the predicted value of the revised Geller radii. The limiting factor in obtaining stoichiometry may be the packing of Nb atoms along the chains of the A15 structure. The Nb-Nb bonds may, due to core-core overlap, no longer be easily compressible for $a_o < 5.12$. If no Nb_3X compounds can be made in the A15 structure with $a_o < 5.12$ (X is a non-tran-sition element) then the route to higher T_c may be by partial substitution on the Si site. This has been attempted many times. The lack of success can be attributed to the fact that there are many more disordered, undesired stable or meta-stable phases that can be reached than the desired ordered

one. It follows that a high degree of control over the
growth parameters is essential. Experimentalists seeking to
synthesize metastable phases have a constantly increasing
arsenal of materials, techniques and characterizations available to help them. The most important ingredients do not have
to be listed – they are the happy combination of inspiration
and hard work.

ACKNOWLEDGMENTS

I thank M. R. Beasley and J. H. Wernick for their help in
putting d–and f–band superconductivity in perspective, and J.
Kwo and R. D. Feldman for helpful discussions concerning their
data.

REFERENCES

Brandt, N. B., Kuvshinnikov, S. V., Rusakov, A. P., and
    Semyonov, M. V. (1978). *JETP Lett. 27*, 37.
Burbank, R. D., Dynes, R. C., and Poate, J. M. (1977). *Int.
    Mtg. on Radiation Effects on Superconductivity, Argonne,
    IL.*
Chu, C. W., Rusakov, A. P., Huang, S., Early, S., Geballe,
    T. H., and Huang, C. Y. (1978). *Phys. Rev. B 18*, 2116.
Collver, M. M., and Hammond, R. H. (1973). *Phys. Rev. Lett.
    30*, 92; (1977) *Solid State Comm. 22*, 55.
Cox, D. E., Sweedler, A. R., Moehlecke, S., Newkirk, L. R.,
    and Valencia, F. A. (1979). This conference.
Dayem, A. H., Geballe, T. H., Zubeck, R. B., Hallak, A. B.,
    and Hull, G. W., Jr. (1978). *J. Phys. Chem. Solids 39*,
    529.
Douglass, D. H. (1972). "Superconductivity in d- and f-Band
    Metals," edited by Douglass, D. H., American Institute of
    Physics, New York.
Douglass, D. H. (1976). "Superconductivity in d- and f-Band
    Metals," edited by Douglass, D. H., Plenum Press, New
    York.
Feldman, R. D., Hammond, R. H., Kihlstrom, K. E., Kwo, J.,
    and Geballe, T. H. (1979). *Bull. Am. Phys. Soc. 24*, 455.
Fertig, W. A., Johnston, D. C., DeLong, L. E., McCallum,
    R. W., Maple, M. B., and Matthias, B. T. (1977). *Phys.
    Rev. Lett. 38*, 987.
Fischer, Ø (1978). *App. Phys. 1*, 16.
Gamble, F. R., Osiecki, J. H., Case, M., Pisharody, R.,
    DiSalvo, F. J., and Geballe, T. H. (1971). *Science 174*,
    493.

Gamble, F. R., and Thompson, A. R. (1978). *Solid State Comm.* *27*, 379.

Gavaler, J. R., Braginski, A. I., Ashkin, M., and Santhanam, A. T. (1979). This conference.

Giaever, T. (1960). *Phys. Rev. Lett.* *5*, 147, 464.

Hammond, R. H. (1975). *IEEE Trans. Magn.* MAG-11, 201; (1978) *J. Vac. Sci. Technol.* *15*, 382.

Hammond, R. H., and Hazra, S. (1972). *Proc. of the Thirteenth Int'l. Conf. on Low Temperature Physics, Boulder, CO.* edited by Timmerhaus, D. K., O'Sullivan, W. H., and Hammel, E. F., Plenum Press, New York (1974) *3*, 465.

Hulm, J. K., and Blaugher, R. D. (1972). "Superconductivity in d- and f-Band Metals," edited by Douglass, D. H., American Institute of Physics, New York, p. 1.

Jorda, J. L., Flükiger, R., and Müller, J. (1978). *Journal of the Less Common Metals 62*, 25.

Klemm, R. A., Beasley, M. R., and Luther, A. (1974). *J. Low Temp. Phys.* *16*, 607.

Kunzler, J. E., Buehler, E., Hsu, F. S. L., and Wernick, J. E. (1961). *Phys. Rev. Lett.* *6*, 89.

Kurti, N., and Simon, F. (1935). *Proc. of the Royal Society Series A 151*, 610.

Kwo, J., Hammond, R. H., and Geballe, T. H. (1979). *Bull. Am. Phys. Soc.* *24*, 455.

Moore, D. F., Zubeck, R. B., Rowell, J. M., and Beasley, M. R. (1979). To appear in *Phys. Rev.*

Orlando, T. P., McNiff, E. J., Jr., Foner, S., and Beasley, M. R. (1979). To appear in *Phys. Rev.*

Prober, D. E., Schwall, R. E., and Beasley, M. R. (1975). *Bull. Am. Phys. Soc.* *20*, 342; and to be published.

Roberts, B. W. (1976). *J. of Physical and Chemical Reference Data 5*, 581-821.

Rowell, J. M., and McMillan, W. L. (1969). "Superconductivity," Parker, R. D., ed., *Chapter 3*, Dekker, New York.

Shen, L. Y. L. (1972). "Superconductivity in d- and f-Band Metals," edited by Douglass, D. H., American Institute of Physics, p. 31.

Shoenberg, D. (1938). "Superconductivity," Cambridge at the University Press.

Stewart, G. R. (1979). This conference.

Varma, C. M., and Dynes, R. C. (1976). "Superconductivity in d- and f-Band Metals," edited by Douglass, D. H., Plenum Press, New York, p. 507.

Waterstrat, R. M., Haenssler, F., and Muller, J. (1979). To be published in *J. Appl. Phys.*

Wolf, E. L., Zasadzinski, J., Arnold, G. B., Moore, D. F., Rowell, J. M., and Beasley, M. R., to be published.

Woollam, J. A., Somoano, R. B., and O'Connor, P. (1974). *Phys. Rev. Lett.* *32*, 712.

THEORETICAL OVERVIEW – EMPHASIS ON NEW MECHANISMS[*]

Marvin L. Cohen[†]

Department of Physics
University of California

and

Materials and Molecular Research Division
Lawrence Berkeley Laboratory
Berkeley, California

Some non-phonon electron pairing mechanisms involving d- electrons are examined. The electron-hole lattice interaction originally proposed for CuCl is shown to give small T_c values. Umklapp or intervalley scattering effects can contribute to raise T_c. The demon (acoustic plasmon) mechanism is explored using model dielectric functions. It is shown that the demon mechanism is capable of reducing the repulsive Coulomb interaction and changing the sign of μ^*.

Recent experimental progress in d- and f- band materials has been significant as is evidenced by the results on new systems and novel properties being presented at this confer-

[*]*This work was supported by the National Science Foundation (Grant DMR7822465) and by the Division of Materials Sciences, Office of Basic Energy Sciences, U.S. Department of Energy (Grant W-7405-ENG-48).*
[†]*Guggenheim Fellow 1978-79.*

ence. Considerable theoretical progress has also been made
since the last conference in this series. Electronic structure calculations for the A15's and other d- band metals have
been done (W. E. Pickett, 1979 and B. M. Klein, 1979), and
despite the complexity of some of these materials, the results
on the whole compare reasonably well with detailed experimental measurements (e.g., Fermi surfaces). The use of these
calculations in analyzing superconductivity properties is just
beginning, and preliminary results are very encouraging.
Theoretical research on phonon spectra has developed in a
similar way (W. Weber, 1979), and contrary to early speculations, the correlations between features in the phonon spectra and superconductivity have been shown to be subtle.
Progress in electronic and phonon calculations will hopefully
lead to reliable calculations of superconducting parameters
in transition metals and hopefully even in A15 compounds.
Some successful attempts will be reported at this conference
(Butler, 1979 and Klein, 1979).

Disorder and transport phenomena have been surprisingly
rich, and theoretical research is progressing well in this
complex area (Allen, 1979). Studies of correlations between
resistivity, electron-phonon interactions, atomic order, and
superconductivity are leading to new points of view and new
schemes for studying superconducting phenomena.

Much recent attention has been given to the interrela-
tionships between superconductivity and magnetism (Varma,
1979 and Enz, 1979). Magnetic field dependent effects, particularly in the rare-earth ternary compounds, are being
examined, and molecular-field theories have been used to
predict some interesting new effects. Other pairing configurations besides the usual s-wave variety have also been
considered, and some detailed estimates of related effects
have been made.

I will not give details of the above theoretical research;
much of it will be discussed here in the next few days.
Rather, I will focus on theoretical work related to pairing
mechanisms with emphasis on non-phonon varieties. Most of
the motivation for research in this area stems from a desire
to increase T_c; however, some of the research attempts to
explain properties of novel superconductors or materials with
unusual normal-state properties which may be related to
superconductivity. A good example of the latter is CuCl
(Brandt *et al.*, 1978 and Chu *et al.*, 1978). Although at this
time it is still unclear whether the anomalous magnetic properties observed in CuCl are signaling the existence of superconductivity in this material, the proposed electron-hole
mechanism (Abrikosov, 1978) still represents a plausible interaction to consider for systems with d- or f- electrons.

HOLE LATTICE MECHANISM

Abrikosov's model is based on an electronic band structure with overlapping conduction and valence bands at different points in the Brillouin zone. An important feature of the model is the requirement that the electron–hole mass ratio $m_e/m_h \ll 1$. It appears unlikely that these conditions are consistent with the electronic band structure of CuCl (Zunger *et al.*, 1979), but electronic configurations of this kind are possible, and for the purposes of exploring the electron–hole mechanism, the above conditions will be assumed.

Abrikosov suggests that if the mass ratio were very small the electron gas parameters for the electrons and holes could satisfy the condition, $r_s{}^e \equiv r_s \lesssim 1$ and $r_s{}^h \gg 1$. Using a suggestion proposed by Herring (1968), it is assumed that the holes crystallize, and electron pairing arises from exchange of the virtual bosons associated with vibrations of the hole lattice. Abrikosov made some estimates of the hole lattice frequency and the effective electron–electron interactions; he estimated $T_c \sim 10^2$–10^3 K. Rajagopal (1979) added renormalization and Coulomb effects and found Abrikosov's estimates were reduced to about 1 K. Since most of the parameters involved can be computed, I will present the results which are based on a more detailed calculation (Cohen, 1979) to explore this proposal in detail.

T_c can be estimated using a BCS type relation, $T_c \sim \omega_0 \exp(-1/K)$; ω_0 is the vibrational frequency of the hole lattice, and $K = \lambda^* - \mu^*$ where $\lambda^* = \lambda/(1+\lambda)$, $\mu^* = \mu/(1+\mu c)$, and $c = \ln(E_F/\omega_0)$. The parameter λ is analogous to the usual electron–phonon coupling constant, but now it represents the coupling to hole lattice excitations; μ is the Coulomb coupling constant. Renormalization effects are included approximately through λ^*. Using Fermi–Thomas screening, the lattice frequency ω_0 at wavevector $2k_F$ and μ can be estimated

$$\hbar\omega_0 = \left(\frac{0.88r_s}{1+r_s/6} \; \frac{m_e}{m_h}\right)^{\frac{1}{2}} E_F \qquad (1)$$

$$\mu = \frac{1}{2} \frac{r_s}{6} \ln\left|1 + \frac{6}{r_s}\right|. \qquad (2)$$

For typical values of r_s, $\mu \sim 0.1 - 0.5$ with a saturation value at 0.5. The Lindhard approximation saturates at 0.65 and the Hubbard approximation at $0.05r_s$.

Without umklapp contributions, stability requires $\lambda = \mu$ (Cohen and Anderson, 1972). This condition gives

$$K = \frac{\mu^2(c-1)}{(1+\mu)(1+c\mu)} \quad . \tag{3}$$

For $r_s \sim 1$, $c = \ln 11.5$, $\mu = 0.16$, and $K \sim 0.023$ giving a
vanishingly small T_c. Even for maximized values: i.e.,
$\mu = 1/2$ and $c \sim \ln 20$, then $K \sim 0.13$ and $T_c < 1~K$.

Abrikosov's high T_c values resulted from his estimates
$\lambda \sim r_s \sim 1$ and $\mu^* = 0$. Rajagopal also took $\lambda \sim 1$; however,
he included μ^* and renormalization. With the choice $\mu = 1$
and $c = \ln 5$, T_c is reduced by orders of magnitude primarily
because of the large μ value.

If the Abrikosov choice of $\lambda \sim 1$ were attainable for
$r_s \sim 1$ and the parameters ω_0 and μ^* were estimated accurately,
$T_c \sim 75~K$. The higher values of T_c require $\lambda > \mu$. This can
be achieved and is achieved in metals through local fields or
umklapp contributions to the electron-phonon interaction.
In the hole lattice model, the electronic density is taken to
be fairly uniform, and local field effects should be small.

Hence, even though ω_0 is large, large λ's are still needed
to yield high T_c's, and this requires umklapp scattering or
intervalley effects. If the hole and crystal lattices are
incommensurate, umklapp processes involving both lattices are
possible. Perhaps these interactions and/or intervalley
scattering can increase λ, but it is not expected that values
larger than those of ordinary metals will be obtained. As λ
increases, renormalization effects will reduce ω_0, and polar-
on coupling will also become important. These effects will
reduce T_c.

In summary, using stability arguments, the electron-hole
mechanism will not produce high values for T_c. Umklapp or
intervalley scattering can increase T_c, but it will be difficult to get $\lambda > \mu$ for this system and, hence, difficult to
obtain large T_c's.

HOLE PLASMA AND "DEMON" MECHANISMS

A two-component plasma with particles of different masses
allows new collective modes, and these, in turn, can be used
to couple particles to form Cooper pairs. In the electron-
hole case considered above, the holes were assumed to crystallize; however, even if the holes were considered to be in the
plasma state, the condition $m_h \gg m_e$ would lead to a sound-
like mode. This mode is analogous to ordinary sound where
the electrons screen the ionic plasma motion producing acoustic phonons. For long wavelengths, the mode would coincide

with the hole-lattice mode with $\omega \sim \upsilon_F \sqrt{m_e/3m_h} \; q$. Hole lattice stability would not be a problem in this case, but interaction strengths and, therefore, the resulting transition temperatures should be comparable to those expected for the hole lattice case.

A variation on the electron and hole plasma configuration is the situation where two types of electrons compose the plasma. The usual case considered is a transition metal with s- and d- electrons and $m_d \gg m_s$ although any system containing electrons satisfying the above condition for the masses will give similar results. In analogy with the ion or hole plasmas, a d- electron acoustic type mode will develop, and the acoustic plasmons or demons (Pines, 1956) can be considered for pairing s- electrons (Garland, 1963; Fröhlich, 1968). Ruvalds *et al.* (1979) have suggested that acoustic plasmons or demons are the dominant interaction causing superconductivity in the A15 compounds and in the alkali tungsten bronzes.

To test the magnitude of the demon mechanism, we take the extreme model of a free electron dielectric function including contributions from s- and d- electrons. Some of the features of this calculation will be given here; for details, see Ihm *et al.* (1979). The principal parameters which can be varied in this model are the mass ratio m_d/m_s, $r_s{}^s/r_s{}^d$, $k_F{}^s/k_F{}^d$, and the background dielectric constant, ε_0. These parameters are not independent, and the choice made here is $k_F{}^s = k_F{}^d$, $r_s{}^s = r_s{}^s$(sodium), $\varepsilon_0 = 1$, and variable m_d/m_s. The RPA or Lindhard form of the dielectric function is assumed even though this approximation breaks down for large r_s. Other important approximations are: the use of a free electron model for the d- electrons and the omission of s-d interactions. Garland (1963) considered s-d interactions and concluded that they did not alter the basic structure of $\varepsilon(q,\omega)$. Although the limits of these approximations will be strained for some choices of the parameters, the object of this calculation is to illustrate the main features of the demon mechanism and to assess the approximate magnitude of this interaction for d- band materials.

The dielectric function for s- electrons alone has a negative (or attractive) region in (q,ω) space. Most of this attractive region is in the small q portion of the plane where the electrons act collectively; i.e., for $q \ll 2k_F$, $5/6\upsilon_F q \lesssim \omega \leq \omega_p(q)$ (See Fig. 1.). To evaluate the Coulomb kernel, K_c, via the total dielectric function,
$$\varepsilon(q,\omega) = \varepsilon_1(q,\omega) + i\,\varepsilon_2(q,\omega),$$

$$K_c \sim \text{Re} \int_{\text{LD}} \frac{4\pi\epsilon^2 \, qdq}{q^2 \epsilon(q,\omega)} \tag{4}$$

where the q-integration is over the region where $\epsilon_2 \neq 0$ or
the Landau damping (LD) region for each ω. Since the attractive region has small overlap with the LD region, K_c is repulsive for all ω although it does have some structure (Cohen,
1964) which under special circumstances can contribute to
superconductivity (Cohen *et al.*, 1967).

     The combined s- and d- system has several important fea-
tures. The d- plasma frequency is renormalized to become the
demon mode as expected, and the demon dispersion curve forms
the upper limit of the d- attractive region while the lower
limit is fixed by the line of zeros which has dispersion
$\omega \approx 5/6 m_s/m_d v_F q$ for small q (Fig. 1). The integration for
the s- Coulomb kernel (Eq. 4) now includes the attractive
region arising from the d- electron contribution. The effec-
tiveness of the attractive region depends on m_d/m_s. As the
ratio increases, the d- attractive region moves down in frequency and concentrates at low ω. The demon mode exists for
$m_d/m_s > 2.26$. For $m_d/m_s \gtrsim 3$, K_c is reduced but not negative.
For large mass ratios, K_c becomes significantly reduced at

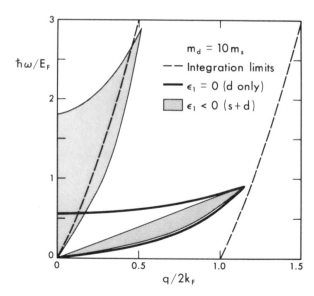

*FIGURE 1. Properties of the Lindhard dielectric function
in the (q,ω) plane. See text for details.*

low ω and is unchanged from the s- only result at higher ω. For very large ratios, i.e., $m_d/m_s \gtrsim 20$, the kernel becomes negative. Hence, for moderate mass ratios, the major effect of the demons is to reduce K_c at low ω enhancing the attractive phonon contribution but not substituting for it. At larger m_d/m_s, K_c can become attractive, and it can yield a superconducting solution without phonons.

Although the above calculations include damping arising from the electron-hole continuum (but not from phonons), the influence of scattering by impurities or defects is not included. The latter effects would average the superconducting gaps in the s- and d- bands (Anderson, 1959). This could significantly reduce T_c; therefore, very pure samples would be the best candidates to investigate for signals of this mechanism.

At present, there is no compelling need for a non-phonon mechanism in transition metals or transition metal compounds. Results will be presented at this conference showing that Nb is apparently well understood using the phonon mechanism (Butler, 1979), and many properties of the A15 compounds can be explained using phonons only (Allen, 1979 and Klein, 1979).

As reported in the last conference in this series and as will be discussed here by Bostock *et al.*(1979), some tunnel-

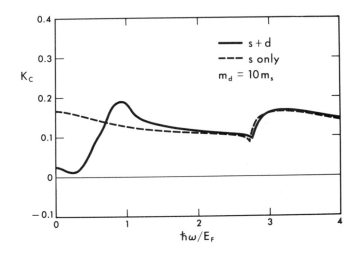

FIGURE 2.   The Coulomb kernel, K_c, of the superconducting gap equation.

ing experiments on Nb have been interpreted in terms of an attractive μ^*. This result is still controversial, and the following calculation does not support or contradict the various interpretations of the tunneling data. It is possible to show, however, that an attractive μ^* is a natural consequence of a mechanism like the demon mechanism which reduces K_c up to frequencies larger than the Debye frequency. We assume that $K_c \sim 0.2$, and it is screened out up to energies $\sim E_F{}^d$. Then, for $E_F{}^s \sim 1\ eV$, $m_d/m_s = 5$, and a Debye energy ~ 0.02 eV, we have $|\mu^*| \sim 0.06$ and it is <u>attractive</u>. For these parameters a $T_c \sim 10\,K$ requires an electron–phonon λ of only $\sim 1/3$. The main purpose of this estimate is to show that attractive μ^*'s are possible and that they are likely to appear if interactions having high frequency cut-offs (e.g., demons) are prominent in the pairing mechanism.

In summary, estimates of the demon mechanism indicate that this interaction can be large enough to contribute to electron pairing, but it is unlikely that it is the dominant mechanism in conventional d- band superconductors.

REFERENCES

Abrikosov, A. A. (1978). JETP Letters *27*, 219.

Allen, P. B. (1979). These proceedings.

Anderson, P. W. (1959). *J. Phys. Chem. Solids 11*, 26.

Bostock, J., and MacVicar, M. L. A. (1979). These proceedings.

Brandt, N. B., Kuvshinnikov, S. V., Rusakov, A. P., and
    Semenov, V. M. (1978). JETP Letters *27*, 33.

Butler, W. H. (1979). These proceedings.

Chu, C. W., Rusakov, A. P., Huang, S., Early, S., Geballe,
    T. H., and Huang, C. Y. (1978). *Phys. Rev. B18*, 2116.

Cohen, M. L. (1964). *Phys. Rev. 134*, A511.

Cohen, M. L. (1979). *Phys. Rev.* (in press).

Cohen, M. L., and Anderson, P. W. (1972). *Superconductivity
    in d- and f- Band Metals*, ed. D. H. Douglass, Amer. Inst.
    of Physics, N. Y., p. 17.

Cohen, M. L., Koonce, C. S., and Au-Yang, M. Y. (1967).
    Phys. Letters 24A, 582.

Enz, C. P. (1979). These proceedings.

Fröhlich, H. (1968). *J. Phys. C. 1*, 544.

Garland, J. W. (1963). *Proceedings of the Eighth Interna-
    tional Conference on Low Temperature Physics*, ed. R. O.
    Davies, Butterworths, N. Y., p. 143.

Herring, C. Y. (1968). This suggestion is discussed in
    Halperin, B. I., and Rice, T. M. (1968). *Rev. Mod. Phys.
    40*, 755.

Ihm, J., Cohen, M. L., and Tuan, S. F. (1979). To be published.

Klein, B. M. (1979). These proceedings.

Pickett, W. E. (1979). These proceedings.

Pines, D. (1956). *Canadian J. of Phys. 34*, 1379. Pines' paper focused on acoustic plasmons in semiconductors, and the word *demon* was associated with the terms "distinct electron motion or D.E.M." In the present paper, *demon* is associated with the d- electron acoustic plasmon.

Rajagopal, A. K. (1979). *Phys. Rev.* (in press).

Ruvalds, J., and Kahn, L. M. (1979). *Phys. Letters* (in press). Also, Kahn, L. M., Ruvalds, J., and Tüttö, I., preprint.

Varma, C. M. (1979). These proceedings.

Weber, W. (1979). These proceedings.

Zunger, A., and Cohen, M. L. (1979). *Phys. Rev.* (in press).

WORKING HYPOTHESES

B. T. Matthias

Institute for Pure and Applied Physical Sciences
University of California, San Diego[*]
La Jolla, California
and
Bell Laboratories
Murray Hill, New Jersey

The behavior of collective phenomena such as superconductivity and magnetism have been well described by theories. In contrast, there are no existing theories that predict the actual materials in which these collective phenomena should occur. At present, the only way to do so is an empirical approach coupled to a more or less sensible working hypothesis. Two of the most recent examples in this direction shall be given below.

For the last two decades, Sc_3In and $ZrZn_2$ were the only truly itinerant ferromagnets known. While they have been described extensively in the literature ever since their discovery, it did not seem possible to predict another one. Now, as the result of some recent theoretical and experimental analysis, it became possible to predict a third itinerant magnet (Enz and Matthias, 1978, 1979). The compound was $TiBe_2$ which indeed showed itinerant antiferromagnetism (Matthias et al., 1978). While the existence of an itinerant antiferromagnetic compound was of great interest, the question arose as to why it had not become ferromagnetic instead. Clearly, the answer was to be found in the Ti–Be bond. This bond was then weakened by replacing part of Be with Cu. And, in fact, this led to itinerant ferromagnetism closely resembling that of $ZrZn_2$. Again, as in the case of $(Zr,Ti)Zn_2$, the highest Curie point was

[*]*Work in La Jolla supported by the National Science Foundation under Grant #DMR77-08469.*

reached at the (C-15) stability limit of Ti(Cu,Be)$_2$. As
pointed out repeatedly in the past, superconductivity is a much
more general phenomenon than magnetism and as a matter of fact,
it emerges as the normal behavior of metals at low temperatures. Therefore, it seems likely to expect superconductivity
in most cubic nonmagnetic compounds. The superconductivity of
the element Eu in its trivalent state, whose ground state is
then nonmagnetic, has not yet been found since the extremely
high pressure necessary to change Eu^{+2} to Eu^{+3} at very low
temperatures are just not available yet. However, it has been
known for quite some time that EuIr$_2$ (van Vleck, 1978), crystal-
lizing with the cubic Laves phase (C-15) structure, contains
europium strictly in its trivalent configuration. Subsequent
experiments (Matthias et al., in press), immediately verified
our assumption of the existence of superconductivity.

Both these compounds, the magnetic TiBe$_2$ and the superconducting EuIr$_2$ jointly illustrate a fascinating feature; that
the combination of two superconducting elements can lead to
magnetism while a compound containing europium, which after all
is situated in the very middle of the strongly magnetic rare
earths, can become superconducting instead.

REFERENCES

Enz, C. P., and Matthias, B. T. (1978). *Science 201*, 828.
Enz, C. P., and Matthias, B. T. (1979). *Z. für Physik B33*, 129.
Matthias, B. T., Giorgi, A., Struebing, V. O., and Smith, J. L.,
     Physics Letters 69A, 221.
Matthias, B. T., Fisk, Z., and Smith, J. L. *Physics Letters*
     (in press).
van Vleck, J. H. (1978). *J. Less-Common Metals 62*, xv.

THIN FILMS AND METASTABLE PHASES[1]

J. R. Gavaler
A. I. Braginski
M. Ashkin
A. T. Santhanam

Westinghouse R&D Center
Pittsburgh, Pennsylvania

I. INTRODUCTION

Theoretical and experimental considerations have led to the
belief that all high-temperature superconducting phases are inherently unstable. This explains the current interest in
growth techniques which might be capable of forming nonequilibrium, metastable phases of potentially high-T_c superconductors.
Among these, attention has centered on techniques which involve
growth from the vapor phase, such as sputtering, evaporation,
and chemical vapor deposition (CVD). This is undoubtedly the
result of the success that these methods have had in preparing
Nb_3Ge. At present this compound remains the highest-temperature
superconductor known with a maximum onset temperature of 23.6 K
(Paidassi et al., 1978).
    Over the past several years, we have investigated the prepa-
ration of thin films of Nb_3Ge and several other apparently non-
equilibrium phases which have high-T_c's (high-T_c is defined
here as being equal or greater than 15 K). In this paper we
review these results and also present our ideas concerning possible mechanisms by which nonequilibrium phases are formed in
thin films.

[1]*Supported in part by the Air Force Office of Scientific
Research Contract No. F49620-78-C-0031.*

II. THIN FILM TECHNIQUES

The ability of sputtering, evaporation and CVD to form phases not obtainable by bulk methods has long been recognized. This ability has generally been attributed to one or a combination of the following factors: (1) These methods can all form phases at much lower temperatures than is possible by growing from the melt. (2) In each of these processes the introduction of impurities into the forming material is likely to occur. (3) Sputtering and evaporation in particular, are often considered to be very high-rate quenching methods by which high-temperature phases can be "frozen in" at lower temperatures.

In the following sections, we discuss the preparation of thin films of NbN, Mo-Re, Nb_3Ge and Nb_3Si and present our views regarding the importance of these and other factors towards stabilizing high-T_c phases in thin films.

III. B1 STRUCTURE FILMS

Historically the first example of an unstable high-T_c superconductor being stabilized by a thin-film growth method is the preparation of NbN films by reactive sputtering. Critical temperature as high as 17.3 K have been reported for these films (Keskar et al., 1971) and also for isomorphic Nb-C-N films (Gavaler et al., 1971). Extensive phase diagram studies (Brauer et al., 1952, 1961; Guard et al., 1967) have shown that the superconducting nearly stoichiometric NbN δ-phase ($T_c \sim$ 16 K), which crystallizes in the cubic B1 (NaCl) structure forms only at high temperatures ($\sim 1400°C$). At lower temperatures, NbN crystallizes into the hexagonal δ' or ε-phases which are not superconducting. Hence, it is generally believed that the δ-phase deposited by sputtering at temperatures as low as 400 to 800°C is metastable. It has been known for a long time, however, that a bulk δ-$NbN_{0.8}O_{0.2}$ phase can be synthesized at 750 to 800°C with a cell parameter of $a_0 = 4.32$ Å (Brauer and Esselborn, 1961b). At lower concentrations of oxygen, a_0 of this phase approaches that of pure δ NbN (4.38 Å). It is also well known that carbonitrides are more stable than the nitrides (Storms, 1975).

In the literature there are almost no analytical data on the O_2- and C-content of sputtered NbN films. Auger analysis of one of our NbN films, (Singer, 1978) which was dc sputtered in a high-vacuum apparatus ($< 10^{-9}$ torr initial pressure) detected 2 at. % carbon and an even higher level of oxygen of up to 9 at. %. The lattice parameter of the film was 4.36 Å. In NbN films reported by Singer et al., (1978), no O_2 was detected

(< 0.2 at. %) but the C-concentration was 5 to 15 at. %. More recent films (Singer 1979) show C-concentrations of 3 to 5 at. %. Recently Gavaler (1979) has been able to form the superconducting δ-phase and attain T_c's in excess of 16 K by annealing amorphous Nb-C-N deposits at temperatures as low as 600°C. This result is considered significant since the annealing process should produce an equilibrium phase. The resistive to superconducting transitions in these annealed films were indeed very sharp ($\sim$ 0.1 K) indicating good phase uniformity. Finally, submicron powders of carbon-free δ'-phase NbN[2] when annealed at 800°C in 1 atm. of N_2, in the presence of O_2 physically adsorbed on the powder surface transformed partly into an oxynitride δ-phase $NbN_{0.9}O_{0.1}$ with a_0 = 4.35 Å and T_c $\sim$ 14 K. Prolonged annealing at 800°C produced the incorporation of additional oxygen resulting in a partial transformation of δ-phase into the tetragonal phase $NbN_{0.6}O_{0.2}$. Based on all of these data, we conclude that the high-T_c δ-phase in thin sputtered films is not metastable. It is rather an equilibrium phase stabilized by the substitution of oxygen and/or carbon impurities thereby forming a stable ternary compound, i.e., an oxy- or carbonitride.

IV. A15 STRUCTURE FILMS

A. Mo-Re System

The first material which was made into a high-T_c superconductor by preparing it into thin-film form was an alloy in the Mo-Re system. Testardi et al., (1971) reported that sputtered films of $Mo_{32}Re_{68}$ had a T_c of almost 15 K, approximately 9 K higher than the corresponding bulk alloy. Later, Gavaler et al., (1972) found another alloy in this system ($Mo_{65}Re_{35}$) which also had a critical temperature of about 15 K. Testardi et al. suggested that sputtering had quenched into the films an unstable Mo-Re cubic phase, and that the low-frequency (soft) phonons associated with this metastable phase were responsible for the enhanced T_c's. The fact that their high-T_c $Mo_{32}Re_{68}$ films could only be obtained over a very narrow temperature range, close to the cubic-to-tetragonal phase transition temperature of 1150°C provided support for the phonon softening idea. However, it was later found by Gavaler (1975) that under certain sputtering conditions this alloy could actually be produced

---

[2]These powders were prepared by the "Herman C. Starck, Berlin" company and characterized by R. Kuznicki of our laboratory.

over a 300°C wide range including temperatures far below the
1150°C phase transition temperature. It was also found that
all of the high-T_c Mo-Re films contained the A15 phase. According to the Mo-Re equilibrium phase diagram, no A15 phase is
known to exist in the bulk. Since it is well known that the
A15 structure is particularly favorable for high-temperature
superconductivity, it was concluded that the enhanced T_c's in
the films were due to the crystallization of the A15 structure.
It has recently been shown (Stewart et al., 1979) that the A15
structure can be stabilized in a bulk Mo-Re alloy by low concentrations of Pt. Other impurities may have a similar effect.
We now know that oxygen was contained in all of our Mo-Re
films; and we believe, therefore, that the stabilization of the
A15 phase in sputtered films is due to the formation of an impurity (oxygen) stabilized ternary phase analogous to the
equilibrium bulk Mo-Re-Pt alloy.

B. *Nb-Ge System*

     Experiments on the Nb_3X series of A15 compounds (where X =
Sn, Al, Ga, Ge, or Si) have established that for a given X, the
highest critical temperature occurs in the compound which is
closest to the ideal 3/1 stoichiometry and which has the highest
degree of order. Secondly, as the X-atom size decreases along
this series, the T_c of the ordered stoichiometric phase becomes
increasingly higher. However, as the X-atom size becomes
smaller, the ordered stoichiometric A15 structure also becomes
increasingly more unstable and thus more difficult to prepare.
The boundary between the stable and unstable lattice appears to
be when a_0 is $\sim$ 5.17 Å. Since the discovery that a very high-T_c
nearly stoichiometric Nb_3Ge phase having $a_0 \simeq$ 5.14 Å could be
obtained in sputtered films, (Gavaler, 1973) there has been a
great deal of effort toward trying to understand how this phase
is stabilized. About three years ago, reports from various
laboratories indicated that oxygen or other impurities have a
beneficial effect on the preparation of the high-temperature
material (Gavaler, 1976; Hallak et al., 1977; Sigsbee,
1977). This led to speculations that high-T_c Nb_3Ge may be stabilized by impurities, perhaps in a manner similar to NbN, or
by some other mechanism involving the distribution of impurities throughout the body of the film. Recent experimental data,
however, make this type of stabilization mechanism improbable
for Nb_3Ge. Unlike NbN, the T_c's of Nb_3Ge films are greatly influenced by the choice of substrate material (Gavaler et al.,
1978). Again, unlike NbN, in Nb_3Ge films several hundred to
1000 Å of growth is required before the film reaches its maximum T_c (Tarutani and Kudo, 1977; Gavaler et al., 1978). Finally, despite efforts by us and various other investigators,

all attempts to crystallize Nb_3Ge from the amorphous phase has
produced material having a T_c at least 5 K lower than obtained
in the directly crystallized films.

We have proposed recently (Gavaler et al., 1978) the following hypothesis to explain the role that impurities can play in
the formation of the stoichiometric or nearly stoichiometric
high-T_c Nb_3Ge phase: During initial film growth, an A15 phase
is formed which is at or close to 3/1 stoichiometry but has an
anomalously large cell parameter, due to the incorporation of
impurities. Cell parameters as high as 5.20 to 5.25 Å have
actually been measured in films that are less than 1000 Å thick.
Because of this large lattice parameter, the stoichiometric
Nb_3Ge phase does not decompose as would otherwise occur. The
films' growth then proceeds by a homoepitaxial process and the
stoichiometric A15 lattice gradually becomes smaller as the impurity level decreases, until it has a cell edge commensurate
with its 3/1 composition and with the size of the Nb and Ge
atoms, i.e., $a_0 \sim 5.14$ Å. Hence, T_c then reaches the very high
values of ~ 23 K. Another means of stabilizing stoichiometric
Nb_3Ge, by epitaxially depositing the high-T_c phase onto the
stable Nb-rich off-stoichiometric material, has been suggested
by Newkirk (1978). In this case, no impurities would be required to stabilize the initially-formed A15 phase. Since it
is known (Gavaler et al., 1976) that the Nb/Ge ratio near the
substrate-film interface in our sputtered films is greater than
3, it is possible that the A15 structure could be stabilized
by an excess of Nb as well as by impurities. Experimentally,
there is no easy way to verify this possibility since chemical
analysis of the material near the interface gives only the
total Nb/Ge ratio in what is known to be a two-phase mixture of
Nb-Ge and Nb-O. In any case, the *crucial* premise on which
these ideas are based is the concept of homoepitaxy. That epitaxial growth may be a means for stabilizing Nb_3Ge was first
suggested by Dayem et al. (1977) based on the deposition of
Nb_3Ge onto A15 structure Nb_3Ir. They concluded that the negative free-energy contribution obtained from the epitaxial
growth is sufficient to allow the metastable stoichiometric
Nb_3Ge phase to start growing on Nb_3Ir. The growth of the metastable phase then continues by homoepitaxy.

Because of the inferred importance of epitaxy in the growth
of Nb_3Ge, we have attempted to reproduce the experimental work
of Dayem et al. using low-energy sputtering as the deposition
method. The sputtering conditions were similar to those used
previously with the exception that a new multiple-target holder
was incorporated into the system. With this target holder, it
was possible to sputter from one target and within a few sec-
onds switch to another thus minimizing the incorporation of impurities at the Nb_3Ir/Nb_3Ge interface. The Nb-Ir and Nb-Ge
targets were constructed so that the exposed surfaces of the

targets had a ratio of Nb to Ge or Ir of $\sim$ 3. The low-energy
conditions used previously to deposit Nb_3Ge (Gavaler, 1973)
were also used to form Nb_3Ir. While sputtering Nb_3Ge on sapphire substrates produced the highest temperature material only
after about 1000 Å of growth, depositing Nb_3Ge on a layer of
freshly sputtered Nb_3Ir produced films with T_c onsets as high
as 20 K even in films less than 200 Å thick. A summary of
these results is shown in Figure 1. X-ray analyses of the
Nb_3Ir layers showed a major A15 phase and a weak second phase
of niobium. The lattice parameters varied between 5.14 to
5.20 Å. Similar to the work of Dayem et al. there seemed to be
no correlation between Nb-Ir lattice parameter and high-T_c's.
As a basis of comparison, to clearly illustrate the influence
of the Nb_3Ir, a second series of experiments was done in which
niobium was substituted for the Nb-Ir target. Using the same
conditions as before, Nb and Nb-Ge layered films were sputtered.
At layer thicknesses of $\lesssim$ 200 Å, where T_c's onsets of nearly
20 K were obtained using the Nb-Ir target, the Nb-Ge films were
found not to be superconducting down to 4.2 K. Finally an experiment was done in which a shutter was used to mask two of
the ten sapphire substrates, during the deposition of the Nb_3Ir
layer. The shutter was then removed and $\sim$ 200 Å of Nb_3Ge were
sputtered on all of the substrates. The results of this experiment are shown in Figure 2. As can be seen, the only nonsuperconducting (above 4.2 K) samples were the Nb-Ge films deposited
on substrates not coated with Nb_3Ir. All of these results are
strong evidence of the efficacy of the Nb_3Ir substrates in pro-
moting the formation of the high-T_c Nb_3Ge phase. The results

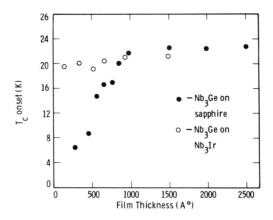

FIGURE 1. *Highest T_c's that have been obtained from a
long series of experiments in which Nb_3Ge films of various
thicknesses were sputtered either on sapphire or Nb_3Ir substrates.*

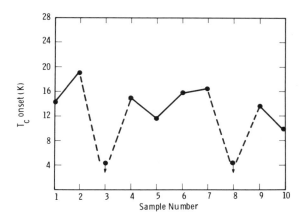

FIGURE 2.   Critical temperature of ∿ 200 Å thick Nb₃Ge films sputtered during a single experiment.   Samples 3 and 8 were deposited on sapphire and the remaining on Nb₃Ir.

however provide no *direct* proof that epitaxy is in fact occurring. In an attempt to obtain such proof, a few of the Nb_3Ir-
Nb_3Ge layers have been successfully removed from their sapphire
substrates and observed by transmission electron microscopy
(TEM). Figure 3 shows a representative electron micrograph of
a combined Nb-Ir and Nb-Ge film. The thickness of the Nb_3Ir is
∿ 1200 Å and the Nb_3Ge ∿ 400 Å. As can be seen, the structure
is characterized by fine, equiaxed (on the plane of observation)
grains. The clean interior of the grains and the absence of any
defect structure such as dislocations and twins suggest little
atomic mismatch between the Nb-Ir and Nb-Ge layers. Furthermore, the lack of evidence of duplex grain structure in the com-
posite film lends credence to the view that Nb_3Ge has grown
epitaxially onto Nb_3Ir. Additional TEM study is in progress.

C.   Nb-Si System

The smallest isoelectronic X atom in the Nb_3X series discussed above is silicon. Since T_c increases with decreasing
X-size and cell parameter in this series, a close-to-stoichiometry,
relatively well ordered A15 Nb_3Si with $a_0 \simeq 5.08$ Å would be expected to have a T_c higher than that of Nb_3Ge. Hence, during
the past few years there has been a great deal of interest in
the study of Nb_3Si. No A15 phase has, so far, been prepared
from the melt. An equilibrium Nb_3Si compound does exist, however; it has a Ti_3P tetragonal structure that is stable above

FIGURE 3. A TEM photograph looking through a layered
Nb_3Ge-Nb_3Ir film. The thickness of the Nb_3Ge is ∿ 400 Å and
the Nb_3Ir ∿ 1200 Å.

∿ 1800°C (Rossteutscher and Schubert, 1965; Deardorff et al.,
1969). Si-deficient and disordered A15 phase can be synthesized
by evaporation, sputtering and chemical vapor deposition in the
temperature range between 450 and 900°C (Hammond and Hazra,
1974; Johnson and Douglass, 1974; Kawamura and Tachikawa, 1975;
Newkirk, 1975; Braginski and Roland, 1975; Paidassi and Spitz,
1978). In most cases the indication of the A15 phase presence
was obtained from X-ray or electron-diffraction rather than
from superconductive measurements, since the T_c values fell in
the range between 6 and 10 K and could be attributed to bcc
niobium phases that could have been present. The cell parameters of the A15 phases were high, from a_0 = 5.14 Å up to a_0 >
5.20 Å. Thus the inferred concentrations of Si were well below
25 at. %. Recently Somekh and Evetts (1977, 1978, 1979) reported T_c onsets of 14 to 17.6 K in Nb-Si specimens prepared by
dc getter sputtering. The presence of A15 phases having a_0 =
5.18 to 5.20 Å was confirmed by X-ray diffraction. These results have not yet been reproduced by other investigators. A
T_c onset of 12 K, significantly higher than that of Nb, was
also obtained by Testardi et al. (1975) when heat pulsing a
sputtered Nb-Si deposit. However, no phase identification was
reported. Haase et al. (1978) also observed $T_c \geq$ 12 K in sputtered Nb-Si with no other evidence for the A15 presence.
    The extreme difficulty in forming close-to-stoichiometry,
high-T_c A15 Nb_3Si has also been indicated by studies of the
pseudo-binary system $Nb_3(Ge_{1-x}Si_x)$. The Si solubility limit in
the high-T_c A15 phase prepared by CVD was only $x \simeq$ 0.2
(Newkirk, 1975; Paidassi, 1979a). By sputtering $x \simeq$ 0.4 was

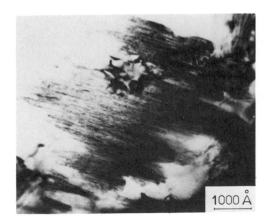

*FIGURE 4.   A TEM photograph of fibrous microstructure in-
side an A15 grain of aged Nb₃Ge₁₋ₓSiₓ film (x = 0.17); before
aging Tc was 22 K (midpoint) and aₒ = 5.135 Å (Paidassi, 1979b).*

attained (Haase et al., 1978). Paidassi (1979b) observed room-
temperature aging of high-T_c ternary specimens with x ≃ 0.2.
Over a period of only a few months, the T_c decreased by 1 to
2 K, accompanied by an increase in the A15 cell parameter from
5.135 to 5.140 Å and the appearance of the hexagonal phase. In
the A15 grains of these samples fibrous striations such as
shown in Figure 4 have been observed by transmission electron
microscopy. These striations could be tentatively associated
with the progressing microscopic segregation of phases culmi-
nating in the formation of larger size precipitates of the
hexagonal phase which is identifiable by the electron-
diffraction. This interesting effect dramatically demonstrates
the instability of the close-to-stoichiometry, Si-containing
A15 phase.

V.   SUMMARY AND CONCLUSIONS

We have reviewed four examples of the preparation in thin-
film form of high-T_c superconducting phases which are not stable
according to bulk equilibrium phase diagrams. We believe that
the high-T_c's in the NbN and Mo-Re films prepared by us are due
to impurity stabilized equilibrium phases, with the stabilizing
impurities being carbon and/or oxygen. In the case of Nb-Ge
films, we hypothesize that the high-T_c stoichiometric Nb₃Ge compound is metastable and can be prepared via a homoepitaxial process starting from either a stable A15 Nb-Ge phase formed near

the film-substrate interface or from another A15 phase such as Nb_3Ir which has a lattice parameter similar to Nb_3Ge. A similar impurity-epitaxy mechanism is probably operative in the stabilization of thin films of sub-stoichiometric, A15 Nb_3Si.

It is our conclusion that the important factors in the growth of unstable phases by thin-film techniques are: (1) the phase formation at relatively low temperatures, (2) the introduction of impurities into the forming material and (3) the free energy contribution from epitaxial growth. We believe that this last factor is particularly important in the growth of stoichiometric Nb_3Ge. There is no reason to believe however that it is not a general phenomenon. Thus, it should be possible to prepare, for example, metastable impurity-free Mo-Re or NbN films by epitaxial growth onto the impurity stabilized phases. We have found no evidence to indicate metastable phase formation by the quenching-in of a high-temperature phase. We conclude therefore that this feature of thin-film deposition has little or no importance in the formation of high-T_c superconducting films.

We also conclude that the stabilization mechanisms discussed become important only in metallurgical systems where the differences in free energies of competing phases of the same composition are quite small. It is becoming increasingly apparent to us that the mechanisms operative during film growth are not energetically sufficient to synthesize a potentially very high-T_c superconductor such as Nb_3Si where the free energy difference between the stoichiometric A15 and the Ti_3P phase is significant. Attempts to obtain the stoichiometric high-T_c phase by thin-film methods will thus remain fruitless. If any further increase in T_c in A15 compounds is obtained by thin-film methods, it will most likely be in the Nb-Ge system where the instability is not so severe.

These conclusions, if valid, have some implications on the current effort to develop a theoretical understanding of the relationship between instabilities and high-temperature superconductivity. However, a detailed discussion in this area is outside the scope of the present paper.

ACKNOWLEDGEMENTS

We gratefully acknowledge the invaluable technical contribution of R. Wilmer, A. L. Foley and P. Yuzawich.

REFERENCES

Braginski, A. I. and Roland, G. W., (1975) unpublished results.
Brauer, G. and Jander, J. (1952), Z. Anorg. Allg. Chem., 270,
    10.
Brauer, G. and Esselborn, R., (1961a), Z. Anorg. Allg. Chem.,
    309, 151.
Brauer, G. and Esselborn R., (1961b), Z. Anorg. Allg. Chem.,
    308, 53.
Dayem, A. H.; Geballe, T. H.; Zubeck, R. B.; Hallak, A. B. and
    Hull, G. W., Jr., (1977) Appl. Phys. Lett. 30, 541.
Gavaler, J. R.; Janocko, M. A. and Jones, C. K., (1971) Appl.
    Phys. Lett. 19, 305.
Gavaler, J. R.; Janocko, M. A. and Jones, C. K., (1972) Appl.
    Phys. Lett. 21, 179.
Gavaler, J. R. (1973) Appl. Phys. Lett. 23, 480.
Gavaler, J. R. (1975) J. Vac. Sci. Technol. 12, 103.
Gavaler, J. R.; Miller, J. W. and Appleton, B. R., (1976) Appl.
    Phys. Lett. 28, 237.
Gavaler, J. R., (1976), Superconductivity in d- and f-Band
    Metals (Proc. of 2nd Rochester Conf.), edited by
    D. H. Douglass (Plenum, New York), p. 421.
Gavaler, J. R.; Ashkin, M.; Braginski, A. I. and Santhanam, A. T.,
    (1978) Appl. Phys. Lett. 33,359.
Gavaler, J. R., (1979) IEEE Trans. on Mag., MAG-15, 623.
Haase, K. L.; Smithey, R. and Meyer, O., (1978) Karlsruhe
    Nuclear Center Progress Report kfK 2670, p. 97.
Hallak, A. B.; Hammond, R. H.; Geballe, T. H. and Zubeck, R. B.
    (1977) IEEE Trans. on Mag., MAG-13, 311.
Hammond, R. H and Hazra, S., (1974) Proc. LT-13, Vol. 3, Plenum,
    New York, p. 465.
Johnson, G. R. and Douglass, D. H., Proc. LT-13, Vol. 3, Plenum,
    New York, p. 468.
Kawamura, H. and Tachikawa, K., (1975) Appl. Phys. Lett. 55A,
    65.
Keskar, K. S.; Yamashita, T. and Onodera, Y., (1971) Japan J.
    Appl. Phys. 10,370.
Newkirk, L. R., (1975), unpublished results.
Newkirk, L. R., (1978), private communication.
Paidassi, S.; Spitz, J. and Besson, J., (1978) Appl. Phys, Lett.
    33, 105.

Paidassi, S. and Spitz, J., (1978) *J. Less Common Metals*, *61*, 213.
Paidassi, S., (1979a) *J. Appl. Phys. (in press)*.
Paidassi, S., (1979b) Ph.D. Thesis (unpublished).
Sigsbee, R. A., (1977) *IEEE Trans. on Mag.*, *MAG-13*, 307.
Singer, I. L., (1978) unpublished results.
Singer, I. L.; Wolf, S. A.; Lowrey, W. H. and Murday, J. S., (1978) *J. Vac. Sci. Technol.*, *15*, 625.
Singer, I. L., (1979) private communication.
Somekh, R. E. and Evetts, J. E., (1977) *Solid State Comm.*, *24*, 733.
Somekh, R. E., (1978) *Proc. LT-15*, *J. de Physique*, *39*, *Suupl. 8*, C6-398.
Somekh, R. E. and Evetts, J. E., (1979) *IEEE Trans. on Mag.*, *MAG-15*, 494.
Stewart, G. R.; Giorgi, A. L. and Flukiger, René (1979) *Bull. Am. Phys. Soc.*, *24*, 426.
Storms, E. V., (1975) *High Temperature Science*, 7, 103.
Tarutani, Y. and Kudo, M., (1977) *Japan J. Appl. Phys.*, *16*, 509.
Testardi, L. R.; Hauser, J. J. and Read, M. H., (1971) *Solid State Commun.*, *9*, 1829.
Testardi, L. R.; Wakiyama, T. and Royer, W. A., (1977) *J. Appl. Phys.*, *48*, 2055.

SUPERCONDUCTIVITY IN THE ACTINIDES*

James L. Smith

Los Alamos Scientific Laboratory of the University of California
Los Alamos, New Mexico

Recent progress has made the pattern of superconductivity in f-band metals clear. Although the concept of valence is unnecessary, considerations of band structures, phonon modes, densities of states, and magnetic effects appear adequate to explain the superconducting properties. The problems of dealing with radioactive superconductors and some current work is reviewed briefly.

I. INTRODUCTION

It has become apparent that f-band superconductors can be understood in a similar manner to d-band superconductors. Rare earth and actinide superconductivity does not appear to require exotic mechanisms and the pattern throughout both series is now obvious. Recent advances in purifying actinide elements have led to the discovery of superconductivity in americium (Smith and Haire, 1978) and to a clean measurement of the superconductivity in protactinium (Smith, Spirlet, and Müller, 1979). This work, coupled with an enlightened view of lanthanum, cerium, thorium, and uranium, and with a recent result on a europium compound make the f-band story simple.

By the actinide elements I am referring simply to actinium and the 5f-electron series following it. They all have unstable nucleii, and are primarily man-made. Since 4f and 5f electrons are fundamentally similar, the rare-earth elements are quite naturally included in this discussion. Figure 1 shows the

Work performed under the auspices of the Department of Energy.

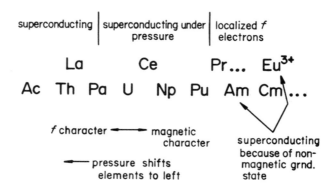

FIGURE 1. *Trends in localization of f electrons are indicated. There is an approximate equivalence in behavior shown by the positioning of lighter elements.*

the pertinent elements schematically with the rare earths spread out to represent the parallel progression through narrowing f bands (or localizing f electrons) across the two series. I have indicated trends in properties, including a few that have not yet been verified experimentally, but the headings and alignments can simply be taken as rough indications of behavior.

There are no essentially new arguments in this paper, but most of them have been previously applied only to *particular* cases. It is not widely appreciated that at the beginning of both of these series f-electrons exist in moderately wide bands, that these band states are bonding, and that they can be non-magnetic and in no way inhibit superconductivity (Hill, 1970; Kmetko and Hill, 1970; Hill, 1971; Freeman and Koelling, 1974; Johansson and Rosengren, 1975; Johansson, 1975; Skriver, Andersen, and Johansson, 1978). The presence of these bonding or nonlocalized (itinerant) f-electrons can be readily recognized (Matthias *et al.*, 1967; Kmetko and Hill, 1976). As the f series further progress, the onset of magnetism naturally affects superconductivity, and finally, localized moments and long range magnetic order set in for the $4f$ and $5f$ electrons. Additionally, Johansson and Rosengren (1975) noted that two elements in the local category, europium and americium, would likely be superconducting. This has now also been essentially demonstrated (Matthias, Fisk, and Smith, 1979; Smith and Haire, 1978).

This paper is concerned primarily with the elemental solids. However, as will become clear, similar arguments apply to compounds and alloys. After putting all f-electron elements on a common, simple basis, I will briefly review both the experimental problems with the actinides and some recent results.

In case the high T_C devotees think they have read enough,
remember that at high pressure, pure lanthanum has a T_C of 13 K
(Probst and Wittig, 1978).

II. NATURE OF f ELECTRONS AT THE BEGINNING OF THEIR SERIES

There are many common features of the electronic structure
of the elements shown in Fig. 1. It should be appreciated that
in general the localization of the $5f$-electron series is intermediate between that of the $3d$ and $4f$ transition metal series.
In the $3d$ series, localized magnetic moments never fully develop
while in the $4f$ series this happens at the beginning. The actinides fall in between. The first two actinides, actinium and
thorium, do not possess strong f character. While we find
protactinium through plutonium to be profoundly affected by
f states at or near the Fermi energy, beyond plutonium, the
actinides have well localized f electrons, and in that sense,
resemble the majority of the rare-earth elements. There are a
few measures of this progress to localization that are extremely
informative. The existence of f states at or near the Fermi
level is demonstrated by depressed melting points, by high
low-temperature heat capacities, by the ability to make the
solids either superconducting or magnetic in various intermetallic compounds, by the extreme sensitivity of most physical
properties to pressure, and by the occurrence of certain unique
crystal structures.

The low-melting points were first correlated to f-electron
bonding by Matthias *et al.* (1967). This melting-point reduction is shown for the actinides in Fig. 2 (Brodsky, 1978) and
is a feature also exhibited by lanthanum and cerium. The low-
melting points occur because the (rather complicated) angular
dependences of the f wavefunctions actually permit better bonding in the liquid than in the more symmetric bcc phase from
which the elements melt (Kmetko and Hill, 1976). It is clear
from Matthias *et al.* (1967) that f electrons *cannot* be passive
(except for a filled shell). That is, they *must* be either
bonding, magnetic, or a mixture of the two. Hence, in the
f-band superconductors their presence must be important.

The electronic heat capacity coefficient γ is a measure of
the density of states at the Fermi energy. The values of γ
(Fig. 2) are seen to peak at plutonium (Mortimer, 1979; Stewart
et al., 1979). This peak, which occurs in the rare earths
at cerium, is due primarily to the contribution from the narrow
(1 eV or so wide) f bands at or near the Fermi energy. The
generally high values of γ in the rare earths (Probst and
Wittig, 1978) must certainly also have a substantial *spd* contribution since, for example, the last rare earth, lutetium,

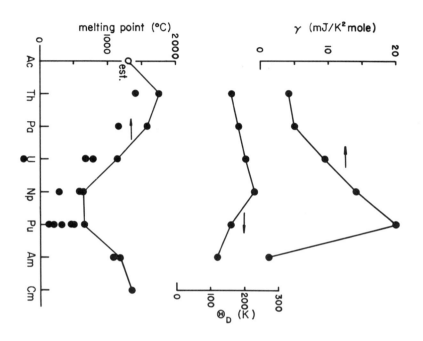

FIGURE 2. The uniform variation of some physical properties of the light actinides are shown. These f electron modified properties are the electronic heat capacity coefficients, the Debye temperatures, and the melting points. The points below the melting line represent structural modifications.

has a γ of 6.8 mJ/mole·K^2 yet its localized closed *f* shell could not be contributing to γ. (Since lutetium is not superconducting at normal pressures, this suggests that a high γ is not a principal cause of the high T_c in lanthanum.)

The Debye temperatures θ_D are also shown in Fig. 2. They are presented primarily to display their smooth behavior in the early actinides and, more importantly, to note their low values (soft lattices) in comparison to other elements (Gschneidner, 1964). This may well result from the symmetry of *f*-bonding which should give rise to more complicated vibrational spectrums than exist in the higher symmetry structures of most elements. It is possible that θ_D is not an adequate measure of such details. The peak in θ_D (stiffest lattice)

at neptunium could be caused by a competition between f-bonding
and f-electron localization or could simply reflect differences
in the phonon modes of the various low-temperature crystal
structures of the actinides. (Between thorium and americium
the structures are all exotic with monoclinic plutonium being
the least symmetric.)

The next aspect of f electron behavior is the question of
how the magnetism first arises in these series. It is the
question of localized versus itinerant electrons. Freeman
(1979) notes that there is only one quantitative approach, the
Hill plot (Hill, 1970). Before 1970 superconductivity or magnetism seemed to occur almost unpredictably in cerium, uranium,
neptunium, and plutonium and their compounds. Hill showed that
this was all understandable in terms of the overlap of atomic
f-electron wavefunctions. If the actinide or rare-earth atoms
are very close together, f-bands must form, and sufficiently
broad bands must destroy local moments and may permit superconductivity. Alternately, if the atoms are far apart, the
atomic-like, partially filled shells will have a magnetic
moment. This picture is obvious, but Hill found that it could
be put on a completely quantitative basis. That is, there is
a critical spacing for the pertinent elements (cerium: 3.37-
3.46 Å; uranium: 3.4-3.62 Å; neptunium: 3.25 Å; and plutonium:
3.4 Å) that delineates the regions for superconductivity (smaller spacing) or magnetism (larger spacing) of the f-electron
atoms in elements, their alloys, or their compounds, and the
value is *independent* of crystal structure. In spite of the
addition of many compounds to Hill's compilation, this plot
approach is still valid today (Brodsky, 1978). It is remarkable that such a simple approach can predict the beginning of
magnetism in f-bands. This beginning can also be called the
onset of electron correlations, of exchange enhancement, of
spin flucuations, of paramagnons, or of whatever we choose to
call our nearly magnetic state. It could be said that
f-electrons cannot be as complicated as all their models.

The Hill plot leads into the last aspects of light rare-
earth and actinide f character, the effects of applying pressure and the existence of a large number of crystal structures.
The simplest way to modify the behavior shown on a Hill plot
is to squeeze the atoms together. This can broaden, shift,
or hybridize bands. Pressure can make magnetic electrons
more itinerant and wash out magnetic behavior altogether, leading to superconductivity. Finally, pressure modifies the
crystal structure. In Fig. 2, the points below the melting-
point line represent various structure modifications. The
large number of phases come about from the low symmetry of
f-bonding (Kmetko and Hill, 1976), and because there are so
many phases, they must be energetically very close together.
So obviously pressure should induce transitions from one to

another. The various structures have somewhat common properties since they are so energetically similar; this partially
explains why the Hill plot is successful although structures
are ignored. Other examples are that the T_c's of the two
structures of lanthanum are comparable and that the energy
band density of states calculations are largely independent of
structure (Hill and Kmetko, 1976). Thus, squeezing the early
rare earths and actinides shifts their properties to resemble
elements to the left because of electron delocalization or
band broadening (Fig. 1) and drives the materials through many
phase transitions. For the rare earths, Probst and Wittig
(1978) have discussed such techniques and effects in detail,
and Johansson and Rosengren (1975) have put this on a remarkably firm theoretical basis.

The foregoing arguments are, in fact, the genesis of Fig.1.
The early *f* electron elements are best viewed in terms of
densities of states, band widths, crystal structures, and the
onset of magnetism. This differs from the *d*-transition elements in the sense of not taking valence as the index of
trends (B. Johansson, 1978). However, the rest of the language
is in fact quite familiar in the field of superconductivity.
I have said rather little about the crystal structures because
this is perhaps the most intrinquing aspect of the next section.

III.   SUPERCONDUCTIVITY AND *f* ELECTRONS

In this section a few significant properties of each *f*-band
superconductor are highlighted. The references should be
consulted for full discussions since they were selected on that
basis. Figure 1 shows lanthanum, actinium, thorium, and
americium as superconductors. They have some *f* character
but certainly exhibit no magnetic effects. These elements have
f bands above the Fermi energy. (Americium with its six well-
localized, *f* electrons in a nonmagnetic, J = 0 state is included in this nonmagnetic category because of the obvious effect
of the additional *f* levels above the Fermi energy.) The
existence of these unfilled bands is shown, for example, by
depressed melting points (Matthias *et al.*, 1967), by the double
hexagonal close-packed (dhcp) structure of lanthanum and americium (Johansson and Rosengren, 1975), and by the need for
including some *f* character in the calculation of the Fermi
surface of thorium (Freeman and Koelling, 1974). These
materials are fairly straightforward superconductors with T_c's
largely understandable in terms of *d*-band superconductivity
but they are still affected to varying degrees by their *f*
character.

Lanthanum ($T_c \sim 5K$, dhcp phase) has been discussed in

great detail by Probst and Wittig (1978). It has a high T_C that is increased to ∿13K by a pressure of 200 kbar. Although it has a high γ, this cannot be an important aspect of its superconductivity since the nonsuperconductors scandium, yttrium, and lutetium also have high γ's. As Probst and Wittig conclude, the high T_C must be phonon related. Lanthanum does have a significantly low θ_D. More specifically, the evidence for several pressure-induced phase transitions shows the f character in the bonding and this is the origin of phonon softenings that can raise T_C's so effectively. The feature of f electron bonding that causes the elements to exhibit so many exotic structures, also means that the typical vibrational modes must be more exotic than for non f-band elements (Weling, 1977). Put more simply, *lattice instabilities raise T_C's*, and if magnetism did not show up so quickly in the f series, the T_C's might be very high.

Actinium has had no reported low-temperature measurements, and the reasons are obvious. It is a β emitter with a half-life of about twenty years. Pure actinium then generates ∿50 mW/g of radioactive self-heating. Its daughters grow in exponentially in days so that for three-month-old actinium, there is ∿18 W/g heating, 1000 ppm lead, and a gamma ray emission from the daughters that is a major health hazard. Furthermore, its high chemical activity makes it very difficult to purify. On the theoretical side, Johansson and Rosengren (1975) predict a T_C for actinium of more than 10K. This is because their pseudopotential calculation lead them to expect it to resemble lanthanum under pressure. On the other hand, if actinium has no significant f character, it should resemble the trivalent nonsuperconductors scandium, yttrium, and lutetium. (An experiment to resolve these arguments is planned.) I point out that this same Johansson and Rosengren paper anticipated the recent demonstration of superconductivity in americium and trivalent europium.

Thorium (T_C ∿ 1.4K, fcc structure) is in many respects the most puzzling of these elements because it appears to be such an ordinary BCS superconductor (Haskell *et al.*, 1972). This is not inconsistent with the views expressed here, but there does appear to be one measurement that begins to raise suspicions of more interesting behavior. Fertig *et al.* (1972) showed that under pressure, T_C drops until 60 kbar in a manner consistent with a BCS superconductor (somewhat unusual for a "transition" metal), but then goes through a weak minimum at 75 kbar, and drops slowly up to the highest measured pressure of 160 kbar. They say this behavior could be associated with a variation of the Fermi surface or a phase change. Either of these changes can signal the presence of f character. We also consider some thermodynamic correlations of entropy in f electron metals. It is clear from a correlation of room temperature entropy to the metallic radius that thorium cannot be completely tetravalent but must contain some f character (Ward and Hill, 1976). Furthermore, Ward (1979) speculates that f electron effects will be seen in the description of superconductivity in thorium hydrides. Thorium is not as simple as it has appeared.

Protactinium superconductivity has been the subject of conflicting reports for many years. Recent work of Smith, Spirlet, and Müller (1979) settles this conflict and finds a T_c of 0.43K (body-centered tetragonal structure) for a high purity single crystal. It is satisfying that this T_c falls smoothly between thorium and uranium as suggested by the trends of properties in Fig. 2. Very little other information on protactinium is available, but it seems likely that its T_c is partially depressed by the onset of electron correlations, an effect that is pronounced in uranium. If so, a positive pressure effect on the T_c is certainly to be expected (as reported by Fowler *et al.*, 1974).

Americium was discovered to be superconducting by Smith and Haire (1978) with a T_c of 0.79K for the dhcp structure; and a T_c of 1.05K for the fcc structure. This latter high-temperature phase had been accidentally retained to room temperature with the addition of 2000 ppm ytterbium as an impurity. We then have a remarkable parallel to the situation of lanthanum with 2000 ppm gadolinium (Levgold *et al.*, 1977). Indeed the parallel between lanthanum and americium is even more complete (Smith and Haire, 1978). This correspondence shows that americium with six localized f-electrons must owe many of its properties to the remaining f bands that are above the Fermi energy. However, its density of states ($\sim \gamma$) is extremely low (Fig. 2) and appears to be the primary cause of its low T_c compared to lanthanum. Americium is also reported to have a rather high critical field (Smith, Stewart, Huang, and Haire, 1979). I speculate that this could be caused by some magnetic nature of the electrons arising from a slightly imperfect J = 0 ground state; that is, the metal might not be exactly trivalent. Determination of the pressure dependence of T_c would also certainly be interesting for a superconductor that has both localized f-electrons and f contributions at the Fermi energy since the separable effects would be mixed. It is clear that sufficient pressure will delocalize the six f-electrons (similar to γ-cerium) and cause the compressed americium to resemble plutonium, as has been suggested by Kmetko (1979) and Johansson (1979).

Closely related to the americium is the recent discovery by Matthias *et al.* (1979) that $EuIr_2$ is a superconductor, similar in *every* respect to $ScIr_2$, YIr_2, $LaIr_2$, and $LuIr_2$. It

appears that if europium metal were trivalent, as is the Eu in
EuIr$_2$, pure europium would also be a superconductor similar to
americium (its actinide analogue) as expected by Johansson
and Rosengren (1975).

The other superconductors in Fig. 1 are cerium, uranium,
neptunium, and plutonium in which the effects of magnetism
must be considered. Cerium and uranium have been discussed
by others in great detail. In contrast, neptunium and plutonium
are so close to being magnetic that their inclusion in the
"superconducting under pressure" category in Fig. 1 is speculation.

Cerium in its room-temperature fcc phase (γ-Ce) is magnetic,
while, its low-temperature fcc phase (α-Ce) and higher pressure
phases are superconducting (Probst and Wittig,1978). There are
three models for the γ-α phase transition that recognize that
this is not a valence change but is simply a loss of the magnetic nature of the f band in γ-Ce. Johansson (1974) views it as
a Mott transition of the f electron, that is, the f electron
goes sharply from a localized, magnetic state to an itinerant,
band state as the pressure is increased. Hill and Kmetko
(1975) see the transition as a hybridization of the d band with
the f band to destroy magnetism. Probst and Wittig (1978)
suggest a simple band broadening picture. These viewpoints
are all in qualitative agreement and the papers contain
extensive information.

Uranium cannot be considered a bulk superconductor above
$\sim$0.1K. The study of its low-temperature behavior is too complicated to discuss here. The Hill plot is comprehensive with
regard to its superconducting and magnetic compounds and alloys
(Lam and Aldred, 1974). The more difficult question of the
behavior of pure uranium is summarized by Bader and Knapp
(1975). They find that the principal effects involved in the
superconducting behavior are large Cooper-pair-weakening
interactions and phonon mode shifting. There is not a more
typical f-electron superconductor.

Neptunium and plutonium are not superconducting at normal
pressure (Smith and Elliott, 1977; Meaden and Shigi, 1964).
In view of their strong similarity to uranium, the possibility
of their superconductivity under pressure was discussed by
Hill (1970). Similarly, in light of the analogy to molybdenum
and niobium stabilized γ-U ($T_c$$\sim$2K), an extrapolated value of T_c
for stabilized bcc neptunium is 30 mK (Hill et $al.$, 1974;
Smith and Elliott, 1977) and 1 mK for bcc plutonium (Hill
et $al.$, 1974). Clearly these elements are so close to being
magnetic (Brodsky, 1978) that study of their superconducting
properties is difficult.

IV. EXPERIMENTAL PROGNOSIS

The experimental progress in actinide superconductivity research is slow for many reasons. The radioactivity of the actinides produces heat which makes cooling difficult, and they can never be truly isothermal. Americium generates the most heat (6 mW/g) of any element yet shown to be a super-conductor. It also accumulates radiation damage at low temperatures that causes its critical field to increase at the rate of a few percent per day (Smith, Stewart, Huang, and Haire, 1979). Hence a proper measurement of actinium (with a heating of 50 mW/g to 18 W/g and a tremendous radiation damage rate) may never be possible. It was reassuring to find a superconducting transition width of only 5 mK at 0.43K in a well-annealed single crystal of protactinium ($\sim$1 mW/g) (Smith, Spirlet, and Müller, 1979). But even this material must remain normal at its center from the heating.

These materials are largely man-made in reactors, are expensive to obtain, and hazardous to work on without special equipment. They are all at least somewhat chemically active, and yet must often be highly purified on a milligram basis.

Nonetheless, work continues (especially now on the magnetism in the heavier actinides). As for current superconductivity efforts, I have made some attempts to measure the critical fields on americium and protactinium. The results are not completely satisfactory and yet the critical field is an important parameter for calculations. The fundamental problem is caused by the heating that maintains the center of the sample normal. In an applied field some outer portions of the sample are also driven normal and this, in turn, raises the thermal conductivity of those portions which then modifies the temperature distribution. Hence, measuring H_{c1} or H_{c2} is a very complex problem of inhomogeneous superconductivity. The largely qualitative results that I have obtained include an H_{c2} of 55 Oe for protactinium and an H_{c1} of 530 Oe and an H_{c2} of 750-1000 Oe for americium. These critical fields, even allowing for improved heat capacity techniques to measure γ (Stewart *et al.*, 1979), still are not precise enough to differentiate clearly between Type I and Type II behavior in the two elements. Certainly more work is needed.

V. CONCLUSION

It seems that the token *f*-electron metal need no longer be included in tables of "transition" metal superconductors. There are now many *f*-electron superconductors, and all we have to do is think about densities of states, lattice heat capacities, band widths, nearly magnetic behavior, and lattice instabilities. This is just what superconductivity has always involved.

ACKNOWLEDGMENTS

I am indebted to the late H. Hunter Hill for starting me on this subject and to B. T. Matthias for assistance with the progress. I want to thank numerous colleagues, particularly B. Johansson, J. W. Ward, and J. H. Wood, for stimulating and helpful discussions.

REFERENCES

Bader, S. D., and Knapp, G. S.(1975). *Phys. Rev. B 11*, 3348.
Brodsky, M. B. (1978). *Rep. Prog. Phys. 41*, 103.
Fertig, W. A., Moodenbaugh, A. R., and Maple, M. B. (1972). *Phys. Lett. 38A*, 517.
Fowler, R. D., Asprey, L. B., Lindsay, J. D. G., and White, R. W. (1974). *In* "Low Temperature Physics-LT13" (K. D. Timmerhaus, W. J. O'Sullivan, and E. F. Hammel, eds.), p. 377. Plenum Press, New York.
Freeman, A. J. (1979). *J. de Phys. 40*, C4-84.
Freeman, A. J., and Koelling, D. D. (1974). *In* "The Actinides: Electronic Structure and Related Properties" (A. J. Freeman and J. B. Darby, Jr., eds.), p. 51. Academic Press, New York.
Gschneidner, Jr., K. A. (1964). *Solid State Physics 16*, 276.
Haskell, B. A., Keeler, W. J., and Finnemore, D. K. (1972). *Phys. Rev. B 5*, 4364.
Hill, H. H. (1970). *In* "Plutonium 1970 and Other Actinides" (W. N. Miner, ed.), p. 2. A.I.M.E., New York.
Hill, H. H. (1971). *Physica 55*, 186.
Hill, H. H., and Kmetko, E. A. (1975). *J. Phys. F: Metal Phys. 5*, 1119.
Hill, H. H., and Kmetko, E. A. (1976). *In* "Heavy Element Properties" (W. Müller and H. Blank, eds.), p. 17. North-Holland, Amsterdam.
Hill, H. H., White, R. W., Lindsay, J. D. G., and Struebing, V. O. (1974). *Sol. State Comm. 15*, 49.
Johansson, B. (1974). *Phil. Mag. 30*, 469.
Johansson, B. (1975). *Phys. Rev. B 11*, 2740.
Johansson, B. (1978). *J. Phys. Chem. Solids 39*, 467.
Johansson, B. (1979). private communication.

Johansson, B., and Rosengren, A. (1975). *Phys. Rev. B 11*, 2836.

Kmetko, E. A. (1979). private communication.

Kmetko, E. A., and Hill, H. H. (1970). *In* "Plutonium 1970 and Other Actinides" (W. N. Miner, ed.) p. 233. A.I.M.E., New York.

Kmetko, E. A., and Hill, H. H. (1976). *J. Phys. F: Metal Phys. 6*, 1025.

Lam, D. J. and Aldred, A. T. (1974). *In* "The Actinides: Electronic Structure and Related Properties" (A. J. Freeman and J. B. Darby, Jr., eds.), p. 109. Academic Press, New York.

Levgold, S., Burgardt, P., Beaudry, B. J., and Gschneidner, Jr., K. A. (1977). *Phys. Rev. B 16*, 2479.

Matthias, B. T., Zachariasen, W. H., Webb, G. W., and Englehardt, J. J. (1967). *Phys. Rev. Lett. 18*, 781.

Matthias, B. T., Fisk, Z., and Smith, J. L. (1979). *Phys. Lett. 72A*, 257.

Meaden, G. T., and Shigi, T. (1964). *Cryogenics 4*, 90.

Mortimer, M. J. (1979). *J. de Phys. 40*, C4-124.

Probst, C., and Wittig, J. (1978). *In* "Handbook on the Physics and Chemistry of Rare Earths" (K. A. Gschneidner, Jr. and L. Eyring, eds.), p. 749. North-Holland, Amsterdam.

Skriver, H. L., Andersen, O. K., and Johansson, B. (1978). *Phys. Rev. Lett. 41*, 42.

Smith, J. L., and Elliott, R. O. (1977). *In* "Proc. 2nd Int. Conf. on the Electronic Structure of the Actinides" (J. Mulak, W. Suski, and R. Troć, eds.), p. 257. Ossolineum, Wrocław.

Smith, J. L., and Haire, R. G. (1978). *Science 200*, 535.

Smith, J. L., Spirlet, J. C., and Müller, W. (1979). *Science 205*, 188.

Smith, J. L., Stewart, G. R., Huang, C. Y., and Haire, R. G. (1979). *J. de Phys. 40*, C4-138.

Stewart, G. R., Smith, J. L., Spirlet, J. C., and Müller, W. (1979). this conference.

Ward, J. W. (1979). private communication.

Ward, J. W., and Hill, H. H. (1976). *In* "Heavy Element Properties" (W. Müller and H. Blank, eds.), p. 65. North-Holland, Amsterdam.

Weling, F. (1977). *Sol. State Comm. 26*, 913.

SUPERCONDUCTIVITY IN IRRADIATED PALLADIUM

B. Stritzker[1]

Institut für Festkörperforschung
Kernforschungsanlage Jülich
Jülich, W. Germany

In pure Palladium films, evaporated between 4.2 K and 300 K, transitions into the superconducting state up to 3.2 K have been achieved by irradiation at low temperatures with He^+-ions. It is shown that the presence of a special kind of defect produced by the irradiation is a necessary precondition for the occurrence of superconductivity. It is argued that the introduced lattice defects, perhaps agglomerations of vacancies and of Pd atoms at interstitial sites, depress the strong spin fluctuations in crystalline Pd. This paramagnetic behavior is thought to be the reason for the absence of superconductivity in crystalline Pd.

I. INTRODUCTION

Palladium is one of the most strongly exchange enhanced materials. The occurrence of the strong spin fluctuation is thought to be the reason for the absence of superconductivity in Pd above 2 mK (Webb

[1]*Present address: ORNL, Solid State Division, Oak Ridge, Tennessee 37830, USA.*

et al., 1978). Theoretical calculations show that Pd
without spin fluctuations should be a superconductor
(Bennemann and Garland, 1973; Papaconstantopoulos
and Klein, 1975; Pinski et al., 1978). The calculated values for the corresponding superconducting
transition temperatures, T_c, vary between 0.3 K and
7 K.

In the present paper it will be shown that Pd
becomes a superconductor by irradiation with He^+-
ions at low temperatures. Furthermore the results
show that a special kind of disorder, introduced by
low temperature irradiation, is necessary for the
occurrence of supercondutivity in Pd. The maximum T_c
obtained is 3.2 K, in good agreement with the predictions for Pd without spin fluctuations. A possible
explanation will be given for the reduction of spin
fluctuations during low temperature irradiation. The
possibility of a "p-wave" pairing mechanism as proposed by Fay and Appel (1977) for Pd can be excluded
in these experiments due to the large number of lattice defects introduced by the irradiation.

II. EXPERIMENTAL

The irradiation experiments at low temperatures
were performed in a modified ^{3}He-^{4}He dilution refrigerator attached to the Jülich 400 kV ion accelerator. The resistance of the sample can be measured
after irradiation down to 0.1 K with a standard
four-point-probe technique. The temperature of the
Pd itself remained below 8 K during the maximum flux
of the He^+-beam. The thickness of the Pd films was
smaller than 1000 Å in order to ensure that nearly
all He^+-ions with an energy of 130 keV pass through
the Pd. Only less than $4 \cdot 10^{-4}$ of the He^+ stick inside the Pd.

Great care was taken to ensure the purity of the
evaporated Pd films. The starting high purity material (5 N Pd from Johnson-Matthey) was evaporated by
different techniques and under various conditions
(Stritzker, 1979). Independent of the special details during evaporation, all investigated Pd films
showed common properties: The as-condensed films do
not become superconducting above 0.2 K, but they become superconducting after He^+-irradiation at low
temperatures. This result leads to the conclusion
that superconductivity is really a property of the
irradiated Pd itself and is not induced by impurities.

III. RESULTS

The maximum transition temperature of 3.2 K can be achieved by He^+-irradiation at low temperatures in Pd films condensed at 4.2 K and immediately irradiated, as well as in films condensed at 300 K and irradiated after cooling down to 4.2 K. However, in the quench condensed case, the maximum T_c value is achieved at a lower He dose (10^{16} cm^{-2}) compared to the sample condensed at room temperature ($2.4 \cdot 10^{16}$ cm^{-2}). Obviously, the superconducting state can be formed more easily by irradiation if the starting material has already a high degree of lattice disorder. This conclusion is in qualitative agreement with the fact that after annealing to 500 K no superconductivity above 0.2 K could be obtained by low temperature irradiation up to a He^+ dose of $5.3 \cdot 10^{16}$ cm^{-2}.

The following experiment was performed to determine whether a special type of lattice disorder as produced by low temperature irradiation is necessary for the occurrence of superconductivity. Different degrees and kinds of lattice disorder were produced as follows. A 400 Å thick Pd-film was quench condensed at 4.2 K. The resulting high degree of lattice disorder was reduced by annealing stepwise to room temperature. The resulting resistivity ρ as a function of temperature is given by the solid curves in Fig. 1. Annealing of the quench condensed Pd leads to an irreversible decrease of resistivity. At various annealing temperatures the Pd is cooled down again to 1.0 K. This leads to a reversible decrease of resistivity with a constant slope above 40 K. Thus the annealing steps end up in a more reduced degree of lattice disorder. In all cases no superconductivity in this non-irradiated but disordered sample could be detected above 1 K. This result demonstrates that Pd cannot be made superconducting by quench condensation.

In contrast, all the Pd films become superconducting after irradiation with He^+ at low temperatures independent of the annealing temperature varying between 4.2 and 300 K. Obviously, only defects introduced by irradiation lead to superconductivity. These defects are essentially different from those achieved by quench condensation. This can be deduced from the different behavior of the defects. The dashed curve in Fig. 1 was obtained for a Pd-film condensed at 4.2 K after subsequent irradi-

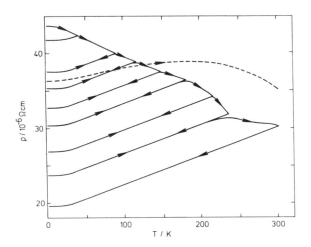

*FIGURE 1. Resistivity ρ of Pd films condensed
at 4.2 K versus annealing temperatures. Solid curve:
as condensed Pd. Dashed curve: after low temperature
irradiation into the superconducting state.*

ation until T_C had reached its maximum value. As can
be seen the temperature dependence of its resistivity is completely different from that of the as-condensed Pd.

IV. DISCUSSION

In the following we will consider first what
type of defects could be consistent with the observed dependence on dose and annealing history of
the sample and second how irradiation induced defects might produce superconductivity. In contrast
to the evaporation, the irradiation at low temperatures can produce vacancies and Pd atoms located at
interstitial sites (Frenkel pairs). We want to propose that these defects are responsible for superconductivity.

The occurrence of interstitial atoms should
widen the lattice. In fact, we observed an increase
of the lattice constant by about 0.4% at 4.2 K in
the irradiated superconducting state of Pd. This
enormous effect (corresponding to about 2% interstitials) is a hint that not only are Frenkel pairs
produced but perhaps also agglomerations of interstitials or vacancies. Such agglomerations could be
achieved due to rather high local temperatures

during the irradiation. These rather stable defects
could explain the fact that the superconducting
phase is rather insensitive to annealing (T_C remains
unchanged after annealing up to 70 K) but very sensitive to the ion flux.

This sensitivity is illustrated in Fig. 2. The
maximum transition temperature, T_{cmax}, (bars, lower
part) as well as the ion dose, $\Phi_{max} \cdot t$, (+, upper
part) necessary to obtain T_{cmax} are presented versus
the ion flux, Φ. The results were obtained for a
400 Å thick Pd-film evaporated at room temperature.
Following each irradiation at 4 K the sample was
annealed at 300 K prior to low temperature irradiation using a different ion flux. The highest value
of T_{cmax}, 3.2 K, was observed by irradiation with
the lowest ion flux, $\Phi = 2.5 \cdot 10^{11}$ cm^{-2}sec^{-1}, at the
lowest dose $\Phi_{max} \cdot t = 9 \cdot 10^{15}$ cm^{-2}. In contrast the T_c
value indicated by the dashed bar was achieved by

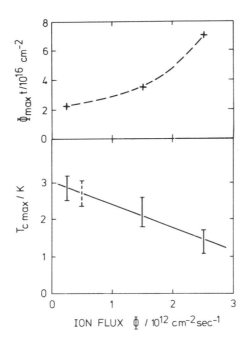

*FIGURE 2. Dependence of the maximum achievable
transition temperature, T_{cmax} , and the corresponding
dose, $\Phi_{max} \cdot t$, on the ion flux Φ. The bars indicate
the width of the superconducting transition.*

reducing Φ from $1.5 \cdot 10^{12}$ cm^{-2} to $0.5 \cdot 10^{12}$ cm^{-2} without intermediate annealing at 300 K. These somewhat puzzling results have not yet been unambiguously explained.

The results of this paper lead to the conclusion that the occurrence of lattice defects introduced by irradiation is necessary for the superconductivity in Pd.[1] In the following a model is proposed explaining the special properties of these lattice defects. From band structure calculations it is known that the Fermi level of crystalline Pd is located in a narrow peak (width 0.1 eV) of the density of states (Anderson 1970; Papaconstantopoulos et al., 1978). The introduction of interstitial Pd atoms changes drastically the symmetry of the nearest surrounding of Pd-atoms. This would lead to enormous consequences in the band structure. According to the special location of the Fermi level in crystalline Pd, it is reasonable to assume, that $N(o)$ would be substantially reduced by the introduced interstitials. In addition the disorder would lead to a smearing of the band structure, and thus to a reduction of $N(O)$. There is experimental evidence for this effect in Ar-sputtered Pd (Hüfner et al., 1974). A reduction of $N(O)$ leads to a decrease of the Stoner enhancement factor. $1/\left[1 - UN(O)\right]$ with exchange parameter U. As a result the strong spin fluctuation would be diminished, and superconductivity might be possible. This assumption does not take into account a possible contribution from enhanced electron-phonon coupling due to weakened phonon modes in disordered Pd or optic phonon modes of interstitial Pd atoms.

---

[1] *In addition preliminary measurements of as-condensed Pd films under uniaxial stress (increase of length 0.13 %) showed no superconductivity above 1.1 K. This indicates that lattice widening alone cannot explain superconductivity.*

ACKNOWLEDGMENTS

Many helpful discussions with J.D. Meyer, F. Pobell
and D. Rainer as well as their intense support
during the progress of this work are gratefully
acknowledged. I would like to thank J. Appel,
W. Buckel, R.W. McCallum, W. Schilling and H. Winter
for stimulating discussions, S. Mesters, K.H. Klatt
and F. Römer for experimental help and W.H.-G.
Müller for performing the stress experiments.

REFERENCES

Andersen, O.K. (1970). *Phys.Rev.B* *2*, 883.
Bennemann, K.H., and Garland, J.W. (1973). *Z.Phys.*
    260, 367.
Fay, D., and Appel, J. (1977). *Phys.Rev.B* *16*, 2325.
Hüfner, S., Wertheim, G.K., and Buchanan, D.N.E.
    (1974). *Chem.Phys.Lett.* *24*, 527.
Papacontantopoulos, D.A., and Klein, B.M., (1975).
    Phys.Rev.Lett. *35*, 110.
Papacontantopoulos, D.A., Klein, B.M., Economou, E.N.
    and Boyer, L.L., (1978). *Phys.Rev.B* *17*, 141.
Pinski, F.J., Allen, P.B., and Butler, W.H. (1978).
    Phys.Rev.Lett. *41*, 431.
Stritzker, B. (1979) to be published in *Phys.Rev.Lett.*
Webb, R.A., Ketterson, J.B., Halperin, W.P.,
    Vuillemin, J.J. and Sandesara, B.B. (1978).
    J. Low Temp.Phys. *32*, 659.

CHARACTERISTICS OF GETTER SPUTTERED THIN FILMS OF Nb$_3$Ge AND MULTILAYERED FILMS OF Nb$_3$Ge/Nb$_3$Ir

P. H. Schmidt
J. M. Vandenberg
R. Hamm
J. M. Rowell

Bell Laboratories
Murray Hill, NJ

The stabilization of A15 Nb$_3$Ge in its high T$_c$ form is of fundamental interest and also technologically important. One factor, the influence of gaseous impurities, has been discussed for some time by various authors. A second, epitaxial growth, was studied on Nb$_3$Ir (Dayem *et al.*, 1978) and an improvement of T$_c$ and phase purification of the Nb$_3$Ge was observed, although T$_c$ did not exceed that of single sputtered films prepared by others. Gavaler *et al.* (Gavaler *et al.*, 1978) have reported that impurities near the substrate-film interface expand the A15 cell and suggest that subsequent deposition results in a homoepitaxial growth with cell size decreasing with increasing thickness. This is consistent with the observed increase in T$_c$ with increasing film thickness (Kudo and Tarutani, 1977). We report the results of similar experiments using the getter sputtering method of film preparation. Of particular interest was examination of (1) the structural and phase relationships of Nb$_3$Ge films prepared with and without an A15 underlayer, (2) how variation of film thickness and deposition temperature affected these properties and T$_c$, and (3) the potential application and usefulness of multilayered films for the stabilization of A15 Nb$_3$Si.

The getter sputter deposition apparatus used previously (Schmidt *et al.*, 1975) for film deposition was modified to permit deposition from two independent targets without breaking vacuum. Substrates were outgassed in vacuum ($\sim 2 \times 10^{-8}$ Torr) to 1100°C prior to a presputter sequence of 1 hour in

600 μm Ar (99.999% purity). Targets were synthesized by arc
melting. Substrates were random orientation single crystal
plates of Al_2O_3. Deposition temperatures (T_D), defined as the
temperature of the Pt heater platform, were monitored continuously with a Pt/Pt 10% Rh thermocouple attached to the under-
side of the heater.

X-ray analysis of the films used a Huber thin film camera
and bulk Pb to calibrate all photographs. Estimates of
second phase components were made by visual comparison of line
intensities and diffuse bands from diffraction photographs at
$2\theta \sim 30°-45°$.

SINGLE LAYER FILMS

The results of Fig. 1 show that T_c increases with increasing film thickness from 4.2 K [midpoint] at 500 Å to 22.2 K at
4000 Å at T_D = 980°C. Resistance ratios for 500 Å films are
< 1, indicative of disorder, while 4000 Å films had a ratio of
2.4. However, when a 500 Å film was deposited at 980°C, and
then maintained at 980°C for the remainder of the time
normally required to deposit 4000 Å (1 hour), T_c was 16.5 K.

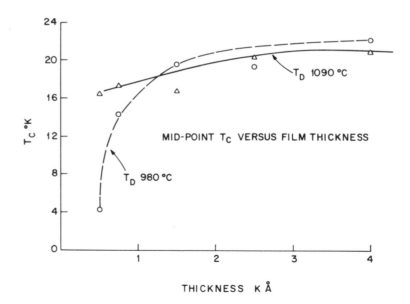

FIGURE 1. *Nb$_3$Ge midpoint T_C versus film thickness in*
K Å. Circles identify samples prepared at T_D = 980°C;
triangles are used for T_D = 1090°C specimens.

At 1090°C the thinner films have improved T_c (16.5 K at 500 Å) but the thick films are somewhat poorer (19.9 K at 4000 Å). Resistance ratios increased from 1.2 to 2.24 from 500 Å to 4000 Å.

X-ray analysis of the above films indicates (Fig. 2) a decrease of the A15 Nb₃Ge cell size with increasing film thickness, in general agreement with the observations of Gavaler et al. (Gavaler et al., 1978). We do not, however, observe an A15 cell size as large as they report ∿ 5.25 Å. (In our experience, such a large cell is only observed when films are deliberately heavily doped with hydrogen (Lanford et al., 1978).) At T_D = 980°C the a_o of our A15 Nb₃Ge decreases from 5.160±.003 Å at 750 Å to 5.149 Å at 4000 Å. At T_D = 1090°C a_o decreases from 5.157 Å at 500 Å to 5.146 Å at 4000 Å film thickness.

The degree of crystallinity (or lack of amorphous component), the relative amounts of A15 Nb₃Ge and hexagonal Nb₅Ge₃ phases present versus film thickness and deposition temperature is pictorially indicated in Fig. 3. For T_D = 980°C, films < 500 Å are amorphous and yield no diffraction lines while in thicker films X-ray lines become sharper and at > 2500 Å thickness there is no remaining amorphous component observable. The hexagonal Nb₅Ge₃ phase was detected in most samples of > 750 Å thickness in amounts of 5-10%. For T_D = 1090°C the amorphous component is absent for

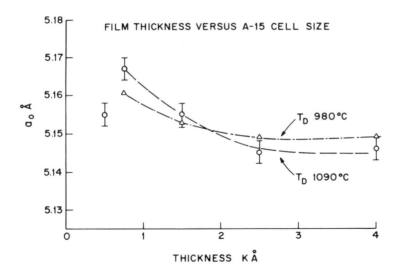

FIGURE 2. *A15 Nb₃Ge cell size is shown versus film thickness in K Å; triangles identify samples prepared at T_D = 980°C and circles show samples prepared at T_D = 1090°C.*

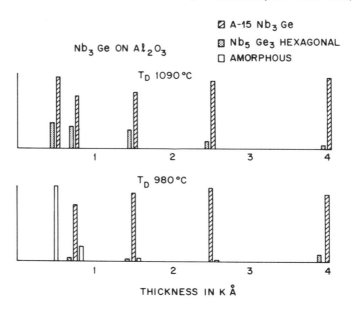

FIGURE 3. *Phase composition of single layer films prepared at* T_D = 980°C *and* 1090°C *is shown versus thickness in K* $\overset{\circ}{A}$.

all film thicknesses investigated (i.e. $\geq$ 500 $\overset{\circ}{A}$). The hexagonal Nb_5Ge_3 phase is observed in all samples with 35% being present at 500 $\overset{\circ}{A}$ and decreasing to $\sim$ 5% at 4000 $\overset{\circ}{A}$ thickness. Thus the hexagonal Nb_5Ge_3 phase is always present in thick samples at T_D of 980°C or 1090°C, i.e. in all the high T_C samples.

Summarizing our results on single layer films, the limited decrease in cell size with thickness is in agreement with the suggestion of homoepitaxy made by Gavaler *et al.* However, at T_D = 980°C it appears that the first 500 $\overset{\circ}{A}$ of film are amorphous (rather than crystalline with a large cell), whereas at 1090°C there is a large amount of Nb_5Ge_3 for the same thickness. There are two other factors worth more detailed study. First is the annealing effect, namely that thicker films generally stay at the deposition temperature for a longer time. Thus the first 500 $\overset{\circ}{A}$ of a 4000 $\overset{\circ}{A}$ film (deposition time 1 hour) is probably much more ordered than a 500 $\overset{\circ}{A}$ film deposited in 12 minutes. Second, it is known that deposition temperature is a critical parameter, but the temperature of the growing surface of the film cannot be measured, and may well change as the film grows from very thin to optically thick, even though the heater below is held at constant temperature. Thus the data of Fig. 1 may simply

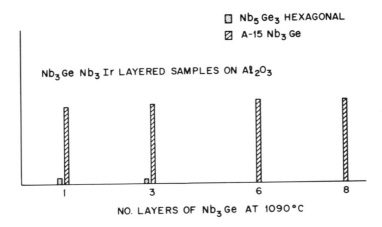

FIGURE 4. *Phase composition of multilayered samples using Nb₃Ge/Nb₃Ir targets, T_D = 1090°C.*

reflect an approach to temperature equilibrium of the film-vacuum interface.

MULTILAYER FILMS

To determine the influence of a similar cell sized A15 material on the structural phase relationships of A15 Nb₃Ge, multilayered films were prepared with a 2000 Å base layer of Nb₃Ir deposited on the Al₂O₃ substrate, followed by alternating layers (750 Å/layer) of Nb₃Ge and Nb₃Ir. The final layer deposited was always Nb₃Ge. The structural and phase relationships of these films are shown in Fig. 4 for T_D = 1090°C. Suppression of the hexagonal Nb₅Ge₃ phase is observed for thick (> 3 layers) multilayered samples. Single phase A15 Nb₃Ge diffraction patterns only were obtained from six and eight layered specimens. The long deposition times of thick multilayered samples may again have permitted annealing and grain growth of the A15 Nb₃Ge phase at the expense of competing phases. Other samples prepared with 1500 Å/layer were structurally identical with our 750 Å/layer specimens. From Fig. 5 we see that T_c is nearly independent of the total film thickness or the number of Nb₃Ge layers deposited. Unfortunately, T_c never exceeds values that can be obtained from single layer specimens. Unlike the results of Fig. 2 for single films the A15 Nb₃Ge cell size (a_0) is also nearly independent of total film thickness or the number of Nb₃Ge layers and is small, 5.136±.003 Å. This is similar

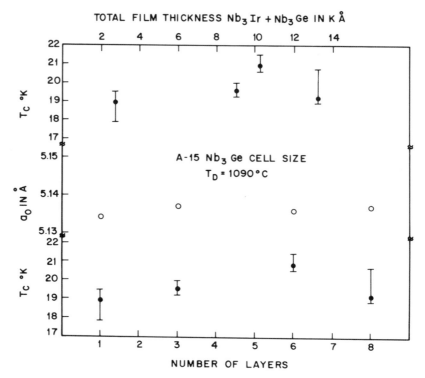

FIGURE 5. T_C *is shown versus total film thickness for multilayered Nb_3Ge and Nb_3Ir samples on the upper third of the figure. The lower portion shows the A15 Nb_3Ge cell size and T_C plotted versus the number of Nb_3Ge layers ($T_D = 1090°C$).*

to the cell size of Nb_3Ir while the lowest value obtained from single films is $a_O = 5.146$ Å. All multilayered films and single layer films showed no evidence of preferred grain growth. Rutherford backscattering analysis of a multilayered Nb_3Ge/Nb_3Ir sample indicated discrete layers, but interdiffusion on the scale of 200 Å cannot be ruled out.

$Nb_{.75}Si_{.25}/Nb_3Ir$ and $Nb_{.8}Si_{.2}/Nb_3Ir$ multilayered samples (3 layers Nb_xSi) of 750 Å/layer were deposited at $T_D = 1090°C$ and 1190°C. Typical results from X-ray analysis yielded A15 Nb_3Ir and tetragonal Nb_3Si; $a_O = 10.21$ Å, $c_O = 5.19$ Å. A15 Nb_3Si diffraction lines were not visible. Maximum transition temperatures (onsets) of 7.6°K were obtained at 1090°C while somewhat lower T_C's (5.9°K) were obtained at high temperatures (1190°C). Resistance ratios of ~ 1.3 were typical and never exceeded the ratio obtained for Nb_3Ir films alone.

Similar resistance ratios and T_C's were obtained without multilayering for films of ~ 4000 Å thickness on Al_2O_3. T_C's of films prepared from $Nb_{.75}Si_{.25}$ targets were $\sim$ 1°$\bar{K}$ higher than those films prepared from our $Nb_{.8}Si_{.2}$ target.

Use of A15 Ti_3Au as a substitute for A15 Nb_3Ir has not improved observed critical temperatures for multilayer samples of Nb_xSi. We found (A15) Ti_3Au more difficult to prepare than (A15) Nb_3Ir. We often obtained a mixed cubic phase Ti_xAu film with a_o = 4.13 Å and 6.94 Å.

In terms of an improvement in T_C, the results on multi-layer samples are disappointing except this does appear to be a way to produce thin layers of Nb_3Ge with relatively high T_C. Epitaxial growth of multilayered films was not observed as we had hoped--rather lattice matching.

REFERENCES

Dayem, A. H., Geballe, T. H., Zubeck, R. B., Hallak, A. B.,
    and Hull, G. W. (1978). *J. Phys. Chem. Solids 39*, 529.
Gavaler, J. R., Ashkin, M., Braginski, A. I., and Santhanam,
    A. T. (1978). *Appl. Phys. Lett. 33*, 359.
Kudo, M. and Tarutani, Y. (1977). *IEEE Trans. Mag. MAG13*, 331.
Lanford, W. A., Schmidt, P. H., Rowell, J. M., Poate, J. M.,
    Dynes, R. C., and Dernier, P. D. (1978). *Appl. Phys. Lett.
    32*, 339.
Schmidt, P. H., Bacon, D. D., Barz, H., and Cooper, A. S.
    (1975). *J. Appl. Phys. 46*, 2237.

LOW TEMPERATURE SPECIFIC HEAT
OF PROTACTINIUM

G. R. Stewart, J. L. Smith[1]

Los Alamos Scientific Laboratory
Los Alamos, New Mexico, USA

J. C. Spirlet, W. Müller

Commission of the European Communities
Joint Research Center
European Institute for Transuranium Elements
Karlsruhe, Federal Republic of Germany

The specific heat of a single crystal of protactinium has been measured between 4.9 and 18 K. The coefficient of the electronic term in the specific heat γ is found to be 5.0 ± 0.5 mJ/mole-K^2, agreeing with the trend in the neighboring elements thorium ($\gamma = 4$) and uranium ($\gamma = 9 - 12$). The question of whether protactinium is a type II superconductor remains unanswered.

I. INTRODUCTION

The lighter actinide elements from thorium on have been subject of renewed interest in recent years, due in part to improved purities and techniques. The present work combined relatively new small sample calorimetry techniques (Bachmann *et al.*, 1972; Stewart and Giorgi, 1978) and the recent production of protactinium of the highest purity to date (Bohet and Muller, 1978). In fact, small sample calorimetry techniques, which eliminate much of the self-heating problem and

---

[1]*Work performed under the auspices of the Department of Energy.*

allow measurement to lower temperatures, have only been applied
to highly radioactive actinides once (americium, Smith, *et al.*,
1979). The present work on protactinium was motivated by the
work by Smith, Spirlet, and Müller (1979) on the superconducting properties of this very high purity protactinium; by the
availability of an 11 mg single crystal of this protactinium;
and by recently published specific heat work on protactinium
(Mortimer, 1979) which yielded a questionable, rather small
value for γ, the coefficient of the electronic term in the
specific heat.

II. EXPERIMENTAL

The protactinium metal was prepared by a Van Arkel process
(Bohet and Müller, 1978) by thermal dissociation of protactinium iodide on a radiofrequency-heated tungsten sphere (Spirlet,
1979). Pelletized protactinium carbide and iodine were sealed
under a pressure of 10^{-6} Torr in the quartz bulb enclosing the
tungsten sphere. During the Van Arkel deposition the bulb was
maintained at 550°C, whereas the metal sphere was heated to
1225°C. The rate of protactinium deposition was 90 mg/hour.
The crystal of protactinium for the present work was selected
from the resultant deposit and sealed under vacuum in pyrex
until use.

The small sample calorimeter (Stewart and Giorgi, 1978) was
used in the earlier work on americium (Smith, *et al.*, 1979).
Important to the accuracy of γ, where

$$C = \gamma T + \beta T^3 \tag{1}$$

is the thermal conductivity of the four 0.08-mm Au-7% Cu alloy
wires that support the sample platform (Stewart and Giorgi,
1978). This conductivity, K, determines the self-heating
temperature rise, $\Delta T = \dot{Q}/K$, where $\dot{Q}$ is the heat produced by
the sample. This ΔT is then added to the helium bath temperature (1.4 K in the present work) to obtain the lowest temperature of measurement (4.9 K). This limit on the lowest temperature affects the accuracy of γ due to the large actinide lattice specific heats, βT^3, or alternately the low Debye temperature, θ_D, where

$$\beta = \frac{1944}{\theta_D^3} . \tag{2}$$

In general, eq. 1 is valid for temperatures below $\theta_D/50$ (sometimes as high as $\theta_D/25$). Above this temperature, additional

terms involving higher powers of T become important. These higher terms, coupled with the relative smallness of the electronic term γT, make determination of γ difficult.

From γ, one determines the electronic density of states, $N(0)$

$$\gamma = (1/3)k^2\pi^2 N(0)(1 + \lambda) \tag{3}$$

where λ, the electron phonon coupling constant, can be determined for superconductors via (McMillan, 1968)

$$\lambda = \frac{1.04 + \mu^* \ln (\theta_D/1.45\ T_c)}{(1 - .62\ \mu^*)\ln(\theta_D/1.45\ T_c) - 1.04} \quad . \tag{4}$$

The transition temperature for protactinium, T_c, is 0.43 K (Smith, Spirlet, and Müller, 1979).

III. SPECIFIC HEAT DATA AND DISCUSSION

The specific heat data between 4.9 and 18 K are presented in Figures 1 and 2 as C/T versus T^2, so that the intercept is γ and the slope is β, from eq. 1. As may be seen in Figure 1, above about 7 K an extrapolation of the data to $T = 0$ becomes very difficult, and in fact would give a spurious γ near zero. In the specific heat results on protactinium between 6.2 and 18.5 K reported upon by Mortimer (1979), their value for γ of protactinium is in fact low, 0.7 ± 1.3 mJ/mole-K^2. However, the lower temperature data of the present work shown in Figure 2 demonstrate conclusively that $\gamma = 5.0\pm0.5$ mJ/mol-K^2. This γ is in fact predictable based on the γ's for the adjoining actinide elements, where $\gamma = 4.08\pm0.03$ for thorium (Luengo, et al., 1972) and $\gamma = 9.1$ to 12.2 for uranium (see the review by Mortimer, 1979).

The Debye temperature found in the present work, $\theta_D = 185 \pm5$ K, agrees well with the value reported by Mortimer (1979), which was $\theta_D = 180$ K.

Using a μ^* of 0.1 in eq. 4 above, λ is calculated to be 0.37. From eq. 3, using this λ and $\gamma = 5.0$ mJ/mole-K^2, $N(0)$ is found to be 1.54 states/ev-atom. This is a relatively high density of states for so low a T_c; however, this merely implies that a great deal of the density of states, probably due to the f electrons, does not contribute to the superconductivity.

The thermodynamic critical field, $H_c(0)$ may also be calculated from γ, using a BCS expression:

$$H_c(0) = 2.24\ \gamma^{1/2}\ T_c \tag{5}$$

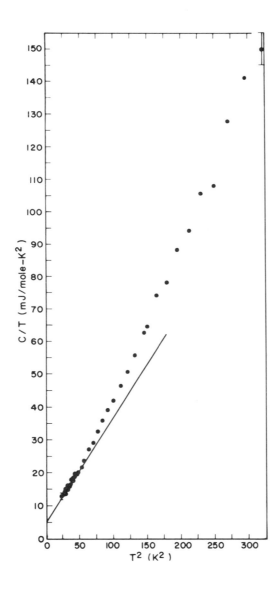

FIGURE 1. Heat capacity data for protactinium.

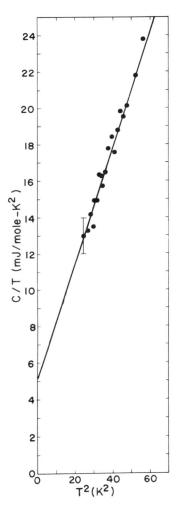

FIGURE 2. Lowest temperature heat capacity for protactinium.

where γ is in ergs/cm³. This gives $H_c(0)$ = 55 Oe. Smith,
Spirlet, and Müller (1979) report an H_{c2} of 56 Oe so that
κ is right at the borderline value of $1/\sqrt{2}$, leaving it unclear
if protactinium is type II or not. Resolution of this question must wait on further measurements.

REFERENCES

Bachmann, R., Disalvo, F. J., Geballe, T. H., Greene, R. L.,
    Howard, R. E., King, C. N., Kirsch, H. C., Lee, K. N.,
    Schwall, R. E., Thomas, H.-U., and Zubeck, R. B. (1972).
    Rev. Sci. Inst. 43, 205.
Bohet, J., and Müller, W. (1978). *J. Less Common Metals 57*,
    185.
Luengo, C. A., Cotignola, J. M., Serini, J. G., Sweedler, A. R.,
    Maple, M. B., and Huber, J. G. (1972). *Solid State Comm.
    10*, 459.
McMillan, W. L. (1968). *Phys. Rev. 167*, 331.
Mortimer, M. J. (1979). *J. de Physique 40*, C4-124.
Smith, J. L., Stewart, G. R., Huang, C. Y., and Haire, R. G.
    (1979). *J. de Physique 40*, C4-138.
Smith, J. L., Spirlet, J. C., and Müller, W. (1979). To appear
    in *Science*.
Spirlet, J. C. (1979). *J. de Physique 40*, C4-89.
Stewart, G. R., and Giorgi, A. L. (1978). *Phys. Rev. B17*, 3534.

THE ADDITIVE PRESSURE DEPENDENCE OF THE SUPERCONDUCTING TRANSITION TEMPERATURE OF Th–Y ALLOYS

John G. Huber[1]

Department of Physics
Tufts University
Medford, Massachusetts

The T_c at zero pressure of solid solution Th-Y alloys increases with Y concentration from the pure Th value to a maximum at 20 a/o Y and then decreases until disappearing at 70 a/o Y. The application of pressure P depresses the T_c of pure Th (negative dT_c/dP), while for pure Y a T_c eventually appears with a dT_c/dP which is positive. For an alloy $Th_{1-X}Y_X$, dT_c/dP at any given pressure seems to vary linearly with composition following the relation

$$\frac{dT_c}{dP} (Th_{1-X}Y_X) = (1-X) \frac{dT_c}{dP} (Th) + X \frac{dT_c}{dP} (Y)$$

independent of T_c itself or even alloy structure (f.c.c. or h.c.p.).

I.   INTRODUCTION

The pressure dependence of the superconducting transition temperature (T_c) of the early transition elements has been a subject of some interest. Finally, Wittig and co-workers (1979) have found even Sc to superconduct. So now the superconductivity of the 3-valent s^2d^1 family of Sc, Y and Lu is confirmed; each

[1]*Formerly supported at the University of California at San Diego by the Air Force Office of Scientific Research; presently supported by the National Science Foundation under grant No. DMR 78-18066.*

eventually superconducts at very high pressures with a positive pressure dependence for T_c. La, the other 3-valent early
transition metal, starts high and soars under pressure to the
highest T_c of any pure element. In the 4-valent s^2d^2 column,
Ti, Zr and Hf - adjacent, respectively, to Sc, Y and Lu in the
periodic table - all superconduct at zero pressure with relatively low T_cs but again with positive pressure dependences.
There remain only 4-valent Ce and Th which exhibit negative
pressure dependences; Ce when it superconducts above about 50
kbar and Th from its moderate zero pressure T_c ($\sim$1.36 $^{\circ}$K). It
has been argued that the d character of Sc, Y and Lu and also
of Ti, Zr and Hf is responsible for their superconducting be-
haviors; whereas an f character, however slight, may explain
what is seen for La, Ce and Th.

   Th, of all the aforementioned elements, has the best metallurgical properties as well as a reproducible T_c; plus, it
alloys extensively with most of the others. We have thus decided to study the low pressure dependence of T_c of Th-X
alloys, and perhaps illuminate the d vs. f character question.
We report here our work with Th-Y alloys.

II. EXPERIMENTAL DETAILS

   All samples were arc-melted alloys of 99.95% pure iodide
bar Thorium and 99.9% pure distilled Yttrium. Weight losses
were minimal, so the quoted compositions are nominal. Most
samples, spheroids approximately 2 mm. in diameter, were measured for superconductivity as cast. Several, however, were
annealed one week at 850 $^{\circ}$C before measurement. Debye-Scherrer
photographs were taken of unannealed filings of selected samples using Cu K radiation. Although considerable cold-work
stress was evidenced by broad and fuzzy high angle lines, the
low angle lines were sufficiently clear to distinguish patterns due to f.c.c. and/or h.c.p. phases.

   All measurements of T_c were made inductively; i.e., the
low field (<0.1 gauss), low frequency (<100 Hz.) a.c. suscep-
tibilities of the samples were monitored via concentric primary and split, counter-wound secondary coils and a mutual inductance bridge - abrupt signal changes were seen with each
advent of superconductivity. To determine T_cs as a function
of hydrostatic pressure, samples were pressurized in beryl-
lium-copper clamps with tungsten-carbide pistons, enclosed
within teflon capsules filled with a 50:50 mixture of n-
pentane and isamyl alcohol. Pressures were inferred from the
T_c of a Sn "manometer" or from that of Th which had been measured against Sn; they are estimated to be accurate to within
0.5 kbar. These measurements were made in He[4], He[3], adiabatic
demagnetization and dilution refrigerator cryostats at both
the University of California at San Diego and Tufts University.

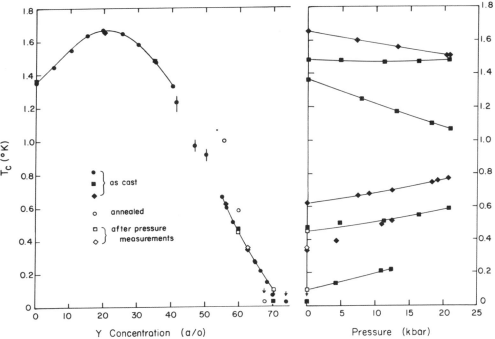

FIGURE 1.   *The superconducting transition temperature* T_c
vs. Y concentration and pressure for Th-Y alloys.

Thermometry was too varied to discuss, but temperatures are
felt to be known within 10 m^{o}K.

III.   RESULTS AND DISCUSSION

        Illustrated in Figure 1 are our T_c measurements on Th–Y
alloys. On the left we show T_c vs. Y concentration at zero
pressure; on the right, T_c vs. pressure for alloys of 0, 20.0,
35.0, 55.9, 59.8, 62.5 and 70.3 atomic percent (a/o) Y concentration. Data for these latter samples have been plotted as
either squares or diamonds, and an open such symbol desig-
nates the zero pressure T_c after the pressure measurements
were made. All closed symbols are for data taken on samples
as cast - arc-melted and unannealed. The open circles show
measurements on annealed samples. Arrows at the bottom of the
figure indicate the temperatures to which measurements were
made with no transition seen for the samples characterized by
the symbols below the arrows. Almost all transitions were on
the order of 10 m^{o}K broad; for those few with a width exceeding the dimension of a symbol, this width is shown by an
"error bar."

The zero pressure T_c results reported here are essentially those of Huber and Maple (1974) with additional data beyond 40 a/o Y. Out to this composition we are confident that the alloys are pure f.c.c.; the X-ray of filings of a 35 a/o Y sample showed only f.c.c. lines, and we confirm the T_c measurements of Gschneidner and Matthias (1961) and Satoh and Kumagai (1973) in this region. After 40 a/o Y we observe a broadening of the T_cs of as-cast samples, and we feel these alloys are mixtures of f.c.c. and h.c.p. phases. There is disagreement in the literature as to the mixed phase portion of the equilibrium Th-Y phase diagram. Eash and Carlson (1959) report it to lie between 38 and 55 a/o Y; Evans and Raynor (1960), between 49 and 69 a/o Y. Undoubtedly, the problem results from the identical size of metallic Th and Y ions. Subtle factors such as impurities must influence which of the close-packed phases might dominate in the center of the phase diagram. Evans and Raynor pointed out that cold-working of Th-Y alloys with Y contents up to 75 a/o can generate the f.c.c. phase. Thus, our X-rays of filings of 56.2 and 59.8 a/o Y samples which show the presence of both f.c.c. and h.c.p. phases mean little. T_cs sharpen again at 55 a/o Y, and we simply believe this to be indicative of the appearance of pure h.c.p. alloys in the as-cast state. The sharp T_cs of annealed samples with Y contents greater than 55 a/o are thought due to f.c.c. portions of the samples with Y compositions less than nominal.

So far we have performed pressure measurements only on as-cast samples which we felt to be pure f.c.c. or pure h.c.p. Most of the h.c.p. samples gave slightly different zero pressure T_cs after being held for varying lengths of time at the pressures under which they were measured (see Figure 1). Although we are uncertain what pressure annealing does, we have accepted the post-pressure T_cs in drawing the T_c vs. pressure curves which are seen in the right-hand side of Figure 1. We have not drawn a curve for the 62.5 a/o Y sample because its behavior seems discontinuous and just doesn't make sense. The slopes (dT_c/dP) which we have taken at 0 and 20 kbar (where possible) from each T_c vs. pressure curve are plotted vs. Y concentration of the sample in Figure 2.

We see in Figure 2 that the 0 and 20 kbar data points invite linear fits, which we have drawn. There exists, then, an apparent empirical relation; dT_c/dP for a single-phase Th-Y alloy at a given pressure equals the sum of the pressure derivatives of T_c for Th and Y at that pressure, scaled by the concentration of each element in the alloy:

$$\frac{dT_C}{dP} \; (\mathrm{Th}_{1-x}\mathrm{Y}_x) \; = \; (1-x) \; \frac{dT_C}{dP} \; (\mathrm{Th}) \; + \; x \; \frac{dT_C}{dP} \; (\mathrm{Y})$$

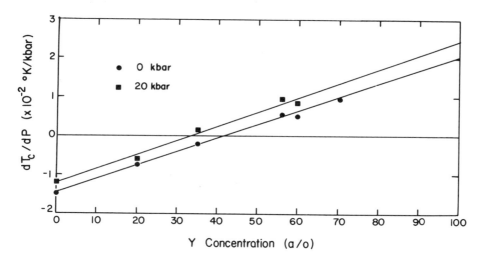

FIGURE 2.   The pressure derivative of the superconducting temperature dT_c/dP vs. Y concentration for Th-Y alloys at 0 and 20 kbar.

Thus, the pressure dependence of the superconducting transition temperature of Th–Y alloys is additive. (Y, of course, does not superconduct at low pressures, but pressure derivatives may be assumed.) From our linear fits in Figure 2 we find that at 0 and 20 kbar the difference between dT_c/dP for Th and that extrapolated for Y is about 3.6×10^{-2} °K/kbar. We note that even at pressures in excess of 120 kbar where Y superconducts with a dramatically increasing T_c (Wittig, 1970) and the T_c of Th is decreasing very gradually (Fertig et al., 1972), the difference between the pressure derivatives of T_c for the two elements is approximately preserved. Finally, we add that for Th–Sc and for Th–Lu alloys in the region where pure f.c.c. samples can be obtained (0 – 35 a/o Sc or Lu concentration), the dependence of the zero pressure T_c s on alloy composition and the pressure dependence of T_c at a given composition are very similar to what we have reported for Th–Y alloys in that region (Huber, 1971).

At this time we offer no interpretation for our observations.

ACKNOWLEDGMENT

We wish to thank John Bulman for making dilution refrigerator measurements.

REFERENCES

Eash, D. T., and Carlson, O. N. (1959). *Trans. Am. Soc. Metals*
    52, 1097.
Evans, D. S., and Raynor, G. V. (1960). *J. Nucl. Mater. 2*,
    209.
Fertig, W. A., Moodenbaugh, A. R., and Maple, M. B. (1972).
    Phys. Letters 38A, 517.
Gschneidner, Jr., K. A., and Matthias, B. T. (1961).   In "Rare
    Earth Research III" (E. V. Kleber, ed.), p.158.   The
    Macmillan Co., New York.
Huber, J. G. (1971). Ph. D. Thesis, University of California
    at San Diego.
Huber, J. G., and Maple, M. B. (1974).   In "Low Temperature
    Physics – LT 13, Vol. 2" (K. D. Timmerhaus, W. J.
    O'Sullivan, and E. F. Hammel, eds.), p. 579.   Plenum
    Publishing Corp., New York.
Satoh, T., and Kumagai, K. (1973).   *J. Phys. Soc. Japan 34*,
    391.
Wittig, J. (1970).   *Phys. Rev. Letters 24*, 812.
Wittig, J., Probst, C., Schmidt, F. A., and Gschneidner, Jr.,
    K. A. (1979).   *Phys. Rev. Letters 42*, 469.

ON THE ELECTRONIC STRUCTURE OF A15 COMPOUNDS

Warren E. Pickett

Department of Physics and Astronomy
Northwestern University
Evanston, Illinois

I. INTRODUCTION

For the past twenty-five years compounds with the A15 crystal structure have dominated the class of high temperature superconductors. The crystal structure of an A15 compound A_3B is cubic (space group O_h^3). However, the site symmetry (D_{2d}) of the A atoms is much lower than cubic, an unusual occurrence in cubic binary compounds. Variations on this theme have supplied the basis of many theoretical models of the anomalous temperature (T) dependence of normal state properties and the low temperature cubic $\leftrightarrow$ tetragonal structural transformations which accompany high values of T_c in A15 compounds (Testardi, 1973). In this paper results of self-consistent pseudopotential band structure calculations (Pickett et al., 1979; Ho et al., 1979) will be used to assess some important aspects of the unique and unusual behavior in A15 compounds: (i) The role of the B atom in determining the overall electronic structure will be shown to be important, (ii) the effect of the low site symmetry of the A atom (e.g. the "linear chains") on the charge density and potential will be assessed and (iii) the bonding will be shown to be metallic—covalent with no significant A-B charge transfer.

II.   LARGE-SCALE ELECTRONIC STRUCTURE

To identify the effects on the band structure due to the B atom, calculations were performed for two hypothetical "elemental" A15 compounds, Nb_3Nb and Nb_3^* (no B atom), as well as for the high T_c compounds Nb3Ge and Nb_3Al. Studying Nb_3^* reveals the band structure due to the Nb chains alone as well as allowing a determination of the distinct effects due to a B atom of sp character (Ge,Al) or d character (Nb). The density of states of each compound is shown in Figure 1.

The density of states (DOS) of a single transition metal chain is the well-known basis of the Labbé-Friedel (1966) model of A15 compounds. From Figure 1 it is clear that the DOS of Nb_3^* is not simply that arising from three degenerate Nb chains, indicating that interchain d-d interactions are important. The DOS is split into a bonding (d^+) complex with energy E≤1eV (measured relative to the Fermi energy E_F), containing states for 40 electrons, and an antibonding (d^-) complex with E≥2eV, separated by a "gap" (low DOS region). The principle effect on the DOS of adding an sp B atom is for the p bands to hybridize strongly with the d^+ complex (1) to produce a strongly delineated DOS in the d^+ region, and (2) to leave E_F in a region of large DOS just below the gap. Both of these effects may be important for high temperature superconductivity (see below).

The main effect of putting Nb on the B site is to create a more strongly delineated DOS in the d^- region. The d^+ region is almost unchanged except for effects due to the 33% increase in the number of bands in the same d band region. In particular, slight smoothing (0.2-0.3eV) of the DOS of Nb_3Nb in the d^+ region leaves a structureless DOS reminiscent of an amorphous system. This arises from the existence of (at least) three distinct d-d interactions of importance: nearest neighbor (intrachain), second neighbor (A-B), and the third neighbor (interchain), which was seen from the DOS of Nb_3^* to be important. These distances are in the ratio $\sqrt{4} : \sqrt{5} : \sqrt{6} = 1 : 1.12 : 1.23$. A chain Nb atom interacts with Nb neighbors at three distances, a situation which begins to approximate what occurs in an amorphous metal. Since the amorphous state is not favorable for high T_c, Nb_3Nb (if it can be formed) should not have high T_c.

[These observations are consistent with other band calculations on A15 compounds (Mattheiss, 1965; Mattheiss, 1975; van Kessel et al., 1978; Klein et al., 1978; Jarlborg, 1979). In fact, they are available in the approximate but realistic results of Mattheiss for V_3Ga, V_3Co and V_3^* reported in 1965(!) however, computational constraints at that time prohibited the calculation of the DOS in enough detail to show the effects clearly. With a single exception, Mattheiss' 1965 conclusions regarding the A15 electronic structure have been verified recently by more accurate calculations. The exception, his conclusion that the d bands "are not greatly affected by the absence of the gallium atoms" in V_3Ga, no doubt resulted from the breakdown of the muffin-tin approximation in V_3*.]

The high values of T_c occur for A15 compounds which have (1) sp B atoms=Al,Si,Ga,Ge,Sn, and (2) E_F lying below the gap (i.e. B≠ P,As,Sb). An important contribution to high T_c is the relatively large DOS at E_F which arises from this combination. However, in addition to the resulting large number of electrons which superconduct, these electrons are in states which have very small dispersion yet are not localized. Qualitatively, it appears that these very flat bands may result from relatively flat d bands being "pushed up" against the bonding-antibonding gap by hybridization with the B atom p states, but a better understanding of these flat bands is needed. The placement of E_F in a region of large DOS immediately below an abrupt d^+-d^- gap also occurs in high T_c members of the C15 structure and the ternary molybdenum chalcogenides (Jarlborg and Freeman, 1979), which suggests that this may be a general inducement to high T_c.

There is another indication, from $(Mo_{1-x}Tc_x)_3Tc$ with $x \simeq 0.5$, that a strongly delineated region of DOS contributes to high T_c. For this alloy, with $T_c \gtrsim 13K$ (Giorgi and Matthias, 1978), the rigid band model on the DOS of Nb_3Nb (Figure 1) puts E_F in the structured region of the DOS near 2eV. A test of this idea would be provided if A15 Nb_3Zr can be made and its T_c measured. Since E_F would lie in the unstructured "amorphous" region of the DOS, Nb_3Zr should be a low T_c material.

III. CHARGE DENSITY AND POTENTIAL

The crystal valence charge density ρ in Nb_3Ge and Nb_3Al, which has been discussed in detail by Ho et al. (1979), is characterized by a prominent covalent-like piling up of charge into a double-hump structure (Ho et al., 1978) between Nb atoms on each of the three chains. Comparison with the charge density ρ_0 due to overlapping spherical atoms reveals, however, that the dominant crystalline charge redistribution is the "flow" of charge from both A and B atoms into the interstitial region, i.e. metallic bonding.

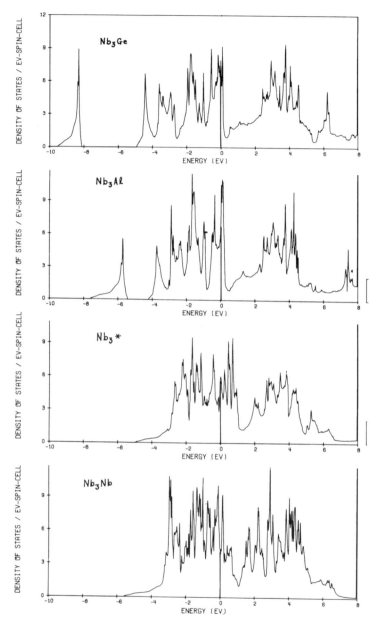

FIGURE 1.   The density of states of Nb₃Ge, Nb₃Al, Nb₃* and
Nb₃Nb, with Fermi levels (taken as the zero of energy) aligned.
Note that the Fermi level falls immediately below the bonding-
antibonding gap in Nb₃Ge and Nb₃Al, but not in Nb₃* and Nb₃Nb.

A contour plot of the difference $\rho-\rho_0$ of the valence
pseudocharge densities in Nb_3Ge is shown in Figure 2. This
difference plot gives the rearrangement in valence charge
when spherical atoms are placed on an A15 lattice, and then
the electrons are allowed to move to achieve a self-consistent
configuration. Both metallic and covalent bonding tendencies
are apparent in Figure 2.

Integrating ρ and ρ_0 over various regions (non-over-
lapping spheres, Wigner-Seitz cells, chain-directed cylinders,
etc.) in the unit cell indicates there is no unambiguous A-B
charge transfer in Nb_3Ge and Nb3Al. Other self-consistent
calculations are in (at least) qualitative agreement with
this conclusion (Klein et al., 1978; Jarlborg, 1979). The
definition of charge transfer is strongly dependent on the
assignment of interstitial charge to A or B atoms, and this
concept is inappropriate in a metallic binary compound with
such markedly different atomic sites as occur in the A15
structure. This accounts for the unreasonably large values
of "charge transfer" which result from Bongi's (1976)
phenomenological model of T_c in A15 compounds.

The only comparison of these charge density results with
experiment is indirect, by comparing ρ of Nb_3Ge with x-ray
data (Staudenmann et al. 1976; Staudenmann, 1978) for the
isoelectronic compound V_3Si. Since the d states in V are
considerably more localized than those in Nb (they peak at
0.85 and 1.4 a.u. respectively in the atom), it is not
evident to just what extent these densities should agree.
On the other hand, the numerous similarities of properties
of V_3X and Nb_3X compounds suggest the similarity of the
important aspects of the electronic structure, so a qualitative comparison is warranted.

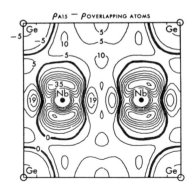

*FIGURE 2. Contour plot of the difference between the
valence charge density in Nb3Ge and the valence charge density
due to overlapping spherical atoms, on a face of the cubic unit
cell. Contours are in steps of 5 electrons/unit cell.*

The x-ray data at both 13K and 300K indicates an excess
of charge along the V chains, compared to overlapping
spherical atoms, in qualitative agreement with our results
for Nb_3Ge. However, the agreement with the low temperature
data ends there. Staudenmann arrives at a "best estimate"
of 1.8-2.4 electrons for the Si-to-V charge transfer,
although part of this may be due to the choice of definition
of charge transfer. Also, no double-hump structure is
observed along the chains on V_3Si. Due to the higher
localization of 3d (compared to 4d) electrons, the double-
hump structure is expected to be less prominent in V_3Si.
His extremely large value of valence charge density at
the V nucleus, ~550 electrons/unit cell (el/cell), reflects
difficulties in subtraction of the V core from the data,
which would then tend to mask a double-hump structure.

More discrepancies between theoretical and experimental
charge densities arise in the interstitial region. The
theory gives a minimum charge density in the (100) face of
25-30 el/cell. This is a very metallic value, about 60-70%
of the average valence charge of 38 el/cell. That the
Fourier representation of the 13K experimental data is
somewhat limited is evident by noting that it gives a
negative value for the valence charge density of -24el/cell
at one point in the (100) face, and, in fact, gives negative
values over most of the intersection of the interstitial
region with the (100) face of the unit cell. The theoretical
charge actually corresponds somewhat more closely to
Staudenmann's 300K data, although substantial differences
remain.

One important implication of Staudenmann's work is that
the charge configuration may have a large temperature
dependence. An example of the charge redistribution which
can occur during vibration of the lattice has been
calculated by Ho et al. (1979). The self-consistent charge
redistribution, which is shown in Figure 3, is that resulting
from a dimerization of a single chain in Nb_3Ge, with each
Nb atom displaced by 1.5% of the lattice constant. Scaling
the Debye-Waller factor of V_3Si (Staudenmann, 1978) by the
mass difference indicates that this displacement is roughly
that expected for the corresponding optic mode at room
temperature. From a study of the charge redistribution in
Figure 3 and the concomitant "fluctuation" in potential, it
is possible to discern some aspects of the strong electron-
phonon interaction in a first principles (model-independent)
manner. A discussion of the effects of chain dimerization
has been given by Ho et al. (1979).

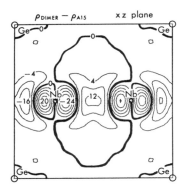

$\rho_{DIMER} - \rho_{A15}$ x z plane

FIGURE 3. Contour plot of the difference in the charge densities of dimerized Nb3Ge and A15 Nb3Ge, in the face of the unit cell containing the dimerized chain. Contours are in steps of 4 electrons/unit cell.

IV. SOME UNANSWERED QUESTIONS ABOUT A15 COMPOUNDS

At the structural transformation in A15 compounds, symmetry requires the tetragonal strain to be accompanied by a frozen-in $\Gamma_{12}^{(+)}$ optical phonon. There have been two prominent models invoking electronic driving mechanisms: the Labbé-Friedel (1967) model, which invokes coupling to the tetragonal strain, and the Gor'kov (1973) model, which assumes coupling to the sublattice distortion to be dominant. All band structure calculations show the specific assumptions about the electronic structure of both models to be unrealistic; however, the general mechanisms of structural instability, i.e. the band Jahn-Teller effect and the Peierls gap-opening respectively, remain possible explanations of the transformation. Band structure calculations do not favor one mechanism over the other and they have not yet provided strong evidence for either. Bhatt (1978) has shown in detail that both are consistent with the available data and that there is little expectation that future experiments can differentiate between the models.

Of course, there are other possible explanations. Testardi's (1972) model of a highly anharmonic lattice remains a viable explanation in terms of a T-dependent phonon, rather than electron, DOS. Other dynamic instabilities also have been proposed (Ngai and Reinecke, 1977). An aspect of strong coupling systems which has not been exploited is the likelihood of strongly T-dependent energies $\varepsilon_k(T)$. For the realistic estimate in a strong-coupling system of $d\varepsilon_k(T)/dT \simeq 5K/K$ for eigenvalues near E_F, this effect could easily dominate the often used (and abused) mechanisms of thermal (Fermi) smearing and disorder broadening.

The significance of placement of E_F is still unclear. It is clear that the placement which occurs in several classes of high-T_c compounds (immediately below the d^+-d^- gap) has the dual advantage of providing a large DOS and large electron-ion scattering $\langle I^2 \rangle$ due to proximity to the d resonance. It has been suggested above that a better understanding of the nearly dispersionless nature of states at E_F may be important for understanding high temperature superconductivity.

Finally, I return to the dominance of the A15 structure among high-T_c compounds: Is there something special about the A15 structure? One feature that seems to emerge is that the A15 structure forces the nearest neighbor A atom separation (along the chains) to be ~10% smaller than occurs elsewhere in molecules or solids. In particular, the V dimer, i.e. the V_2 molecule, has a nuclear separation (Harris and Jones, 1979) similar to the nearest neighbor separation in bcc V rather than that occurring in A15 compounds. A preliminary survey of various high T_c V-, Nb- and Mo- based compounds seems to confirm this distinctive feature. A precarious (in)stability of the closely-packed chain is a natural, if ill-understood, occurrence from this point of view.

ACKNOWLEDGMENTS

I am indebted to many fruitful conversations with T. Jarlborg and A.J. Freeman, and for communication of unpublished work. I gratefully acknowledge communication with A.R. Williams on the vanadium dimer and on the chain instability problem, and collaboration of K.M. Ho and M.L. Cohen on the band structure calculations which provided the basis of these ideas. This work was supported in part by the Air Force Office of Scientific Research Grant 76-2948.

REFERENCES

Bhatt, R.N. (1978). Phys. Rev. B17, 2947.
Bongi, G.H. (1976). J. Phys. F 6, 1535.
Giorgi, A.L. and Matthias, B.T. (1978). Phys. Rev. B17, 2160.
Gor'kov, L.P. (1973). JETP Lett. 17, 379.
Harris, J. and Jones, R.O. (1979). J. Chem. Phys. 70, 830.

Ho, K.M., Pickett, W.E. and Cohen, M.L. (1978). Phys. Rev.
     Lett. 41, 580.
Ho, K.M., Pickett, W.E. and Cohen, M.L. (1979). Phys. Rev.
     B19, 1751.
Jarlborg, T. and Freeman, A.J. (1979). (Unpublished).
Jarlborg, T. (1979). J. Phys. F 9, 283.
Klein, B.M., Boyer, L.L., Papaconstantopoulos, D.A. and
     Mattheiss, L.F. (1978). Phys. Rev. B18, 6411.
Labbé, J. and Friedel, J. (1966). J. Phys. Radium 27,
     153, 303, 708.
Mattheiss, L.F. (1965). Phys. Rev. 138, A112.
Mattheiss, L.F. (1975). Phys. Rev. B12, 2161.
Ngai, K.L. and Reinecke, T.L. (1977). Phys. Rev. B16, 1077.
Pickett, W.E., Ho, K.M. and Cohen, M.L. (1979). Phys. Rev.
     B19, 1734.
Staudenmann, J.L. (1978). Solid State Commun. 26, 461.
Staudenmann, J.L., Coppens, P. and Muller, J. (1976).
     Solid State Commun. 19, 29.
Testardi, L.R. (1972). Phys. Rev. B5, 4342.
Testardi, L.R. (1973). Physical Acoustics, ed. W.P. Mason
     and R.N. Thurston (Academic, New York), Vol. I, p. 193.
van Kessel, A.T., Myron, H.W. and Mueller, F.M. (1978).
     Phys. Rev. Lett. 41, 181.

FERMI SURFACE MEASUREMENTS IN A-15 COMPOUNDS

*A. J. Arko**
*G. W. Crabtree**

Argonne National Laboratory
Argonne, IL 60439

Z. Fisk[†]

Institute for Pure and Applied Physical Sciences
University of California at San Diego
La Jolla, CA 92093

I. INTRODUCTION

The electronic structure of compounds based on the A-15 crystal structure continues to be the subject of intense investigation by theorists and experimentalists alike (Clogston and Jaccarino, 1961; Fisk and Webb, 1976; Hardy and Hulm, 1954; Matthias et al., 1954; Testardi, 1977; Weger and Goldberg, 1973). The basis for this is the widely held belief that the underlying electronic and/or phononic structure which is responsible for the unusual normal state properties is also connected with the occurrence of high T_c superconductivity (Testardi, 1977, and references therein) in this group of materials. This connection is illustrated by the structural distortion in V_3Si, which grows with decreasing temperature below the phase transformation at ~ 23 K, until it is suddenly arrested at T_c (Testardi et al., 1965; Batterman and Barrett, 1964; 1966). Like much of the A-15 behavior, the explanation of this effect depends on the interplay of many aspects of the problem; e.g., electrons, phonons, crystal structure and defects (Testardi, 1972; Varma et al., 1974). For such complicated systems, a clear picture of the basic electronic structure is essential to understanding the experimental behavior.

[*]*Work supported by the U.S. Department of Energy.*
[†]*Work supported by NSF grant No. NSF/DMR77-08469.*

87

All of the electronic structure models proposed to explain A-15 properties have stressed the importance of sharp structures near the Fermi level. In addition, various other features have been emphasized as being particularly relevant: narrow d-bands in the vicinity of E_F (Clogston and Jaccarino, 1961), the A-site one-dimensional interpenetrating chain structure (Labbe and Friedel, 1966), flat planar pieces in the Fermi surface (Berko and Weger, 1972), and electronic instabilities at special points in the Brillouin zone which could drive the structural transformation (Abrikosov, 1975; Gorkov, 1974; Ho et al., 1979; Lee et al., 1977).

A number of band structure calculations have now appeared (Klein et al., 1978; Mattheiss, 1975; Pickett et al., 1979; van Kessel et al., 1978). High precision experimental studies of the k-space electronic structure in A-15's are required in order to evaluate the importance of these various models. The de Haas-van Alphen (dHvA) effect provides the most direct and highest precision method of Fermi surface determination presently known. Such measurements have now been carried out on a few A-15 materials (V$_3$Ge (Graebner and Kunzler, 1969), Nb$_3$Sn and V$_3$Si (Arko et al., 1978), Nb$_3$Sb (Arko et al., 1977)]. Though not yet developed to the same degree as for pure metals (see, for example, Crabtree et al., 1979), these studies have already yielded valuable information. It is a difficult experiment in that good stoichiometric single crystals are difficult to obtain, and, in the case of high-T_c materials, magnetic fields H > 20 T are needed to reach the normal state (Foner and McNiff, 1976). This has precluded a thorough experimental examination in the first three materials where the strongest conclusion that could be made was that the observed surfaces are probably located at M. The situation is much better in the case of Nb$_3$Sb where crystals with resistance ratios ~100-150 are available and the critical field is low. Because a complete detailed study of its Fermi surface can be done with present technology, Nb$_3$Sb can be used as a test case for evaluating different approaches to electronic structure calculations in A-15 materials.

In this paper we present new dHvA data in Nb$_3$Sb (now nearly complete) and compare them with the various band calculations, while at the same time reviewing the progress of Fermi surface measurements in A-15 compounds in general. Some definite statements can finally be made regarding the band structure. It appears that indeed the various theorists are converging on a more or less correct description in A-15 compounds, with those using a generalized potential having somewhat greater success than those making a muffin-tin sphere approximation.

II.   REVIEW OF FERMI SURFACE DATA TO DATE

A. V_3Ge

This material was the first A-15 in which quantum oscillations (magnetothermal) were seen (Graebner and Kunzler,
1969). Three separate pieces of Fermi surface were observed
which are consistent with cylindrical surfaces (possibly
ellipsoids) centered at the M point in the Brillouin zone.
Both the initial calculations of Mattheiss (1975) and subsequent work of Klein et al. (1978) predict structure at M
as does the most recent calculation of Pickett et al. (1979)
for isoelectronic Nb_3Ge. In all cases however, it appears
that even the smallest predicted surface is at least twice as
large as the smallest observed surface. Moreover, more structure is predicted than observed at M even though the expected
effective masses (m*), from a visual inspection of the bands,
are not excessively high. It is difficult to carry the comparison further without either a set of predicted dHvA frequencies to compare with the data, or a more complete set of
experimental frequencies. The theoretical Fermi surface is
very complicated and possibly open in some directions. There
is the additional complication of probable magnetic breakdown
due to numerous degenerate bands which, we will see later, is
very important in Nb_3Sb. What little experimental data exists
is in qualitative agreement with theory, although no definitive statement can be made without a great deal more work.
Using modern Fourier transform and field modulation techniques,
much more can undoubtly be learned about the V_3Ge Fermi
surface.

B. Nb_3Sn and V_3Si

These data were obtained by direct inductive pickup
of the magnetization utilizing the 40 T slow-pulsed magnet at
the University of Amsterdam (Arko et al. 1978). The results
of the Nb_3Sn investigation are shown as the data points in
Fig. 1 superimposed on a background of frequencies predicted
by van Kessel et al. (1978) and attributed to nested ellipsoids centered at M. Because the data are sparse, the apparent excellent agreement does not unambiguously confirm the
band structure. Much more data will be needed for verification. (It might be pointed out that no adjustment of the
calculation was needed to obtain this fit.) Only two orientations ([100] and [110]) were investigated in V_3Si with results
nearly identical to those obtained in Nb_3Sn.

Further comparison of experiment and theory will be
fruitless without experimental data on more of the Fermi surface. Especially in the case of Nb_3Sn, with its complex

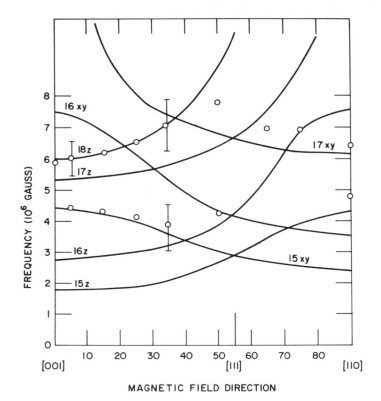

FIGURE 1. *Nb₃Sn dHvA data (circles) superimposed on a*
set of predicted dHvA frequencies (solid lines) from Arko et
al. (1978). The numbers (15, etc.) refer to theoretical orbit
labels (band 15, (π/a, π/a,0) orbit center).

Fermi surface and the magnetic breakdown that almost certainly
occurs between nearly degenerate bands, a meaningful comparison would be very difficult. Perhaps the A-15 wars are
better fought over a simpler material like Nb_3Sb, where a
complete set of data is now becoming available.

III.   RESULTS AND DISCUSSION OF Nb_3Sb

As already stated, a substantial amount of data for
Nb_3Sb were first published (Arko et al., 1977) in 1977.
Three pieces of Fermi surface were established on the basis
of that data, the α_i, β and γ_i frequencies of Fig. 3. Only
one family of frequencies (α_i) could be observed at all angles.

In addition, fragments of the η and δ branches were observed
but were too short to make any definitive statements. It was
obvious that the rather complex Fermi surface would require a
band structure to assist in identifying the experimental data.

Until very recently, no band structure calculation existed specifically for Nb₃Sb. Van Kessel et al. (1979), using
the same technique as in their previous approach to Nb₃Sn
(i.e. nonself-consistent APW, overlapping charge density model,
but using extremely high precision) have now produced a set of
bands complete with predictions of dHvA frequencies. This
enormously simplifies comparison with experiment and allows
quantitative rather than merely qualitative comparisons. Their
predicted dHvA frequencies are shown in Fig. 2. The very close
agreement between the α, β and γ frequencies and bands 18, 20
and 24, respectively, prompted us to re-examine Nb₃Sb in an
effort to observe the missing bands 19, 21, 22 and 23, and the
missing branches in band 20.

A complete new investigation was performed in the (100)
plane utilizing the full power of the Fourier transform technique and the frequency discrimination capability of the field
modulation technique. The new data was taken at slightly higher fields (to 75 kG) and lower temperature (0.35 K) than that
already published. The experimental difficulties were enhanced by the complexity of the Fermi surface and the fact
that amplitudes varied by several orders of magnitude even
when the frequencies themselves were nearly identical.
In addition, Fourier transforms of previous data were reexamined and found to confirm many of the new features we
present here (new points in the (110) plane reported below
were found in this way). Figure 3 contains the sum total of
the frequency data obtained in Nb₃Sb.

A. High Frequencies $\approx 10 - 25 \times 10^6$ Gauss

Three unrelated sets of frequency branches have now been
confirmed to exist (Fig. 3) above $\approx 10 \times 10^6$ Gauss. In keeping with our previous labeling, we call these γ, δ and η. If
we compare this to Fig. 2, we find a great deal of correspondence even though the absolute values of the cross-sectional
areas differ from predictions by as much as 20%. We can
identify the γ frequencies with band 20, the δ frequency with
band 21, and the η frequency with band 22. The γ-surface
would then be an M-centered hole surface which is larger than
predicted, while η and δ are R-centered electron surfaces,
both smaller than predicted. A lowering of the Fermi energy
by about 4 milliRydbergs would bring theory in line with experiment for these three surfaces, but, as we will see later,
may cause problems for other pieces of the Fermi surface.

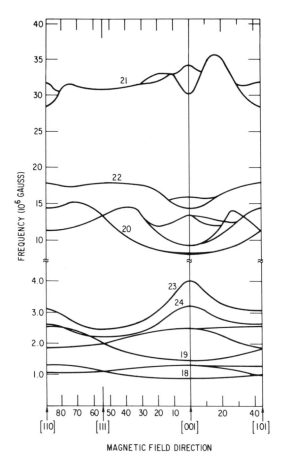

FIGURE 2.   Predicted dHvA frequency for Nb₃Sb from van Kessel, et al. this conference.

The γ_3 branch, previously observed only near [101] has now been found to connect with our previously labelled χ-frequency at [001] and the noncentral ε_2 branch connects with γ_3 as predicted by van Kessel et al.   We have not observed the connection between γ_2 and χ and ε_1, the signals in the region in question being just too weak.   The splitting of γ_1 around [001] is easily understood as due to the slight, dumbbell shape of the band 20 surface (see Fig. 4) with γ_1 being the noncentral orbit.

The dashed-in portions of the δ and η frequencies in the (110) plane are done on the basis of the (100) plane observations and are meant only as an aid to the reader.   We

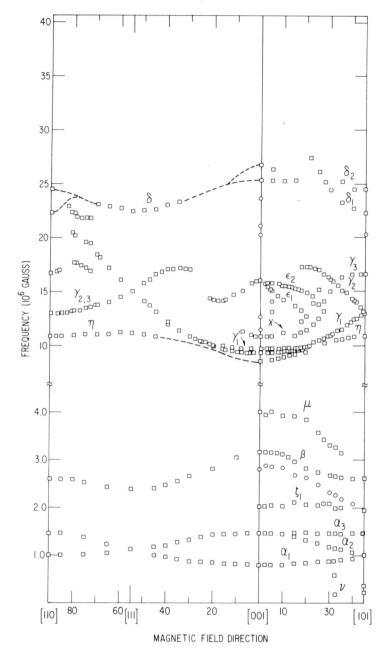

FIGURE 3. Experimental dHvA frequency spectrum for Nb₃Sb. Dashed lines in (110) plane are meant to show the expected connectivity.

believe, however, that these will prove reasonably accurate
when we examine the (110) plane in detail.

Theory fails to predict the multiple splitting of γ_1 near
[110] in the (110) plane which may be due to magnetic breakdown involving band 19. Magnetic breakdown in fact is the
most likely explanation for the very weak amplitudes of η, δ_i,
γ_3 and χ frequencies. The inset in Fig. 4 shows a plot of the
predicted orbits for H//[001]. Note the large number of de-
genercies where magnetic breakdown may occur on all surfaces
except band 18. The effective masses [see van Kessel et al.,
this conference, for a detailed discussion of m*] are in general not too large (max. m* = 1.7 for band 21) so that one
would normally not expect any of the amplitudes to be especially weak. Because of the weak signals and the dense bunch-
ing of frequencies, it is not possible to do an amplitude
plot vs 1/H to look for breakdown effects in the usual manner.
We would point out, however, that the η frequency increased in
amplitude by one order of magnitude as we tipped H from [101]
by a mere 5°. In this angular range the number of possible
breakdown switching junctions would have decreased from 8 to
2 for η. Furthermore, at both [100] and [101] we observe
numerous additional Fourier peaks in the transform (see Fig.
4) which we cannot follow for more than $\sim$5°. Some of the
stronger peaks have been included as circled data in Fig. 3.
We believe these are magnetic breakdown orbits.

In general, there is excellent topological correspondence
and rough quantitative agreement between theory and experiment
for the higher frequencies.

B.   Low Frequencies $< 5 \times 10^6$ Gauss

The picture is somewhat less neat and tidy at lower frequencies. It is not a lack of Fourier transform peaks that
creates the problem, but rather their profusion. The problem
arises from the low masses of frequency branches α_i ($\sim$0.3 m_0)
which result in numerous harmonics which often tend to obscure
real data. There may also be some confusion due to magnetic
breakdown. Figure 3 shows the data which we cannot account
for as being due to harmonics of α_i except for the circles
which we include for reasons stated later.

A comparison of Fig. 3 with Fig. 2 shows that frequency
β can be identified with the band 24 surface and frequency
μ with the band 23 surface. Both of these are R-centered
electron surfaces and the agreement between theory and ex--
periment is better than 2%. Thus, a lowering of the Fermi
energy can only make matters worse for these R-centered pieces.

A similar phenomenally good agreement is found for the
frequency branches α which would seem to correspond to band
18, an M-centered hole ellipsoid. There are, however, several

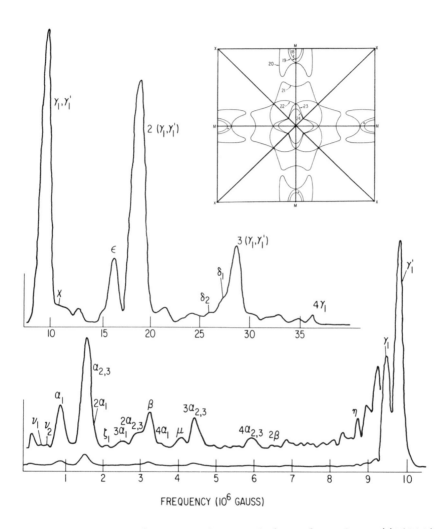

FIGURE 4. Fourier transforms of dHvA data for H//1(001). Lower frequencies (0-10 × 10^6 Gauss) are shown in the bottom traces with frequencies below ≈ 9 × 10^6 Gauss magnified × 10. Upper trace shows frequencies above 10^7 Gauss. All unidentified peaks are observed only near the symmetry axes and are probably a result of magnetic breakdown. Inset shows the R- and M-centered orbits for H//(001) from van Kessel et al., this conference. Note the numerous degeneracies.

problems in the comparison. The theory of van Kessel et al. predicts a second M-centered hole ellipsoid from band 19 which we are not certain that we have observed. The reason for the uncertainity is that while we do observe a very weak frequency branch (ζ_1 in Fig. 3) that cannot be accounted for as being due to harmonics of α_1, and which is approximately of the correct cross-sectional area to correspond to the lower branch of the band 19 ellipsoid, the middle branch of band 19, ξ_2, falls precisely on the second harmonic of α_2 (= $2\alpha_2$) and may not be distinguishable as an independent frequency. Moreover, the upper branch of band 19 is <u>not</u> observed. It can be argued that this is due to magnetic breakdown since band 19 is degenerate with band 20 along the R-M line thus affecting the upper branch in the entire (100) plane. We find amplitude of the $2\alpha_2$ branch to be considerably larger than $2\alpha_1$ or $2\alpha_3$, indicating there may indeed by a separate orbit at this frequency. (These are the circles in Fig. 3). This question is yet to be resolved.

Another unexplained feature is a weak frequency (ν) at $\sim$0.4 x 10^6 Gauss. This very small frequency cannot be optimized with our apparatus and hence cannot be mapped out in detail. We know that it does <u>not</u> have the symmetry of Γ or R since multiple frequencies are observed. It appears to have a heavier m* than the α_i frequencies. A preliminary magnetoresistance experiment yielded quantum oscillations of this frequency. Thus ν is possibly a switching orbit controlling magnetic breakdown, but the small signals and the low frequencies again do not allow a detailed mapping. It is also possible that by lowering the Fermi energy 3-4 mRy we obtain an additional piece of Fermi surface either from band 20 along the Q-M line or from band 17 along the R-M line.

IV. DISCUSSION AND CONCLUSIONS

We have seen that there is not only qualitative but also a great deal of quantitative agreement between experiment and the band structure calculation of van Kessel et al. Very small pieces of Fermi surface are always difficult to determine in any calculation so it is gratifying to see that the problems occur with these pieces. It must also be emphasized that the bulk of bands 21, 22, 23 and 24 were discovered only <u>after</u> being predicted by van Kessel et al., as well as some rather crucial connections on band 20.

We have confined our comparisons of the data to the calculations of van Kessel et al., primarily because their theoretical dHvA frequencies (Fig. 2) allow quantitative comparison. But we can also make qualitative comparisons with other calculations.

There are two other <u>preliminary</u>, unpublished calculations
for Nb_3Sb, one using the LMTO method (Jarlborg, private communication) with no nonspherical corrections and the other using
self-consistent APW again with no inside muffin-tin corrections (Klein, private communciation). In both cases a correct
description is obtained around the R-point, but shifts of
$\sim$30 mRy at the M point are needed to obtain the small hole
pockets.

One can also gain some insight by using a rigid band
shift of Nb_3Sn or Nb_3Ge bands to correspond to Nb_3Sb. Since
the wave function character at E_F is primarily d-like and
coming from Nb sites, a rigid band shift can be useful if
one recognizes that results are only qualitative. Using such
a shift for the Nb_3Sn bands of van Kessel et al. (1978) and
Nb_3Ge bands of Pickett et al. (1979) we obtain a qualitatively
correct description at both M and R while a similar shift of
the corresponding bands of Klein et al., (1978) results in a
correct description at R but approximately the same 30 mRy
failure at M. The calculations of van Kessel et al. and
Pickett et al. both use a generalized potential and their
resulting bands are strikingly similar.

It is curious that nonspherical corrections inside the
muffin-tin sphere should result primarily in an upward shift
of some bands at the M point. Perhaps there is some important
physics to be learned from this. We ask our theoretical
friends to clarify this point.

ACKNOWLEDGEMENTS

We gratefully acknowledge many helpful discussions with
Drs. D. D. Koelling, F. M. Mueller, H. W. Myron, A. T. van
Kessel, A J. Freeman, B. W. Veal, T. Jarlborg and W. E. Pickett.
We wish to thank T. Jarlborg and B. M. Klein for sending us
their preliminary calculations for Nb3Sb, and to A. T. van
Kessel for sending us the Nb3Sb calculations prior to publica-
tion.

REFERENCES

Abrikosov, A., (1975) *JETP Lett. 43*, 217.
Arko, A.J., Lowndes, D.H., Muller, F.A., Roeland, L.W.,
    Wolfrat, J., van Kessel, A. T., Myron, H.W., Mueller,
    F.M., and Webb, G. W., (1978). *Phys. Rev. Lett. 40*,
    1590.
Arko, A.J., Fisk, Z., and Mueller, F.M., (1977). *Phys. Rev. B
    16*, 1387.

Batterman, B.W. and Barrett, C.S., (1964). *Phys. Rev. Lett. 13*,
    290; (1966). *Phys. Rev. 199*, 296.
Berko, S. and Weger, M. (1972). *Phys. Rev. B 4*, 521.
Clogston, A.M. and Jaccarino, V., (1961). *Phys. Rev. 121*,1357.
Crabtree, G., Dye, D.H., Karim, D.P., and Ketterson, J.B.,
    (1979). *J. Mag. and Mag. Materls. 11*, 236.
Crabtree, G., Dye, D.H., Karim, D.P., and Ketterson, J.B.,
    (1979). *Phys. Rev. Lett. 42*, 390.
Fisk, Z. and Webb, G.W., (1976). *Phys. Rev. Lett. 36*, 1084
    and references therein.
Foner, S. and McNiff, Jr., E.J., (1976). *Phys. Lett. 58A*,
    318.
Gorkov, L.P., (1974). *Sov. Phys. JETP 38*, 830.
Graebner, J.E. and Kunzler, J.E., (1969). *J. Low. Temp. Phys.
    1*, 443.
Hardy, G.F. and Hulm, J.K., (1954). *Phys. Rev. 93*, 1004.
Ho, K. M., Pickett, W. E., and Cohen, M. L., (1979). *Phys.
    Rev. B 19*, 1751.
Klein, B.M., Boyer, L.L., Papaconstantopoulos, D.A., and
    Mattheiss, L.F., (1978). *Phys. Rev. B 18*, 6411.
Labbe, J. and Friedel, J., (1966). *J. Phys. Radium 27*, 153
    and 303.
Lee, S.K., Birman, J. L. and Willaimson, S. J., (1977). *Phys.
    Rev. Lett. 39*, 39.
Matthias, B. T., Geballe, T. H., Geller, S., and Corenzwit,
    E., (1954). *Phys. Rev. 95*, 1435.
Mattheiss, L.F., (1975). *Phys. Rev. B 12*, 2161.
Pickett, W. E., Ho, K. M., and Cohen, M. L., (1979). *Phys.
    Rev. B 19*, 1734.
Testardi, L.R., Reed, W.A., Bateman, R.B., and Chirta, V.G.,
    (1965). *Phys. Rev. Lett. 15*, 250.
Testardi, L.R., (1972). *Phys. Rev. B 5*, 4342.
Testardi, L.R., (1977). *Cryogenics 17*, 67.
van Kessel, A.T., Myron, H.W., and Mueller, F.M., (1978).
    Phys. Rev. Lett. 41.
van Kessel, A.T., Myron, H.W., and Mueller, F.M., (1979).
    This conference.
Varma, C.M., Phillips, J.C., and Chui, S.-T., (1974). *Phys.
    Rev. Lett. 33*, 1223.
Weger, M. and Goldberg, T.B., (1973). *Solid State Phys. 28*,
    1, eds. F. Seitz and D. Turnbull (Academic Press, New
    York).

THEORETICAL EVIDENCE FOR AN
ELECTRONICALLY DRIVEN ISOSTRUCTURAL
PHASE TRANSITION IN FCC LANTHANUM[1]

Warren E. Pickett
Department of Physics and Astronomy
Northwestern University
Evanston, Illinois

Arthur J. Freeman
Department of Physics and Astronomy
Northwestern University
Evanston, Illinois
and
Argonne National Laboratory
Argonne, Illinois

Dale D. Koelling
Argonne National Laboratory
Argonne, Illinois

I. INTRODUCTION

The observation in lanthanum of a high superconducting
transition temperature, T_c, and its large pressure derivative,
dT_c/dp, has led to wide speculation that a pairing mechanism
involving f electrons is responsible. Under pressure, T_c
rises sharply from 6K at ambient pressure and saturates at a
value of nearly 13K around 200 kbar. In addition, at 53 kbar
and low temperature, a phase transition occurs which is

[1]*Supported in part by the Air Force Office of Scientific
Research (Grant 76-2948), the Department of Energy and the
National Science Foundation (through the Northwestern University Materials Research Center, Grant 76-80847).*

apparently an isostructural fcc-fcc transition (Balster and
Wittig, 1975). The development of reliable tunnel junctions
has led to the unambiguous assignment of La as a relatively
strong coupling d electron superconductor, with electron-
phonon coupling constant $\lambda \simeq 0.8$, but there remains a question
whether dT_c/dp is enhanced by f electrons. There is also the
possibility that the structural transition represents another
facet of the intriguing relation between structural instability and high temperature superconductivity.

We describe below results of careful band structure
calculations which provide insight into the origin of the
unusual behavior of La under pressure. The method of band

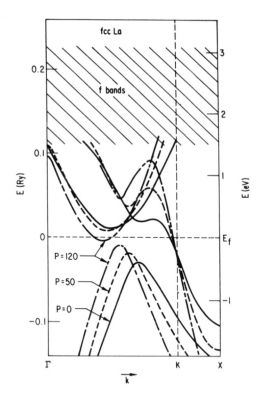

FIGURE 1. The behavior under pressure of bands 1 and 2 in
fcc La, along the (110) direction. The Fermi energies have
been aligned and taken as the zero of energy. Pressures P are
in kbar.

structure calculation is the Linearized Augmented Plane Wave
method of Koelling and Arbman (1975) in the semi-relativistic
approximation (Koelling and Harmon, 1977). Both the eigenvalues and the charge density have been carried to self-
consistency, using stringent convergence criteria, on a grid
of 128 k points in the irreducible Brillouin zone. Both the
potential and the charge density were allowed to have a
general (i.e., non-muffin-tin) form. Consequently, we believe
the results to be quite accurate. All the results described
here were obtained using the Kohn-Sham-Gaspar (α=2/3) local
density approximation for exchange.

II. THE ELECTRONIC STRUCTURE OF fcc LA

Calculations have been carried out for lattice constants
of 10.0378, 9.4760 and 9.0340 a.u., corresponding approximately to ambient (0), 50 and 120 kbars pressure, respectively.
At ambient pressure the center C_f of the f band lies 2.2eV
above the Fermi level E_F, and C_f increases monotonically to
2.7eV at 200 kbar. The f bandwidth W_f increases by roughly
a factor of two (from ~1 to 2eV) in this pressure range,
depending somewhat on how it is defined.

The density of states (DOS) $N(E_F)$ <u>decreases</u> from 27.5 to
21.4 states/Ry-atom in going from 0 to 120 kbar pressure, a
direct result of the increase by 40% in the d bandwidth. The
angular-momentum-projected partial DOS shows $N(E_f)$ to be
dominated by d states, with a weakly pressure dependent
f contribution of ~5%. The corresponding f electron charge
is 0.1 electron at ambient pressure, increasing to nearly
0.2 electron at 120 kbar. Examination of the ℓ=3 radial
function reveals that only about one-half of the "f electron
charge" lies in the atomic 4f state region (in the
outer core), with the remaining one-half reflecting the d
states overlapping from neighboring atoms. The large
separation C_f-E_F and the small amount of f character at E_F
suggests that La should be considered a true transition
metal rather than a rare-earth, or f band, metal.

Our calculations also show that, in the range 50-120
kbar, band 2 along the Σ=(110) direction passes through the
Fermi level, as shown in Figure 1. Including the spin-orbit
interaction in a non-perturbative fashion (MacDonald et al.,
1979) does not qualitatively alter this result. The minimum
of this band along the (110) direction is actually a saddle
point arising from strongly hybridized states (50%d, 20%f,
30%sp). Implications are discussed in the following section.

III. SUPERCONDUCTIVITY AND THE STRUCTURAL TRANSITION

We have used the McMillan (1968) equation for T_c, with λ
calculated in the rigid-muffin-tin approximation (Gaspari and
Gyorffy, 1972) to provide a theoretical estimate of T_c vs.
pressure. Assuming a pressure independent Debye temperature
of 140K, the calculated values of electronic "stiffness" η,
coupling constant λ, and T_c, respectively, increase from
$(2.24eV/Å^2$,   0.90,6.9K) at ambient pressure to $(3.37eV/Å^2$,
1.52, 13.6K) at 120 kbar, in spite of the decrease in $N(E_F)$
noted above. This rather good agreement of T_c with experiment could be improved somewhat if the degree of stiffening
of the lattice were known. These energy band results are
consistent with those of Glocker and Fritsche (1978).
Consequently a localized f electron mechanism is not necessary
to explain the magnitude of either T_c or its pressure
derivative.

Lifshitz (1960) observed that a change in Fermi surface
topology (a band minimum, maximum or saddle point crossing
E_F) leads to an electronically induced isostructural transition of "5/2 order". Dagens (1976) noted that coupling of the
electronic system to the lattice leads to a phonon induced
transition of "3/2 order". Although the latter transition
is stronger, in the sense that singular behavior occurs in
lower derivatives of the free energy, in general, either
mechanism may be the driving force for the transformation
(since the transformation may not occur exactly at the
singularity of free energy functional). Although the
Lifshitz instability has been suggested to occur in various
systems, for the 53 kbar transition in La there are good
reasons for invoking an instability of the Lifshitz-Dagens
type: (1) since $\lambda(50$ kbar$) \simeq 1$ the electronic system is
known to be strongly coupled to the lattice, (2) the band
which crosses E_F is strongly hybridized, which implies
stronger than average coupling of this band to the lattice,
(3) the transition should be accompanied by appreciable
phonon softening (Dagens, 1978) which could account for the
"excess" increase in T_c (Balster and Wittig, 1975) in the
vicinity of the transition.

REFERENCES

Balster, H. and Wittig, J. (1975). J. Low Temp. Phys. **21**, 377.
Dagens, L. (1976). J. Physique Lett. **37**, L37.
Dagens, L. (1978). J. Phys. F **8**, 2093.

Gaspari, G.D. and Gyorffy, B.L. (1972). Phys. Rev. Lett.
    28, 801.
Glocker, R. and Fritsche, L. (1978). Phys. Stat. Sol. 88, 639.
Koelling, D.D. and Arbman, G.O. (1975). J. Phys. F 5, 2041.
Koelling, D.D. and Harmon, B.N. (1977). J. Phys. C 10, 3107.
Lifshitz, I.M. (1960). Sov. Phys. JETP 11, 1130.
MacDonald, A.H., Pickett, W.E. and Koelling, D.D. (1979)
    (unpublished).
McMillan, W.L. (1968) Phys. Rev. 167, 331.

SUPERCONDUCTIVITY AND CHARGE-DENSITY WAVES

C.A. Balseiro[1]
L.M. Falicov

Department of Physics[2]
University of California
Berkeley, California

I. INTRODUCTION

Superconductivity (SC) and charge-density waves (CDW) are the two most well known cases of broken symmetry in a solid which require an effective attractive electron-electron interaction; this is normally provided by a phonon-mediated mechanism (Bardeen, Cooper and Schrieffer, 1957, hereafter referred to as BCS; Parks, 1969; Falicov, 1979 and references therein; Wilson *et al.*, 1975; Bilbro and McMillan, 1976; Chan and Heine, 1973). It is therefore to be expected that systems in which CDWs have been experimentally observed (Falicov, 1979) exhibit also SC at very low temperatures (Di Salvo *et al.*, 1971; Morris and Coleman, 1973; Morris, 1975).

The tendencies towards the formation of SC and CDW are to certain degree opposing one another: while the SC state has infinite conductivity and a Meissner effect, the CDW state, for large enough interactions, produces a semiconductor gap in the spectrum and a non-conducting state. From the microscopic point of view on the other hand the SC state arises from electron-electron coupling into Cooper pairs, the CDW state from electron-hole coupling and charge redistribution: two effects which are in principle independent of one another.

[1]*On a fellowship from CNICT, Argentina, on leave from Centro Atómico Bariloche, Argentina*
[2]*Work supported in part by the National Science Foundation Grant DMR78-03408.*

Although considerable uncertainty exists, it has been mentioned in the literature (Wilson and Yoffe, 1969) that some CDW layered compounds, e.g. $1T-TaS_2$, despite the apparent lack of metallic properties are reported to be superconducting below a given temperature T_{SC} ($T_{SC} \sim 0.8$ K for $1T-TaS_2$).

It is the purpose of this work to study the interdependence of the SC and CDW states. In particular we are interested in setting up a model hamiltonian which allows us to test both effects on the same footing. We thus expect to clarify several points: (a) the difference in order of magnitude of T_{CDW}, the CDW transition temperature, and T_{SC}, the SC transition temperature; (b) the range of relevant parameters over which a CDW exists; (c) the nature of the competition between SC and CDWs; and (d) the possibility of coexistence of both effects, i.e. a non-uniform charge distribution with its attendant CDW gap and a superconducting state also with its characteristic gap.

II. THE MODEL HAMILTONIAN

In the spirit of the BCS theory, a hamiltonian describing a system of crystalline electrons moving in a single tight-binding band and interacting with lattice phonons

$$H_0 = H_e + H_p + H_{ep} \tag{1}$$

is transformed into an effective electron hamiltonian

$$H_{eff} = H_e + H_{SC} + H_{CDW} \quad . \tag{2}$$

Here

$$H_e = \sum_{k\sigma} \varepsilon_k \, a_{k\sigma}^{\dagger} a_{k\sigma} \quad , \tag{3}$$

$$H_{SC} = \sum_{kq} V_{kq} \, a_{(k+q)\uparrow}^{\dagger} \, a_{(-k-q)\downarrow}^{\dagger} \, a_{-k\downarrow} \, a_{k\uparrow} \quad , \tag{4}$$

$$H_{CDW} = \sum_{\substack{kk' \\ \sigma\sigma'}} \tfrac{1}{4}(V_{kQ} + V_{k'Q}) \, a_{(k+Q)\sigma}^{\dagger} \, a_{k\sigma} \, a_{(k'-Q)\sigma'}^{\dagger} \, a_{k'\sigma'} \, \tag{5}$$

Q is the wave-vector of the CDW and, following BCS once again, we choose

$$V_{kq} = \begin{cases} -\lambda & \text{if } |\varepsilon_k - \varepsilon_F| < \hbar\omega_D \quad , \\ & |\varepsilon_{k+q} - \varepsilon_F| < \hbar\omega_D \quad , \\ 0 & \text{otherwise.} \end{cases} \qquad (6)$$

where ε_F is the Fermi energy and ω_D is a characteristic (Debye) frequency.

The above hamiltonian H_{eff}, in the mean-field approximation yields three order parameters, a superconducting one

$$\Delta \equiv \lambda \sum_k{}' < a_{-k\downarrow} a_{k\uparrow} > \qquad , \qquad (7)$$

and two related to the CDW

$$G_0 \equiv \lambda \sum_k{}'' < a^\dagger_{(k+Q)\uparrow} a_{k\uparrow} > \qquad , \qquad (8)$$

$$G_1 \equiv \lambda \sum_k < a^\dagger_{(k+Q)\uparrow} a_{k\uparrow} > \qquad . \qquad (9)$$

In the above equations the primed summation is over all states k such that $|\varepsilon_k - \varepsilon_F| < \hbar\omega_D$, the doubly primed summation is over states which satisfy the two conditions $|\varepsilon_k - \varepsilon_F| < \hbar\omega_D$ and $|\varepsilon_{k+Q} - \varepsilon_F| < \hbar\omega_D$, and <...> indicates thermodynamic mean values. For the sake of definiteness we now restrict ourselves to a well defined model. We choose a two-dimension-al structure in a square lattice of constant a with first and second neighbor couplings, where the band energies are given by

$$\varepsilon_k = -2t_0 [\cos(k_x a) + \cos(k_y a)] - 4t_1 \cos(k_x a) \cos(k_y a).$$
$$(10)$$

The structure given by (10) has two degenerate saddle points of energy $\varepsilon_{sp} = 4t_1$ and located at the midpoint X of the side of the square Brillouin zone. For $t_1 = 0$ the saddle point energy coincides with the Fermi level of the half-filled band. We choose our Q vector to be that which connects the two saddle points, i.e. $Q = (\pi/a, \pi/a)$.

III. RESULTS

1. Figure 1 shows the region of stability of the CDW for our model. It gives the region in parameter space (inter-action strength λ; Fermi energy ε_F) measured in units of t_0

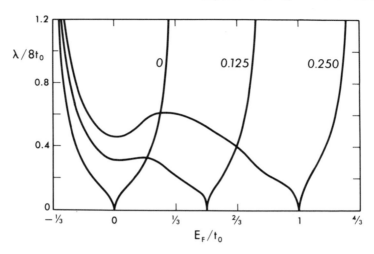

FIGURE 1. Phase diagrams (interaction strength λ versus Fermi energy ε_F) for CDW formation. Various values of (t_1/t_0) are shown. For ε_F at the saddle point of the band structure a CDW is stable at any value of λ.

for various values of (t_1/t_0). At $T=0$ and half-band occupation for $t_1=0$ the CDW is stable at all positive values of λ. This is easily understood in terms of the features of the Fermi surface, which is a perfect square with perfect nesting (Lomer, 1962; Penn, 1966; Falicov and Penn, 1967), and even an infinitesimal perturbation is sufficient to open a complete gap in the spectrum and stabilize the CDW.

For $t_1=0$, $\varepsilon_F \neq 0$ a minimum value of λ is required to stabilize the CDW. This value is shown in Figure 1.

2. The CDW transition temperature T_{CDW}, for $t_1=0$ and a half-filled band has a particularly simple form, similar to T_{SC} in the BCS theory

$$T_{CDW}(t_1=0, \varepsilon_F=0) = 1.14\hbar\omega_D \exp[-1/\lambda_{eff}\rho_0] \quad , \tag{11}$$

where, to a very good approximation,

$$\rho_0 \equiv (8t_0)^{-1} \quad , \tag{11}$$

and $\lambda_{eff} \equiv \lambda[2-\lambda\rho_0 \ \ell n(2\rho_0\hbar\omega_D)]$. $\tag{12}$

A comparison of T_{CDW} and T_{SC} can now be made. If in temperature-energy units we choose $t_0=1500$ K (a bandwidth of 1.2×10^4 K ~ 1eV), $\hbar\omega_D=300$ K and $\lambda=2000$ K we obtain T_{CDW} ~ 31 K,

$T_{SC} \sim 0.8$ K. If we increase λ to 3000 K we obtain $T_{CDW} \sim 80$ K and $T_{SC} \sim 6.3$ K. In both cases T_{CDW} is more than one order of magnitude larger than T_{SC}.

3. Figure 2(a) shows the value of the CDW energy gap parameter $G \equiv G_0 + G_1$ in the absence of SC (i.e. $\Delta \equiv 0$) as a function of λ for $T=0$, $(t_1/t_0)=0.125$, $(\hbar\omega_D/t_0)=\frac{1}{3}$ and a half-filled band (the bandwidth is in this case $8t_0$, independent of t_1). The CDW state is stable only for values of $\lambda > \lambda_c$; in our example $\lambda_c = 1.67t_0$. As λ increases beyond λ_c, G increases rapidly. For $\lambda_c < \lambda < \lambda^*$, where $\lambda^* = 2.23t_0$ in the example of Figure 2(a), a CDW state is stable, but the system exhibits no gap in the quasiparticle spectrum, i.e. we have a metallic CDW state. For $\lambda > \lambda^*$ a gap appears in the spectrum; for a half-filled band the system now becomes a semiconductor. The value of the gap in the spectrum is also shown in Figure 2(a) for $\lambda > \lambda^*$. For the sake of completeness we also show in Figure 2(a) the value of the SC parameter Δ corresponding to the same model and in the absence of CDWs ($G \equiv 0$). It should be noted that $\Delta \neq 0$ for any value of λ, and that the SC and CDW gaps are of the same order of magnitude for $\lambda \sim \lambda^*$.

4. The solution of the complete problem has been carried numerically; particular results follow. For $t_1 = 0$ and $\varepsilon_F = \varepsilon_{sp} = 0$ there is no stable SC solution; the stable solutions at all temperatures is $\Delta = 0$, with $G \neq 0$ for $T < T_{CDW}$ and $G = 0$ for $T > T_{CDW}$.

5. For $t_1 \neq 0$, even when CDWs are stable, if the nonsuperconducting state is metallic then a SC state is always present and stable at low enough temperatures.

6. In Figure 2(b) we show the results for the same parameters of Figure 2(a), i.e. $T=0$, $(t_1/t_0)=0.125$, $(\hbar\omega_D/t_0)=\frac{1}{3}$ and a half-filled band.

7. For $\lambda < \lambda_{c1}$ ($\lambda_{c1}=1.75t_0$ in the example of Figure 2) the CDW is not stable and an ordinary BCS superconductor appears, with its characteristic gap parameter Δ and transition temperature T_{SC}. It should be noted that λ_{c1} is larger than the value λ_c obtained when SC correlations were not included.

8. For $\lambda_{c1} < \lambda < \lambda^{**}$ ($\lambda^{**} \sim \lambda^* = 2.23t_0$ in Figure 2) the ground state of the system is a state in which SC and CDW coexist, i.e. both order parameters are non-vanishing; the value of Δ remains practically unchanged throughout the interval because of the competing effects of increasing λ and decreasing density of states at the Fermi level. It should be mentioned that the pure CDW state is metallic in this range.

9. For $\lambda^{**} < \lambda < \lambda_{c2}$ ($\lambda_{c2}=2.46t_0$ in Figure 2) the CDW induces a (semiconductor) gap in the quasiparticle spectrum, but SC does exist and the ground state of the system is SC. In this range of λ-values, Δ decreases drastically with increasing λ. We expect in this case to have a transition as a function of increasing temperature between a SC and a semiconducting CDW state.

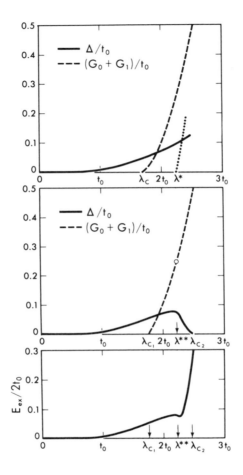

FIGURE 2. Various order parameters for the square lattice and a half-filled band as a function of the interaction
strength λ for $(t_1/t_0)= 0.125$. (a) The CDW parameter (G_0+G_1)
in the absence of SC for $\lambda>\lambda_c$ (dashed line); the semiconducting half-gap for $\lambda>\lambda^*$ (dotted line); and the SC order parameter Δ in the absence of CDWs (full line). (b) The CDW
order parameter (G_0+G_1) (dashed line) and the SC order parameter Δ (full line) when the two effects are considered;
$G_0+G_1=0$ for $\lambda < \lambda_{c1}$ and $\Delta=0$ for $\lambda>\lambda_{c2}$. Coexistence occurs for
$\lambda_{c1} < \lambda < \lambda_{c2}$. (c) The excitation half-gap (half the excitation energy of a pair) for the case shown in (b). All order
parameters are energies in units of t_0.

10.   For $\lambda_{c2} < \lambda$ no SC state is stable, and the ground state of the system is an insulating CDW state.

11.   In Figure 2(c) we give the minimum quasiparticle excitation energy (one half of the pair excitation energy) as a function of λ. For $\lambda < \lambda^{**}$ this energy corresponds to the breaking up of a Cooper pair; for $\lambda > \lambda_{c2}$ it is the excitation across the CDW gap; at intermediate values $\lambda^{**} < \lambda < \lambda_{c2}$ it is a combined effect. Comparison of the gaps in Figure 2(c) with the various gaps in Figure 2(a) is instructive.

12.   In the region $\lambda^{**} < \lambda < \lambda_{c2}$, as the temperature increases there is a transition at T_{SC} between SC and a semiconductor solid. The SC exhibits an ordinary Meissner effect, but the order parameter Δ has a strong dependence on the current j and consequently the critical current j_c, in response, is much smaller than it would be in an ordinary SC with the same gap.

13.   In general terms the order parameters interfere destructively, with CDW tending to suppress SC and vice versa. A large enough CDW which produces a semiconductor state may completely destroy SC. But all metallic CDW states become SC at low enough temperatures. Not all semiconducting CDWs on the other hand show a lack of SC at low temperature: if the CDW gap is small enough, smaller than an "ideal" SC gap, as the temperature decreases the system makes a transition from a non-uniform semiconductor to a non-uniform SC. This semiconducting-SC state shows persistent currents and a Meissner effect, but the dependence of the SC parameters on the total current is different and more pronounced than in the metallic SC state.

14.   Experimental data on the SC properties of the transition-metal dichalcogenides show that:

(i) all metallic CDW systems are SC at low enough temperature;

(ii) those layered compounds which exhibit no SC show a non-linear low temperature specific heat, i.e. a lack of ordinary metallic properties;

(iii) it has been reported, although uncertainties still remain unsolved, that some systems which are SC at low enough temperature are probably semiconducting at temperatures higher than T_{SC}.

All these findings are in agreement with our theory.

REFERENCES

Bardeen, J., Cooper, L.N. and Schrieffer, J.R. (1957). *Phys. Rev. 108*,1175.

Bilbro, G. and McMillan, W.L. (1976). *Phys. Rev. B 14*,1887.

Chan, S.K. and Heine, V. (1973). *J. Phys. F: Metal Phys. 3*,795.

DiSalvo, F.J., Schwall, R.,Geballe, T.H., Geballe, F.R. and Osiecki, J.H. (1971). *Phys. Rev. Lett. 27*,310.

Falicov, L.M. (1979). *In* "Physics of Semiconductors 1978" (J.A. Wilson, ed.), p. 53. The Institute of Physics Conference Series Number 43, Bristol and London.

Falicov, L.M. and Penn, D.R. (1967). *Phys. Rev. 158*,476.

Lomer, W.M. (1962). *Proc. Phys. Soc. (London) 80*,489.

Morris, R.C. (1975). *Phys. Rev. Lett. 34*, 1164.

Morris, R.C. and Coleman, R.V. (1973). *Phys. Rev. B 7*,991.

Parks, R.D. (1969) editor "Superconductivity". Dekker, New York, (in two volumes).

Penn, D.R. (1966). *Phys. Rev. 142*,350.

Wilson, J.A., DiSalvo, F.J. and Mahajan, S. (1975). *Adv. Phys. 24*,117.

Wilson, J.A. and Yoffe, A.D. (1969). *Adv. Phys. 18*,193.

ANISOTROPY OF ELECTRON-PHONON INTERACTION AND SUPERCONDUCTING ENERGY GAP IN NIOBIUM[*]

G. W. Crabtree, D. H. Dye, D. P. Karim

Argonne National Laboratory
Argonne, Illinois 60 439

J. B. Ketterson
Northwestern University
Evanston, Illinois 60201

Despite a great deal of effort devoted to the study of the electron-phonon interaction, our knowledge of anisotropy effects is rather meager. Much of our experimental information comes from tunneling experiments which measure the electron-phonon spectral function $\alpha^2F(\omega)$ and from inversion of empirical T_c equations which give λ directly from T_c. As a result, most theoretical effort has gone into calculation of these quantities, neither of which is sensitive to k-space anisotropy. In transition metals the comparison of theory with λ and $\alpha^2F(\omega)$ has, to a large extent, reached its natural limit. Severe experimental problems have prevented tunneling from yielding much hard information about $\alpha^2F(\omega)$, and the empirical values of λ are not inherently accurate enough for close quantitative comparison. On the other hand, calculations based on both the rigid muffin tin and tight binding schemes are capable of reproducing the observed qualitative trends across the transition series fairly well.

A much more comprehensive test of theory is provided by the k-dependence of the coupling. Such information is more fundamental than the average quantities $\alpha^2F(\omega)$ and λ because it allows the coupling strength for individual electron states to be examined. Any disagreement between

[*]*Work supported under the auspices of the U.S. Department of Energy.*

113

theory and experiment in the k-dependence of the electron-
phonon interaction cannot be attributed simply to a uniform
over- or underestimate of the coupling strength for all
electrons, but must be recognized as a more fundamental
shortcoming of theory.

In this paper we present data on the anisotropy of the
electron phonon coupling in Nb. We define a quantity $\lambda(\vec{k})$
which describes the interation of an electron in a particular
k state with the system of all phonons. Through a combination
of accurate, complete de Haas-van Alphen (dHvA) measurements
and sophisticated band theory we are able to derive $\lambda(\vec{k})$ on
all 3 sheets of the Fermi surface. Furthermore, the aniso-
tropies in $\lambda(\vec{k})$ can be related to anisotropies in $\Delta(\vec{k})$,
so that a rather complete picture of both the electron-phonon
interaction and superconductivity is obtained. This represents a new source of information based largely on experiment
which is much more complete and detailed than that previously
available.

We obtain $\lambda(\vec{k})$ from a comparison of Fermi velocities
derived from dHvA experiments, $|\vec{v}_{exp}(\vec{k})|$, with that derived
from band structure $|\vec{v}_{band}(\vec{k})|$ according to the relation

$$\frac{|\vec{v}_{band}(\vec{k})|}{|\vec{v}_{exp}(\vec{k})|} = 1 + \lambda(\vec{k}) \qquad .$$

Defined in this way, $\lambda(\vec{k})$ includes contributions from both
the electron-phonon and the electron-electron interaction.
Because of its strong electron-phonon coupling and high
superconducting T_c, we expect that the former contribution is
dominant. However, our knowledge of electron-electron effects
in transition metals is at present very weak, so that we cannot rule out a significant contribution from the latter.

The dHvA areas and masses for Nb were taken from an
earlier study (Karim et al., 1978) in fields up to 130 kG and
temperatures as low as 0.3 K. On-line Fourier transforms
were used to recover weak signals in the presence of much
stronger ones, and to separate the many closely spaced frequency branches. The full power of the experimental system
was required to observe the complete Fermi surface, and it is
quite clear that still higher fields and lower temperatures
will be needed to study materials in which many body enhancements or flat energy bands produce even higher effective
masses.

The orbital areas and masses were inverted using a phase
shift parametrization scheme based on the KKR method of band
calculation (Crabtree, et. al. 1979 a,b). Briefly, the
method consists of adjusting a set of phase shifts (which,
together with a lattice constant and a Fermi energy, com-

pletely determines the detailed Fermi surface geometry) to
obtain the best agreement between the areas of certain orbits
in the parametrization with experiment. This set of phase
shifts then gives a definitive model for the Fermi surface
geometry. For Nb, a non-relativistic, 5 parameter phase
shift fit was made to 10 symmetry direction areas (see
Crabtree, et al., 1979b, for details) The quality of the fit
is indicated by the 0.6% rms error in fitting the 10 orbital
areas, and by the agreement of the occupied volume of the
Brillouin zone with the requirement of charge neutrality to
better than 0.3% of one electron.

To obtain Fermi velocites $v_{exp}(\vec{k})$, the energy derivatives
of the phase shifts were adjusted to give the best agreement
of 8 effective masses with experiment. The rms error of the
fit was 3.5%, (just outside experimental error of the mass
measurements) and the enhanced density of states predicted
by the fit agreed with specific heat experiments (Ferreira
da Silva, et. al., 1969) to within 1.5%. This excellent
agreement of both orbital and Fermi surface average quan-
titites is strong evidence that the phase shift model is a
correct description of the experimental Fermi surface
properties.

The one-electron velocities $v_{band}(\vec{k})$ were taken from the
work of Elyashar and Koelling (1977), who did a relativistic,
self-consistent, general potential calculation with $\rho^{1/3}$
exchange. Their published results for the Fermi surface geometry are in quite good agreement with our experiment, differing by about 7% rms error for the 10 symmetry direction
dHvA orbits. To improve the agreement still further, the p
states in their calculation where shifted down by 50 mRyd
and the Fermi energy recalculated. With this single empirical
adjustment, the rms error was reduced an order of magntiude
to about 0.7%. The detailed k_F from the band theory and
the parametrization consistently lie within 0.3% of each
other on all three sheets of the Fermi surface and show no
systematic error.

Figure 1 shows $v_{exp}(\vec{k})$, $v_{band}(\vec{k})$ and the values of $\lambda(\vec{k})$
derived from them, on the intersection of the (100) and (110)
planes with all 3 sheets of the Fermi surface. There are
several features to note. First, the bumps in the $\lambda(k)$
curve for the octahedron and jungle gym in the (100) and (110)
planes are an artifact of our fitting procedure and should be
ignored. In our non-relativistic fit, $v_{exp}(\vec{k})$ is allowed to
be discontinuous where the two sheets are degenerate, introducing unphysical jumps at these isolated angles which are
amplified in the $\lambda(\vec{k})$ curve. Second, the most remarkable
feature of $\lambda(\vec{k})$ is that it does not vary much within any
one sheet. Despite large changes in $v_{exp}(\vec{k})$, $\lambda(\vec{k})$ changes
by no more than about 10-15% on any one sheet. However,

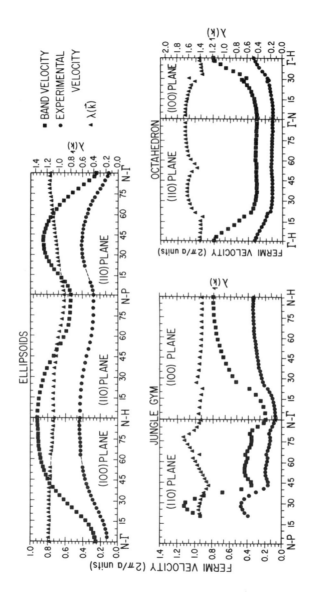

FIGURE 1. $|\vec{v}_{band}(\vec{k})|$, $|\vec{v}_{exp}(\vec{k})|$ and $\lambda(\vec{k})$ for the symmetry planes inter-secting all three sheets of the Fermi surface of Nb.

there are significant changes in λ between any two sheets.
By integrating over each sheet separately, we obtain the
average values of λ for each sheet shown in Table I.
There is a definite trend: the ellipsoids are lowest, the
jungle gym intermediate, and the octahedron the largest.
This makes the k dependence particularly easy to characterize.
The three average values give a concise yet accurate picture
of the λ anisotropy in Nb.

By doing a surface average, the overall value of λ = 1.33
is obtained. This agrees perfectly with the value inferred
from comparison of specific heat data with band theory density
of states, since our phase shift model reproduces the specific
heat to within 0.3%. The agreement with tunneling values of
λ is less easy to characterize. For Nb, the problems of
producing a good barrier while maintaining a metal surface
characteristic of the bulk (Arnold et. al., 1978) have impeded
experimental work, while conflicting results for $\alpha^2 F(\omega)$ and
unphysical values of μ^* required to invert the data (Bostock
et al., 1977) have clouded the interpretation. Current work
seems to put λ at about 1.0, (Arnold et al., 1978; Robinson
and Rowell, 1978) but it is probably too early to use that
value to draw hard conclusions. In contrast to tunneling,
our work is based entirely on bulk measurements of well-
characterized single crystals, and the phase shift fitting
scheme and band calculations used to analyse the data predict
accurate experimental numbers whenever comparison can be
made.

Of the many theoretical calculations of electron-phonon
effects in Nb, there are three which specifically examine

Table I: Average values on each sheet of the Fermi surface
of the one electron density of states from the adjusted band
theory, n_{band}, enhanced density of states from the phase
shift model, n_{exp}, electron-phonon interaction parameter
λ<sheet>, and superconducting energy gap Δ<sheet>.

	Ellipsoids	Octahedron	Jungle Gym	Total
n_{band}	0.596	0.202	0.648	1.446
n_{exp}	1.249	0.547	1.573	3.360
λ<sheet>	1.1	1.7	1.4	1.3
$\dfrac{\Delta\text{<sheet>}}{\Delta_o}$	0.9	1.0	1.1	

Table II: Theoretical calculations of λ <sheet>.

	Ellipsoids	Octahedron	Jungle Gym	Total
Harmon and Sinha Rigid muffin tin	1.85	1.28	1.37	1.58
Pinski, et. al. Rigid muffin tin	1.17	1.09	1.08	1.12
Peter et. al Tight Binding	1.35	1.92	1.90	1.69

anisotropy effects. The work of Harmon and Sinha (1977) and
of Pinski, Butler, and Allen (1979) is based on the rigid
muffin tin approximation, while Peter, Ashkenazi, and
Dacorogna (1977) use a tight binding scheme. Average values
of λ on each sheet of the surface are shown in Table II
for these calculations. The two rigid muffin tin calculations
predict trends opposite to our results, while the tight
binding work is more nearly in agreement. The qualitative
disagreement with rigid muffin tin results is very striking,
in view of its popularity and success in dealing with transition metals generally. To find the underlying causes for
the discrepency, a detailed analysis of the roles of charge
deformation within the muffin tins and of Coulomb interaction
of charge outside the muffin tins needs to be made. Perhaps
the frozen phonon approach discussed by Harmon, et. al.
(1979) will shed more light on these points.
    In addition to information on the electron-phonon inter-
action, our results can be used to infer a great deal about
anisotropic superconducting properties. The reason for this
is the direct connection between the energy gap $\Delta(\vec{k})$ and
the electron-phonon coupling strength $\lambda(\vec{k})$. In the weak
anisotropy limit, this connection can be made explicit (Allen,
1978)

$$\frac{\Delta(\vec{k})}{\Delta_o} = 1 + \frac{(1+\mu^*)}{(1+\lambda)(\lambda-\mu^*)}\,[\lambda(\vec{k})-\lambda]$$

where Δ_o and λ are average values of $\Delta(\vec{k})$ and $\lambda(\vec{k})$.
Using this relation, our model for $\lambda(\vec{k})$ translates directly
to a model for $\Delta(\vec{k})$. All of the comments about $\lambda(\vec{k})$
being characterized by the average values on the three sheets
apply equally to $\Delta(\vec{k})$. Using our average values of $\lambda(\vec{k})$
shown in Table I and $\mu^* = 0.13$, the gap values shown in
Table I are obtained. Of course the gap values are distributed smoothly throughout some range, but the distribution

will be peaked about the three values shown.

Our data support a multiple gap model for Nb. That is, for most directions in k-space gaps associated with each of the three sheets exist simultaneously. However, very little orientation dependence to the gaps is expected, since within any sheet the gap is fairly constant. Whether this picture of the gap structure in Nb could be confirmed in a tunneling experiment is a very interesting question.

Suggestions that two gaps of widely different magnitude exist in Nb have occasionally been made to explain various tunneling, heat capacity, ultrasonic attenuation, and thermal conductivity data (see Almond, et. al., 1972 and references therein). Such suggestions require a second gap either a factor of 10 smaller or a factor of 3 larger than the main gap. Our data show all the gaps to be within about 10-15% of the average, with no indication that gaps of the suggested size exist.

REFERENCES

Allen, P. B. (1978), Phys. Rev. B17, 3725

Almond, D. P., Lea, M. J., and Dobbs, E. R. (1972), Phys. Rev. Letters 29, 764.

Arnold, G. B., Zasadzinski, J. and Wolf, E. L. (1978) Physics Letters 69A, 136.

Bostock, J., Diaduk, V., Cheung, W. N., Lo, K. H., Rose, R. M. and MacVicar, M. L. A. (1977), Ferroelectrics 16, 249.

Crabtree, G. W., Dye, D. H., Karim, D. P. and Ketterson, J. B. (1979a) Phys. Rev. Letters 42, 390.

Crabtree, G. W., Dye, D. H., Karim, D. P. and Ketterson, J. B. (1979b) J. Magnetism and Magnetic Materials 11, 236.

Elyashar, N. and Koelling, D. D. (1977), Phys. Rev. B15, 3620.

Ferreira da Silva, J. Burgemeister, E. A. and Pokoupil, Z. (1969), Physica (Utrecht) 41, 409.

Harmon, B. N. and Sinha, S. K. (1977), Phys. Rev. B16, 3919.

Harmon, B. N. and Ho, K. M. (1979), this conference.

Karim, D. P., Ketterson, J. B., Crabtree, G. W. (1978), J. Low Temp. Phys. 30, 389.

Peter, M., Ashkenazi, J., and Dacorogna, M., (1977), Helv. Physica Acta 50, 267.

Pinski, F. J. Butler, W. H. and Allen, P. B. (1979) Phys. Rev. B19, 3708.

Robinson, B. and Rowell, J. M. (1978), Inst. Phys. Conf. Ser. 39, 666.

FERMI SURFACE OF Nb₃Sb: MICROSCOPIC COMPARISON
OF THEORY AND EXPERIMENT

A. T. van Kessel, H. W. Myron and F. M. Mueller

Physics Laboratory and Research Institute for Materials
University of Nijmegen, Toernooiveld, Nijmegen
The Netherlands

A. J. Arko and G. Crabtree

Material Science and Solid State Divisions
Argonne National Laboratory
Argonne, Illinois

Z. Fisk

Institute for Pure and Applied Physical Sciences
University of California, San Diego
La Jolla, California

*An ab initio band structure calculation of Nb₃Sb is pre-
sented which yields a Fermi surface model of this A-15 compound. Direct comparison between the measured dHvA frequencies
can be made. The predicted and measured dHvA extremal cross
sectional areas shows near perfect agreement for seven sheets
of the Fermi surface. This represents the first Fermi surface
model of an A-15 compound which shows quantitative agreement
with a dHvA measurement.*

INTRODUCTION

For the last twenty years the electronic structure of the
A-15 compounds has been studied in order to understand the
origins of the various anomalous physical, phononic and electronic properties.[1] Density of electron states models[2] for the
A-15 compounds, although differing in detail and motivation,

has been proposed which qualitative demonstrates temperature
dependent anomalies. A common thread throughout these models
is a high value of the density of electron states at or near
the Fermi energy dropping off rapidly to a background value.
This common feature of a high density of states at the Fermi
level for the high T_c materials has also been recently verified
by various groups who have presented *ab initio* electronic structure calculations.[3-7] Although these various calculations do
have general structure in common, they differ substantially on
a microscopic level.

In order to evaluate the relative merits of these calculations, experiments which probe the Fermi surface are essential.
To date there have been a number of experimental investigations
of the Fermi surface by a variety of experimental techniques.
Graebner and Kunzler[8] measured the magnetothermal oscillations
of V_3Ge finding three orbits of very small extremal areas.
Arko, Fisk and Mueller[9] using the de Haas-van Alphen (dHvA)
effect studied Nb_3Sb finding sheets of the Fermi surface which
seemed to be consistent with energy surfaces surrounding the
points M and R of the Brillouin zone. Arko et al.[10] observed
dHvA oscillations in Nb_3Sn which were interpreted as a series
of nested ellipsoids centered about M.[6] These results were
quite preliminary because of the inherent difficulty of the
experiment.[10]

The positron annihilation technique has also been applied
to the study of the Fermi surface in V_3Si.[11] More recently,
two-dimensional angular correlation of angular radiation[12,13]
have yielded additional information concerning V_3Si. Because
of the complexity of the Fermi surface and the limited resolution of the technique a definitive model has not yet emerged.
However, some features of the Fermi surface from these experimental studies on V_3Si seem to be consistent with the band
structure of V_3Si and the isoelectronic Nb_3Sn.[6]

Because of all the inherent experimental difficulties
mentioned above, we have decided to investigate the Fermi surface of Nb_3Sb[14] since crystals are now available with resistivity ratios of 100 to 150, the critical field (H_{c2}) is low,
and band structure calculations for this specific material are
available. In this paper we present new dHvA data of Nb_3Sb
and compare them to a band structure in which a quantitative
comparison can be made. In addition, some results on the
effective mass and electron phonon coupling constants are made.

SAMPLE PREPARATION, EXPERIMENTAL AND THEORETICAL
TECHNIQUES

The single crystals of Nb_3Sb used in this study were prepared by means of an iodine vapor transport technique.[9,15] The
resistivity ratios of the actual samples used in the experiments were between 100-150; the samples were cut to 1 mm^3 with
a spark erosion machine and surface damage was etched away.

The dHvA techniques and methodologies utilized in this investigation have been described previously.[16] The oscillatory
component of the magnetic susceptibility was measured by means
of field modulation techniques for fields up to 75 kG and temperatures of 0.35 K, higher fields and lower temperatures than
previously reported.[9] Fourier transforms of previous data
were reexamined and new points in the (110) plane were found in
this way.

The band structure was calculated along the lines previously outlined.[6] Briefly, the electronic structure was calculated via the APW method (non-self consistently) using the
Slater local exchange approximation. The non-spherical corrections to the muffin tin (MT) potential were incorporated both
outside and inside the MT spheres. Since a large number of $\vec{k}$
points in the vicinity of E_F are required in order to calculate
dHvA frequencies, a generalized $\vec{k} \cdot \vec{p}$ interpolation method for
the electronic structure was developed.[17] This model scheme is
three orders of magnitude faster than the APW method and produces bands in the vicinity of E_F to within 1 mRy of their *ab
initio* values. This is not severe since the *ab initio* eigenvalues have been calculated to within 3 mRy of convergence.

DISCUSSION AND RESULTS

The electronic structure of Nb_3Sb in the vicinity of the
Fermi energy is shown in Fig. 1 along high symmetry lines and
planes. The Fermi energy falls above the states at Γ_{12} and has
intersections with the band structure surrounding the M point
(holes) and the R point (electrons). The box like Fermi surface sheet surrounding the Γ point previously found in Nb_3Sn[6]
is absent in the Sb by the addition of one extra electron per
non transition metal. This box like Fermi surface sheet has
been experimentally confirmed for isoelectronic V_3Si.[11-13] In
addition, one may expect that $Nb_3Sb_{1-x}Sn_x$ forms a rigid band
system, but this is not the case. The ordering of the states
at M from below to above E_F are M_9, M_{10}, M_2, and M_8 for Nb_3Sn
and M_2, M_9, M_{10}, and M_8 for Nb_3Sb. In all other respects, the
$Nb_3Sb_{1-x}Sn_x$ system behaves rigid band like in the vicinity of

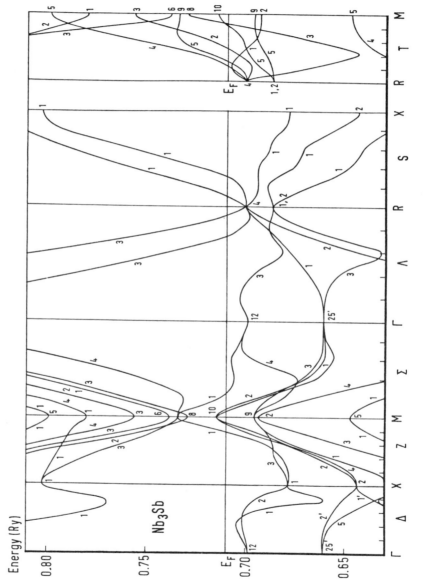

FIGURE 1.   The energy bands of Nb$_3$Sb *in the vicinity of the Fermi energy.*

E_F. The relative shift downward of the M_2, predominantly p non metal state, with respect to the d band manifold is due to the increasing electronegativity of Sb with respect to Sn. Thus a rigid band interpretation of the Fermi surface of Nb₃Sb using the Nb₃Sn band structure can lead to a false interpretation of some M centered sheets.

In Fig. 2 the Fermi surface of Nb₃Sb (intersecting bands 18-24) is shown in the principle symmetry planes. The principle structures are oscillated ellipsoidal holes (bands 18, 19 and 20) surrounding the M point and a confluence of electron sheets (bands 21 to 24) surrounding the R point. Other calculations[5,7] for Nb₃Sb yield a similar description of the Fermi surface around R, but shifts of approximately 30 mRy are required to obtain small hole pockets about the M point.

The measured and calculated dHvA frequencies are shown in Fig. 3a and 3b, respectively. We should note at this point that the theoretical curves have been shifted non uniformly in order to achieve an optimal fit to the measured dHvA frequencies. The shifts relative to the *ab initio* Fermi level are in mRy 2.5, 1.9, 1.0, −1.2, −3.6, 2.6, and 2.5 for bands 18 to 24, respectively. In the (100)plane, for frequencies greater than 9 MGauss, the 21st band (R centered) corresponds to the δ frequency, the 20th band (M centered) corresponds to the γ frequencies and the 22nd band (R centered) corresponds to the η frequency. Theory predicts the connectivity of the ε branches to the γ branches. The connection between γ_2 and χ and ε_1 have not been observed, the signals in this region being too weak.

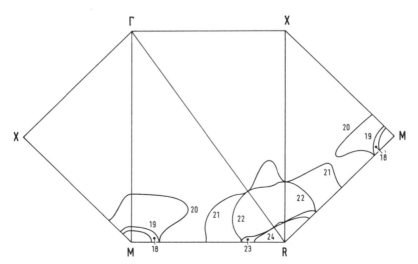

FIGURE 2. *The Fermi surface of Nb₃Sb along the principle symmetry planes.*

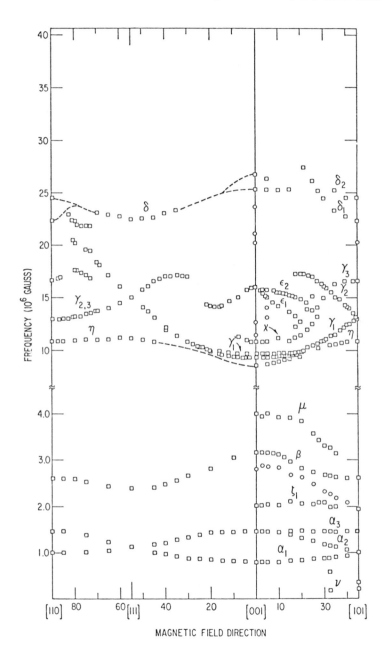

FIGURE 3a. Experimental dHvA frequencies of Nb₃Sb
plotted along high symmetry directions.

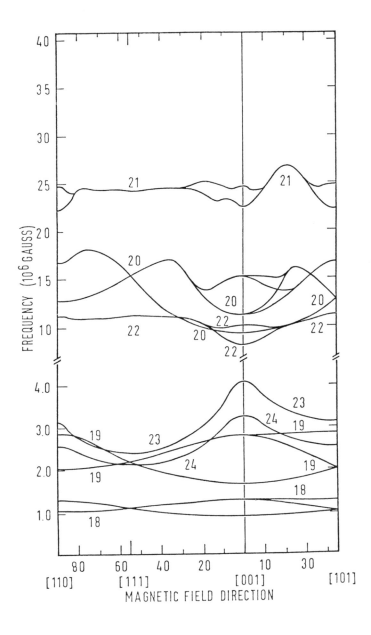

FIGURE 3b. Calculated dHvA frequencies of Nb₃Sb plotted along high symmetry directions.

For the low frequency dHvA orbits, band 18 (M centered) corresponds quite nicely to the α orbit. Two other Fermi sheets which correspond quite well to the field dependent dHvA frequencies are 23 (μ) and 24 (β), both of which are R centered. However, the higher M centered frequency, denoted by the calculation as 19, is not seen in the (100) plane. These orbits are seen (ξ_1) in the (100) plane, but the signal is quite weak and the upper branch of band 19 is not observed. This disappearance of the upper branch in the (110) plane and all the branches in the (100) plane can be explained by magnetic breakdown of this orbit. Preliminary experimental evidence indicates that this is the case.[18] The very low dHvA frequencies labelled by ν near the [101] axis on the (101) plane cannot be explained on the basis of the band structure calculation.

The effective masses have been determined experimentally for certain orbits and have been compared to the band masses. From this we have extracted some orbital electron-phonon coupling constants (λ). We see from Table I, that the values of λ

TABLE I. *Effective masses and orbital γ's for Nb_3Sb*

Frequency	Band	Orientation	m(exp)	m(theory)	
M centered					
α	18	[111]	0.26	0.21	0.24
γ_1	20	[001]	0.70	0.64	0.09
γ_1	20	[110]	1.75	1.35	0.30
$\gamma_{1,2}$	20	[101]	0.89	0.84	0.06
ε	20	[001]	1.34	1.12	0.20
R centered					
δ	21	[111]	1.58	1.30	0.22
δ	21	A	1.70	1.61	0.06
η	22	B	1.10	1.05	0.04
β	24	[101]	0.64	0.50	0.28
β	24	[111]	0.52	0.38	0.37

A - 27.5° *from* [001] *to* [101].

B - 40.0° *from* [001] *to* [101].

range from 0.04 to 0.35. However, what clearly emerges is an average λ which is a factor of 10 smaller than found for Nb₃Sn. Independent measurements[19] using heat capacity techniques yield an average λ of 0.3.

In conclusion we should like to mention the following points: 1) the Fermi surface of Nb₃Sb as calculated without the adjustment yields qualitative agreement with a majority of the dHvA branches, 2) many parts of the dHvA branches denoted as 21 (δ), 22 (η), 23 (μ), and 24 (β) were initially not measured[9] and were predicted by the band model before the data was reexamined, 3) after remeasuring the frequencies, the dHvA areas were recalculated by shifting the Fermi energy of the various branches by a maximum of 3.6 mRy in order to obtain an optimal fit. We believe this represents the first successful marriage of an *ab initio* calculation to the experimental FS of any A-15 compound.

This work was supported as part of the research program of the Stichting voor Fundamental Onderzoek der Materie with financial support from the Nederlandse Organisatie voor Zuiver Wetenschappelijk Onderzoek, the Department of Energy and the National Science Foundation.

REFERENCES

1. For a review, see M. Weger and I. B. Goldberg, in *Solid State Physics*, edited by H. Ehrenreich, F. Seitz and D. Turnbull (Academic, New York), Vol. 28; and (b) L. R. Testardi, in *Physical Acoustics*, edited by W. P. Mason and R. N. Thurston (Academic, New York), Vol. 10.

2. A. M. Clogston and V. Jaccarino, Phys. Rev. *121*, 1357 (1961); J. Labbe and J. Friedel, J. Phys. Radium *27*, 708 (1966); R. W. Cohen, G. D. Cody and J. J.Halloran, Phys. Rev. Lett., *19*, 840 (1967).

3. L. F. Mattheiss, Phys. Rev. B*12*, 2161 (1975).

4. T. J. Jarlborg and G. M. Arbman, J. Phys. F*7*, 1635 (1977); T. J. Jarlborg, J. Phys. F*9*, 283 (1979).

5. B. M. Klein, L. L. Boyer, D. A. Papaconstantopoulos and L. F. Mattheiss, Phys. Rev. B*18*, 6411 (1978).

6. A. T. van Kessel, H. W. Myron and F. M. Mueller, Phys. Rev. Lett. *41*, 181 (1978).

7. W. E. Pickett, K. M. Ho and M. L. Cohen, Phys. Rev. B*19*, 1734 (1979).

8. J. E. Graebner and J. E. Kunzler, J. Low Temp. Phys. *1*, 443 (1969).

9. A. J. Arko, Z. Fisk and F. M. Mueller, Phys. Rev. B*16*, 1387 (1977).

10. A. J. Arko, D. H. Lowndes, F. A. Muller, L. W. Roeland, J. Wolfrat, A. T. van Kessel, H. W. Myron, F. M. Mueller and G. W. Webb, Phys. Rev. Lett. *40*, 1590 (1978).
11. S. Berko and M. Weger, Phys. Rev. Lett. *24*, 55 (1970).
12. A. A. Manuel, S. Samoilov, M. Peter and A. P. Jeavons, Helv. Phys. Acta *52*, 37 (1979); A. A. Manuel, S.Samoilov, R. Sachot, P. Descouts and M. Peter, Solid State Commun. *31*, 955 (1979).
13. W. S. Farmer, F. Sinclair, S. Berko and G. M. Beardsley, Solid State Commun. *31*, 481 (1979).
14. A preliminary report of this work has been presented by A. J. Arko, G. W. Crabtree and Z. Fisk, *Conf. on Superconductivity in d- and f-Band Metals*, La Jolla, CA, June 1979 (this volume).
15. H. Schafer and W. Fuhr, J. Less Common Metals *8*, 375 (1965).
16. R. W. Stark and L. R. Windmiller, Cryogenics *8*, 872 (1968); A. J. Arko, M. B. Brodsky, G. W. Crabtree, D. Karim, D. D. Koelling and L. R. Windmiller, Phys. Rev. B*12*, 4102 (1975).
17. This procedure has been applied to paramagnetic bcc Fe by H. J. F. Jansen and F. M. Mueller, Phys. Rev. B*20*, 1426 (1979); A. T. van Kessel (unpublished).
18. D. J. Sellmeyer (private communication).
19. F. Y. Fradin, G. S. Knapp, S. D. Bader, G. Cinader and C. W. Kimball, *Superconductivity in d- and f-Band Metals*, edited by D. H. Douglass (Plenum, New York, 1976).

RECENT CALCULATIONS OF PHONON
SPECTRA OF TRANSITION METALS
AND COMPOUNDS

Werner Weber[1]

Kernforschungszentrum Karlsruhe
Institut für Angewandte Kernphysik I
7500 Karlsruhe, Postfach 3640
Federal Republic of Germany

*A new formulation of the lattice dynamics is presented,
based on the nonorthogonal tight binding scheme. The method
is applied for calculations of the phonon dispersion curves in
various transition metals and compounds. All anomalous features in the phonon spectra of these crystals are reproduced.
The physical origin of the anomalies and their relation to
superconductivity are elucidated.*

I. INTRODUCTION

Over the last decade, the intensive experimental and theoretical studies of the lattice dynamics of transition metals
and their compounds (TMC) have been spurred[2] by the observation that anomalies appear in the phonon spectra of those TMC
which exhibit high superconducting temperatures T_c and are absent in materials with vanishing T_c. The microscopic theory of
lattice dynamics could provide an understanding of this empirical correlation. However, first principles calculations of
the phonon spectra of TMC are very intricate, when based on

---

[1]*Part of this work was performed in collaboration with C.M.
Varma at Bell Laboratories, Murray Hill, N.J.*
[2]*See for instance the proceedings of the previous "Rochester"
conferences (Douglass, 1972 and 1976)*

131

the customary dielectric function method (Sham, 1974). So far
only simplified model calculations have been performed (Sinha
and Harmon, 1975, Hanke et al, 1976).

In recent years, we have developed a relatively simple, alternate approach of lattice dynamics, especially suited for
TMC with their tightly bound d-electrons (Varma and Weber, 1977
and 1979). In the following - after a brief outline of our
method - we will present results based on this theory and will
discuss the origin of the anomalies in TMC as well as their relation to superconductivity.

II. OUTLINE OF THE THEORY

Our theory is based on two ideas: i) In TMC the change of
the electronic charge caused by the displacements of the ions
is best described using orbitals which move with the ions[1].
Therefore, a nonorthogonal tight-binding scheme (NTB) has been
adopted. ii) The terms of the dynamical matrix are rearranged
such that the ion-ion and electron-electron Coulomb interactions V_{i-i} and V_{e-e} cancel each other as much as possible[2].

In harmonic approximation, the dynamical matrix D of the
phonons is obtained from the second derivatives of the total
crystal energy E_{tot} with respect to ion displacements

$$D = \nabla\nabla E_{tot} = \nabla\nabla(V_{i-i}+E_e)$$

$$= \nabla\nabla(V_{i-i}-V_{e-e}+E_\Sigma) \tag{1}$$

Here, E_e is the total electronic energy, treated in the adiabatic approximation as a parametric function of the ionic coordinates. Further is

$$E_\Sigma = \sum_{k,\mu} \epsilon_{k\mu} f_{k\mu}$$

the sum of the energies $\epsilon_{k\mu}$ of all occupied one electron states
(k,μ are wavevector and band indices, respectively, and $f_{k\mu}$ is
the Fermi occupation factor).

---

[1]*This idea goes back to Fröhlich (1966) and Friedel (1969)*
[2]*This rearrangement has also been suggested by Pickett and
Gyorffy (1976)*

We combine

$$D_o = \nabla\nabla(V_{i-i} - V_{e-e})$$

In D_o, the two large Coulomb terms cancel each other and we are left with essentially short range forces.
The remaining part

$$\nabla\nabla E_\Sigma = D_1 + D_2$$

consists of the first and second order perturbation theory "band structure" contributions. In elemental crystals, D_1 contains only short range forces and is not discussed here in detail. There remains D_2 which reads as

$$D_2(\lambda\alpha,\lambda'\beta|q) = -\sum_{\substack{k\mu\mu' \\ k'=k+q}} \frac{f_{k'\mu'}-f_{k\mu}}{\varepsilon_{k\mu}-\varepsilon_{k'\mu'}}\ g_{k\mu,k'\mu'}^{\lambda\alpha}\ g_{k'\mu',k\mu}^{\lambda'\beta} \tag{2}$$

Here, λ,λ' and α,β are sublattice and cartesian indices, respectively, and q is the phonon waverector.
The electron-ion form factors g are understood as fully renormalized and are given as

$$g_{k\mu,k'\mu'}^{\lambda\alpha} = \sum_{\kappa,m'} A^*(\kappa m|k\mu)\ [\gamma_\alpha(\kappa m,\kappa'm'|k')\delta_{\kappa\lambda}$$
$$- \gamma_\alpha(\kappa m,\kappa'm'|k)\delta_{\kappa'\lambda}]\ A(\kappa'm'|k'\mu') \tag{3}$$

with $A = S^{-1/2}\ U$. Here, U are the eigenvectors originating from the eigenvalue equation $S^{-1/2}\ H\ S^{-1/2}\ U = UE$ for NTB Hamiltonian. Finally,

$$\gamma_\alpha(\kappa m,\kappa'm'|k) = N^{-1} \sum_{ll'} [\nabla_\alpha H(l\kappa m,l'\kappa'm')$$
$$\varepsilon\nabla_\alpha S(l\kappa m,l'\kappa'm')]\ \exp[ikR(l\kappa,l'\kappa')] \tag{4}$$

where $\nabla_\alpha H$ and $\nabla_\alpha S$ are the real space derivatives of energy and overlap matrix elements between sites $l\kappa$ and $l'\kappa'$ involving orbitals m and m'. Further we define $\varepsilon = (\varepsilon_{k\mu} + \varepsilon_{k'\mu'})/2$.

The phonon linewidth[1] is given by

$$\gamma_{qj} = \pi h/2 \sum_{\substack{\lambda\lambda' \\ \alpha\beta}} (m_\lambda m_{\lambda'})^{-1/2} e(\lambda\alpha|qj)e(\lambda'\beta|-qj)$$

$$x \iint_{FS} d\sigma_{k\mu} \, d\sigma_{k'\mu'} \, g^{\lambda\alpha}_{k\mu,k'\mu'} \, g^{\lambda'\beta}_{k'\mu',k\mu} \, \delta(k-k'-q) \tag{5}$$

with $e(\lambda\alpha|qj)$ being the phonon polarization vector and m_λ an atomic mass. The integrals are performed over the Fermi sur-face.

The computational advantages of our approach are as follows: i) The changes in the Hamiltonian due to ion motion are ex-pressed in terms of derivatives of matrix elements taken at discrete lattice points. We do not explicitly require the knowledge of the wave functions and potential gradients at all points in space. ii) The effects of screening, exchange and correlation are included in the terms ∇H and ∇S, as long as these quantities are determined from self-consistent band-structure calculations at different lattice configurations. iii) In TMC, the significant terms ∇H and ∇S are confined to a few nearest neighbors.

III. METHOD OF CALCULATION

So far, in our calculations we have focussed on D_2 which, as we will show, is the source of the long range oscillatory force fields manifest in the phonon anomalies. To compute D_2, we use a NTB Hamiltonian with an orbital basis sufficiently large to yield good agreement with APW band structure calculations. The radial derivatives in ∇H and ∇S are determined as indicated above, provided that APW calculations for different lattice constants have been available. Otherwise, we have calculated them from Herman-Skillman atomic wave functions and potentials. In general, the results of the two methods agree within $\sim 20\%$.

At present, the short range terms $D_0 + D_1$ are parametrized by nearest neighbor force constants which are fitted such that the eigenvalues of D yield best agreement with experimental phonon curves.

[1]*Only contributions second order in displacements are con-sidered.*

IV. RESULTS AND DISCUSSION

A. Nb-Mo

Calculations for the phonon dispersion curves of Nb($T_c \sim 9K$), $Nb_{0.25}Mo_{0.75}$ ($T_c < 0.4K$) and Mo ($T_c \sim 1K$) are shown in Fig. 1. Apart from the good overall agreement with experiment we note that all anomalous features in the phonon curves are correctly reproduced; e.g. the dip near (0.7,0,0), the deep minimum near

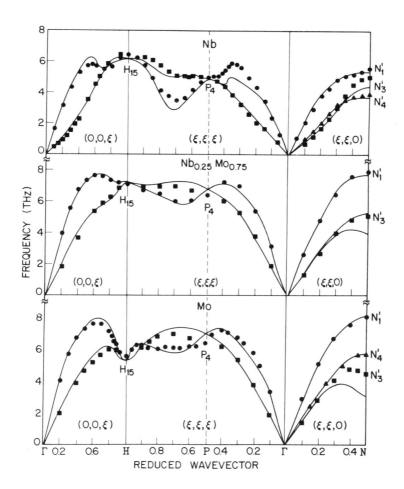

FIGURE 1. Calculated phonon dispersion curves for Nb, $Nb_{0.25}Mo_{0.75}$, and Mo. The circles, squares, and triangles are the experimental results of Powell et al. (1968).

(0.7,0.7,0.7) in the longitudinal branches of Nb, the sharp minimum at the H point in Mo, concomitant with a band crossing
along (ξ,ξ,ξ) near H. No anomalous features are seen in
$Nb_{0.25}Mo_{0.75}$, where the electronic density of states at the
Fermi energy $N(E_F)$ has a minimum value (Mattheiss, 1970).

All the structure in the dispersion curves of Nb and Mo
arises from the term D_2. As we have used a rigid band model to
calculate D_2 in Nb and Mo and thus have only changed E_F according to the different band-filling, it is evident that the
anomalous features in D_2 arise from scattering between states
near $E_F (\sim \pm 0.5eV)$. However, the anomalies are not caused by
structure in the energy denominator (the "bare susceptibility")
of Eq. 2; i.e., they are not Fermi-surface anomalies in the
Kohn-Overhauser sense. Instead, they mainly stem from the
strong q-dependence of the electron-ion form factors g, roughly
approximated as the difference of the electron velocities of
states (k,μ) and $(k'\mu')$

$$g^{\alpha}_{k\mu,k'\mu'} \;\; \propto \;\; (V^{\alpha}_{k\mu} - V^{\alpha}_{k'\mu'}) \; ; \;\; V^{\alpha}_{k\mu} = \partial \varepsilon_{k\mu}/\partial k^{\alpha}$$

Thus the form factors g^{α} are large when they couple states
with large and opposite band dispersion along the α-th components of wave vectors k and k'. Along other components of k and
k', flat bands are advantageous to produce a large phase space
for the sum over k in Eq. 2.

Therefore, very anisotropic electronic band topologies, and
as a consequence, very anisotropic Fermi surfaces are favourab-
le for the occurence of the phonon anomalies. In Nb this is the
"jungle gym", three perpendicular, interpenetrating Fermi surface pieces of cylindrical shape; in Mo these are the octahedra
around the Γ and H points. We note that, of course, a large
number of states near E_F should be available; i.e., $N(E_F)$ has
to be relatively large.

B. *Refractory Compounds*

Very anisotropic electronic bands are also prevailent in the
refractory compounds ZrC, TiC, NbC, TiN, VN, NbN, etc. all of
rocksalt structure (Schwarz 1977, Neckel et al, 1976). To a
good approximation, a rigid band model applies to these materials too. For ZrC or TiC, $N(E_F)$ lies close to a minimum of the

electronic density of states, we have $T_c < 1K$, and the phonon dispersion curves look very smooth, similar to those of alkali halides (Smith 1972, Pintschovius et al., 1978). With one more valence electron to fill the bands, NbC ($T_c \sim 10K$) or TiN($T_c \sim 6K$) have much higher values of $N(E_F)$. Furthermore, their Fermi surfaces exhibit very typical "jungle gym" structure, even more distinct than in elemental Nb (Gupta and Freeman, 1976, Klein et al 1976). In the phonon dispersion curves of NbC or TiN, the very pronounced minima (Smith, 1972, and Kress et al. 1978) roughly correspond to diameters of the jungle gym cylindrical rods.

A phonon calculation similar to the one for Nb-Mo reproduces all the observed structure (see Fig. 2). Again the anomalies stem from D_2, from scattering between states near E_F. This is again caused mainly by the form factors g. These are very large for the reasons discussed in the case of Nb-Mo, but there is further substantial enhancement of the form factors due to σ-type interaction between Nb-d_{xy} and C-p orbitals (Weber,1979). We also note that, in contrast to Nb, there is some structure in the bare susceptibility alone, as was pointed out by Gupta and Freeman (1976) and by Klein et al (1976).

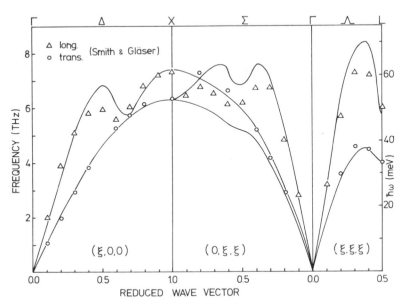

FIGURE 2. Acoustic phonon dispersion curves for NbC. Experimental results from Smith and Gläser (1971).

A challenging problem to experimentalists has been the study of the phonon spectra of NbN or VN, because of the enormous difficulties to grow sufficiently large single crystals. As NbN ($T_c \sim 16K$) and VN ($T_c \sim 9K$) have 10 valence electrons, one more than TiN or NbC, the bands are further filled and the Fermi surface changes drastically. The bandfilling also causes big changes in D_2 as is evident from Fig. 3. The maximum in $-D_2$ along ($\xi,0,0$) shifts from (0.65,0,0) to the X point at the zone boundary. In addition, this maximum becomes stronger and

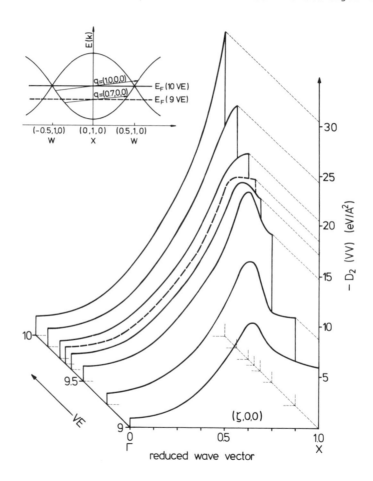

FIGURE 3. Plot of the longitudinal component of $-D_2$ in VN as functions of the wave vector along ($\xi,0,0$) and of the valence electron concentration. The dashed curve is for VE = 9.65. The insert shows schematically the transitions most relevant for D_2.

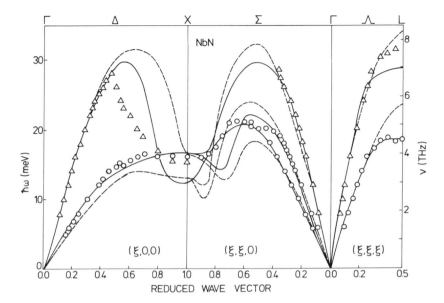

FIGURE 4.   *Acoustic phonon dispersion curves of NbN.*
Experimental values for NbN$_{0.84}$ from Ref. 4.   Dashed lines
show predicted curves for fully stoichiometric NbN.   Solid
lines represent calculations which take into account the
non-stoichiometry.

very sharp. As a consequence, a pronounced minimum at X should
appear in the phonon dispersion curves (see Fig. 4).

This result has been confirmed by very recent neutron scattering studies at NbN$_{0.84}$ single crystals (Reichardt and Scheerer, 1979, Kress et al. 1979), which agree well with the theoretical curves (see Fig. 4). Some discrepancies remain, we believe that they are caused by the N vacancies in the NbN$_{0.84}$
sample. Indeed, when we decrease the bandfilling corresponding
to ∿9.6 valence electrons, we are able to further improve the
agreement between theory and experiment. Very similar results
are found for VN (Weber et al. 1979).

V. ANOMALIES AND HIGH T$_c$

To elucidate the problem, in how far the phonon anomalies
are related to high T$_c$ values we have also studied the phonon

linewidths γ_{qj} (eq.6) for Nb and NbC. The γ_{qj} are connected to the electron-phonon coupling parameter λ by (Allen 1972)

$$\lambda = (2\pi hN(E_F))^{-1} \sum_{qj} \gamma_{qj}/\omega^2_{qj}$$

and λ essentially determines T_c as

$$T_c \underset{\sim}{\sim} <\omega> \exp(-(1+\lambda)/\lambda) .$$

D_2 and γ can be interpreted as the real and imaginary part of the phonon self-energy; thus there have been speculations that the two quantities are closely related (Allen and Cohen 1972).

Results of γ_{qj} and D_2 for NbC are shown in Fig. 5. Very clearly, their q-dependences are almost identical. In particular, very large peaks of γ_{qj} are found for the anomalous phonons, just where $-D_2$ exhibits large maxima, too. In Nb, $-D_2$ and γ_q also show this parallelism; however the peaks are less pronounced. We note that, in Nb, our γ_{qj} results agree well with those of Butler et al (1977) who used the rigid-muffin-tin approximation.

In NbC, the effect of the anomalies is indeed essential for the magnitude of λ, as the anomalous phonons contribute not only via the frequency softening - this has been found to be by far not enough to account for the change of λ between, for instance, ZrC and NbC - but to an even larger extend via the very big linewidths.

The very anisotropic nature of the electron-phonon coupling in NbC is manifest not only in the finding that only the few anomalous phonons couply strongly between electronic states at E_F, but also that relatively few electronic states are involved in this strong coupling.

In conclusion, as γ and D_2 are indeed closely related, the anomalies and the high T_c's both result from the same strong electron-phonon coupling.

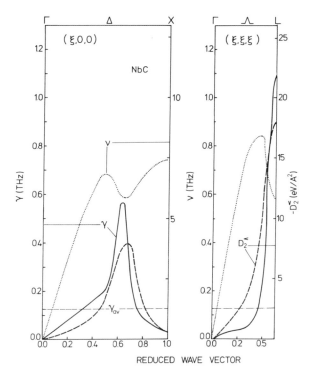

FIGURE 5. Longitudinal acoustic phonons ν, $-D_2^<(Nb-Nb)$
and phonon linewidths γ along two symmetry directions in
NbC. $D_2^<$ includes scattering processes $\geq \pm 0.5$ eV around E_F.
Also shown is γ_{av}, the average value of all linewidths γ_{qj}
in the Brillouin zone.

REFERENCES

Allen, P.B. (1972). *Phys. Rev.* B6, 2577
Allen, P.B., and Cohen, M.L. (1972). *Phys. Rev. Lett.* 29, 1593
Butler, W.H., Smith, H.G., and Wakabayashi, N. (1977). *Phys.*
    Rev. Lett. 39, 1004
Douglass, D.H., Ed. (1972). "Superconductivity in d- and f-Band
    Metals", lst Rochester Conference. Am. Inst. Phys., New
    York
Douglass, D.H., Ed. (1976). "Superconductivity in d- and f-Band
    Metals", 2nd Rochester Conference. Plenum Press, New York
Friedel, J., (1969). In "The Physics of Metals" (J.M. Ziman,
    ed.). Cambridge Univ. Press, Cambridge, U.K.
Fröhlich, H., (1966). In "Perspectives of Modern Physics",
    (Marshak, R.E., ed.) Interscience, New York

Gupta, M., and Freeman, A.J. (1976). in Douglass (1976)

Hanke, W., Hafner, J., and Bilz, H. (1976). *Phys. Rev. Lett.37*, 1560

Klein, B.M., Papaconstantopoulos, D.A., and Boyer, L.L. (1976). in Douglass (1976).

Kress, W., Roedhammer, P., Bilz, H., Teuchert, W.D., and Christensen, A.N. (1978). *Phys. Rev. B 17*, 111

Kress, W., Teuchert, and Christensen, A.N. (1979). to be published

Mattheiss, L.F. (1970). *Phys. Rev. B1*, 373

Neckel, A., Rastl, P., Eibler, R., Weinberger, P., and Schwarz K. (1976). *J. Phys. C 9*, 579

Pickett, W.E., and Gyorffi, B.L. (1976). in Douglass (1976)

Powell, B.M., Martel, P., and Woods, A.D.B. (1968). *Phys. Rev. 171*, 727

Reichardt, W., and Scheerer, B. (1979). to be published

Sinha, S.K., and Harmon, B.N. (1975). *Phys. Rev. Lett. 35*, 1515

Sham, L.J. (1974). In "Dynamical Properties of Solids" (G.K. Horton and A.A. Maradudin, ed.s), North Holland, Amsterdam.

Smith, H. in Douglass (1972).

Smith, H.G., and Gläser, W. (1971). In "Phonons", Proc. Int. Conf. in Rennes (Nusimovici, M.A., ed.). Flammarion, Paris.

Schwarz, K., (1977). *J. Phys. C 10*, 195

Varma, C.M., and Weber, W. (1977). *Phys. Rev. Lett. 39*, 1094

Varma, C.M., and Weber, W. (1979). *Phys. Rev. B19*

Weber, W., Roedhammer, P., Pintschovius, L., Reichardt, W., Gompf, F., and Christensen, A.N. (1979) to be published.

Weber, W. (1979). to be published.

SOFT MODES AND LATTICE INSTABILITIES IN α-U

H. G. Smith
N. Wakabayashi
W. P. Crummett[1]
R. M. Nicklow

Solid State Division
Oak Ridge National Laboratory[2]
Oak Ridge, Tennessee

G. H. Lander
E. S. Fisher

Materials Science Division
Argonne National Laboratory
Argonne, Illinois

I. INTRODUCTION

The physical properties of α-U have been extensively studied by many investigators. Anomalies in the elastic constants (Fisher and McSkimin, 1961; Fisher and Dever, 1968) and lattice constants (Barrett et al., 1963) near 43 K and 22 K have been reported, suggesting some type of structure transformation. There is an overall volume expansion below 43 K. However, careful x-ray (Barrett et al., 1963) and neutron diffraction (Lander and Mueller, 1970) studies did not reveal any changes from the orthorhombic space group Cmcm or any significant internal distortion from the room temperature

[1]*Present address: Baptist College of Charleston, Charleston, South Carolina.*
[2]*Operated by Union Carbide Corporation with the U.S. Department of Energy.*

structure. Actually, dilatation measurements (Steinitz et al., 1970) indicated three transitions, two first-order transitions at 23 K and 37 K, and a second order transition in the vicinity of 43 K. Specific heat studies (Crangle and Temporal, 1973) on a "single" crystal specimen revealed similar anomalies, but studies on polycrystalline samples indicated only a second-order type of anomaly near 43 K.

The superconducting properties of uranium have also attracted much attention (Geballe et al., 1966; Fisher et al., 1968) and here again the results seem to depend on the crystalline state of the samples; for example, T_c is much less in a single crystal than in some polycrystalline specimens (e.g. ≤ 0.1 K vs ≈ 1.0 K) (Bader et al., 1975). However, at 10 kbar both types of samples exhibit bulk superconductivity with a T_c of the order of 2 K (Smith, 1972; Smith and Fisher, 1973). Modest pressures also tend to inhibit the elastic constant anomalies (Fisher and Dever, 1970). It has been suggested (Fisher and Dever, 1968; Bader and Knapp, 1975; Wittig, 1975) that soft modes exist in α-U and are related in some way to the above mentioned anomalies.

Extensive measurements (Crummett et al., 1979) of the phonon dispersion curves were carried out along the principal symmetry directions, and the results, indeed, revealed that the frequency of one of the Σ_4 branches becomes very low near $q = (\frac{1}{2},0,0)$. This mode has the eigenvector corresponding to the in-phase y-displacements and the out-of-phase x-displacements of the two basis atoms in the unit cell, namely a mixture of the TA and LO modes. The observed neutron scattering intensities indicated that these low frequency Σ_4 phonons are mainly LO-like. Preliminary low temperature measurements showed a pronounced softening of this mode and the formation of a superlattice incommensurate with the main lattice. This report represents a more detailed study of the lattice softening and the superlattice formation. A very brief account has been given elsewhere (Smith et al., 1979).

II. EXPERIMENTAL RESULTS

The neutron scattering measurements were carried out on two single crystals of uranium, each with a volume of approximately 0.1 cm^3, on the HB-3 triple-axis neutron spectrometer at the Oak Ridge HFIR. One crystal was used for measurements in the (hk0) zone and the other for the (h0ℓ) zone. For the low temperature measurements the crystals were mounted in a closed-cycle He-gas refrigerator. Be monochromators were used throughout the investigation, whereas, both Ge and Be were employed as analyzers, occasionally with a graphite filter.

A. Inelastic Scattering

Most of the inelastic scattering measurements were made
under rather coarse resolution conditions in order to permit
measurements with adequate intensity within a reasonable
length of time, primarily because of the small sample size and
the effect of the Boltzman factor at low temperatures. Data
were collected in either the constant-Q or constant-E mode
with scattered energy E' of the order of 6-8 THz. The intensity of the Σ_4-LO mode was highest at Q vectors near (2.5,0,1)
and (2.5,2,0) confirming the short wavelength optic-like
character of the atomic displacements. The temperature dependence of this mode is shown in Fig. 1. It was impossible to
precisely determine the soft mode behavior near and below T_c
due to the growth of a central peak. In order to separate the
inelastic scattering from the elastic scattering it was
necessary to incorporate finer collimation and a lower E'
(3.6 THz). At this energy the inelastic scattering could only

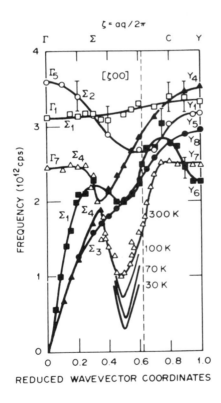

FIGURE 1. *Dispersion curves in* $[100]$ *direction.*

be studied at $Q = (1.5,0,1)$, resulting in a significant loss
of intensity since the intensity is proportional to Q^2.
Furthermore, there are fewer incident neutrons under these
conditions of finer resolution. Nevertheless, the indications
are that the dip in the Σ_4-LO branch does not collapse to zero
energy, but remains finite at an energy of about 0.4 THz, well
separated from the elastic superlattice peaks which are described below.

B. Elastic Scattering

As mentioned in the introduction, the softening of the
Σ_4-LO branch was accompanied by the formation of superlattice
reflections. They occurred in the vicinity of $(h+\frac{1}{2},k,\ell)$. A
careful study of the temperature dependence of the extra reflections revealed a small but finite intensity at 60 K which
increases slowly until 42 K is reached. Then the intensity
increases at a more rapid rate reaching a maximum at 10.4 K,
the lowest temperature studied, with a small but abrupt change
observed at 18 K. A high resolution study of the intensity of
the (1.5,0,1) reflection is shown in Fig. 2. Figure 3 shows a
slight deviation from the exact periodicity of a commensurate
superlattice which cannot be attributed to instrumental resolution effects. However, these are not satellites about the Γ
points, as is typical of incommensurate charge density waves
(ICDW), for all the new reflections $(h+\frac{1}{2}-h\delta,k,\ell)$ are shifted
a very small amount toward the origin parallel to h, somewhat
reminiscent of the ω-phase structure in the low concentration
Zr-Nb alloys (Lin et al., 1976). On the basis of 28 measurements of 18 superlattice reflections $(1-\delta) = 0.9963 \pm 0.0003$.

III. INTERPRETATION OF THE DATA

In the α-phase the four uranium atoms are in the positions
$4(c)$: $\pm(0,y,\frac{1}{4}) + (\frac{1}{2},\frac{1}{2},0)$, $y = 0.1024$. The extra reflections possibly represent a new phase α' growing in the α-
matrix and with a lattice constant slightly larger than the
bulk lattice constant. It can, perhaps, be considered a commensurate CDW state in this phase. Below 43 K the peaks are
as narrow as the main Bragg reflections indicating appreciable
long range order. The eigenvector corresponding to this distortion is $(u,v,0; \bar{u},v,0)$ where (as can be shown from intensity considerations), the v component is very small. The
crystallographic description requires a doubling of the a_0
lattice constant with eight atoms now in the positions

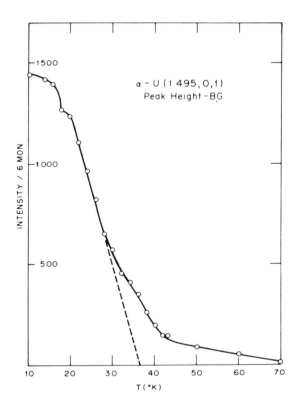

FIGURE 2.   Intensity of (1.495,0,1) reflection vs T.

+(ε_x,y,¼); +(¼,½+y,¼); ±(¼,½-y,3/4); ±(½-ε_x,y,¼) (see
Fig. 4). This arrangement may be thought of as a simple sinu-
soidal modulation in the x-direction, with the actual phase
angle undetermined; we arbitrarily choose the phase to make
the displacement zero at x = ¼. A comparison of $|F_0|^2$ vs
$|F_c|^2$ for the observable extra reflections based on $\varepsilon_x \cong 0.001$
gives satisfactory agreement for the (h0ℓ) and (hk0) zones of
both crystals (Table I). The induced magnetization density
studies of α-U by Maglic et al. (1978), and the angular dependence of the superlattice intensities observed in the present
study precludes the possibility that the extra reflections are
of magnetic origin.
     The possibility of two similar phases of α-U coexisting
is not a new concept. It was suggested by Fisher and Dever
(1968) in their interpretation of the elastic constant
studies. However, an independent coexistence of these two

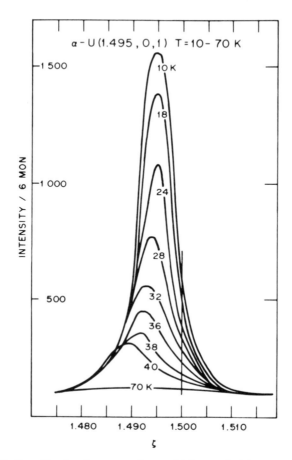

FIGURE 3.   Peak shape and position of (1.495,0,1) reflection vs T.

states should also produce temperature dependent peaks at the main Bragg positions slightly displaced from the intense peaks of the α-phase, but such peaks were not detected. Nor were they seen in the earlier high resolution x-ray experiments (Barrett et al., 1963).

The shifts in the weak reflections may be due to discommensurations (McMillan, 1976), and preliminary calculations with one such model in the form of a so-called 'stacking soliton' (Horovitz et al., 1978; Bak and Timonen, 1978; Pynn, R., 1978) remove some of the difficulties. Horowitz et al. (1978) qualitatively explained the rather unusual diffuse scattering observed in some Zr-Nb alloys, in which a diffuse

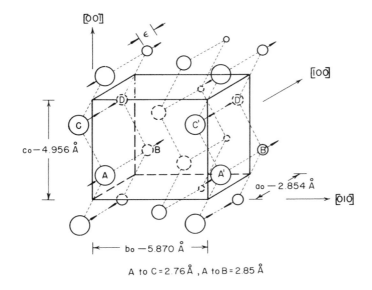

FIGURE 4.   Modulated structure of α-U at 10.4 K.

ω-phase coexists with the bcc β-phase, in terms of <u>narrow</u>
'stacking solitons.' The ω-phase is thought to 'feel' the
perturbing potential of the β-phase.

Apparently the uranium atoms want to be displaced slightly
from the crystallographic mirror planes at x = 0 causing a
lattice expansion incommensurate with the α-lattice and
producing discommensurations between the α and α' regions.
Similarly, the α' regions 'feel' the perturbing potential of
the α lattice (and perhaps vice versa) and produce, in this
case, <u>wide</u> 'stacking solitons', which cause the weak superlattice peaks to be 'density-shifted' to lower angles and the
stronger fundamental α' peaks to be 'phase shifted' to higher
angles, but still below the α peaks.

The one-dimensional formalism that Horovitz et al. (1978)
used for the ω-phase, was applied to our conception of the
state of the uranium lattice. The value of the parameter λ we
used (λ=.98) is very different and produces vastly different
effects from that used in the ω-phase problem, (λ=0.2). The
results for uranium are shown in Fig. 5. There is semi-
quantititative agreement with the observed data. The calculations omit any coherent relations between the α and α' regions
and, therefore, predict two (200) peaks slightly separated,
contrary to observation. An alternative explanation (unproven
here) is that coherent relations exist between the α and α'

TABLE I. Observed and Calculated Superlattice Reflections
at 10.4 K

h	k	ℓ	h^2	$|F|^2$	KI_{calc}	I_{obs}
0.5	1	0	0.25	0.36	1.5	vw
1.5	1	0	2.25	0.36	12	18
2.5	1	0	6.25	0.36	36	32
0.5	2	0	0.25	0.92	3	5
1.5	2	0	2.25	0.92	32	37
2.5	2	0	6.25	0.92	89	90
0.5	3	0	0.25	0.88	3	9
1.5	3	0	2.25	0.88	31	30
2.5	3	0	6.25	0.88	84	69
1.5	5	0	2.25	0	0	vw
1.5	6	0	2.25	0.44	15	w
0.5	7	0	0.25	0.96	3	w
1.5	7	0	2.25	0.96	34	M
0.5	0	1	0.25	1	4	3.4
1.5	0	1	2.25	1	35	43
2.5	0	1	6.25	1	98	92
0.5	0	2	0.25	0	0	-
1.5	0	2	2.25	0	0	-
2.5	0	2	6.25	0	0	-
1.5	0	3	0.25	1	4	5.6
1.5	0	3	2.25	1	35	33

regions such that they contribute coherently to the fundamental peaks. Such a model would probably involve a more complex
arrangement of the discommensurations than has been proposed
here. Additional diffuse scattering at low temperatures also
indicates that the structural distortion may be somewhat more
complex than has been described above. There appears to exist
a general background scattering plus some diffuse scattering
near the $(h+\frac{1}{2},k,0)$ superlattice peaks elongated in the b-
direction. These very tentative results suggest a possible
modulation of the y-parameter which is still incomplete even
at 11 K. This would be consistent with the small thermal expansion of the b_0 lattice constant. The thermal contraction
of the c_0 lattice constant is thought to represent a relaxation of the lattice in this direction and not a modulation of
the z-parameter from the value z = $\frac{1}{4}$. A careful search was
made for evidence of such a modulation, but none was found.
However, a very small distortion could have been masked by the

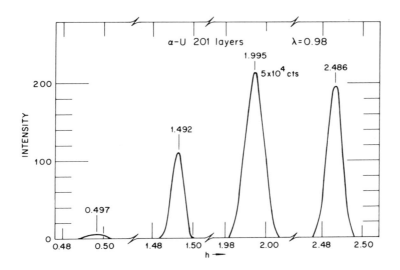

FIGURE 5. Shifts and intensities of superlattice reflections based on a 1-dimensional model of wide 'stacking solitons' (Horowitz et al., 1978).

large Renninger-type scattering at (h,k,ℓ/2) reflections from the λ/2 component in the incident neutron beam, even when a graphite filter was in the scattered beam.

CONCLUSIONS

The many predictions of structure transitions and soft modes in α-U have been verified by elastic and inelastic neutron scattering. This would not have been possible except for the discovery of the pronounced anomaly in the Σ_{4-LO} branch by Crummett et al. (1979), for the identification of the eigenvector associated with the anomalous mode led to the regions in reciprocal space where one would expect to observe the superlattice reflections. The discovery of the unusual shift in the superlattice reflections toward the origin is somewhat puzzling. However, the timely paper by Horovitz et al. (1978) suggests a possible mechanism due to discommensurations in terms of <u>wide</u> 'stacking solitons.' Further elucidation of the structure transformations in α-U should be very beneficial in understanding the chemical bonding and electron-phonon interactions in this unusual material. They undoubtedly all have their origin in subtle changes in the electronic band structure at low temperatures. Similar studies as a function of pressure should be very illuminating and will be attempted in the near future.

ACKNOWLEDGMENTS

The authors wish to thank A. F. Zulliger, III and Francisco Li-Aravena for technical assistance.

REFERENCES

Bader, S. D. and Knapp, G. S. (1975). Phys. Rev. B 11, 3348.
Bader, S. D., Phillips, Norman E., and Fisher, E. S. (1975). Phys. Rev. B 12, 4929.
Bak, Per and Timonen, J. (1978). J. Phys. C 11, 4901.
Barrett, C. S., Mueller, M. H., and Hitterman, R. L. (1963). Phys. Rev. 129, 625.
Crangle, J. and Temporal, J. (1973). J. Phys. F 3, 1097.
Crummett, W. P., Smith, H. G., Nicklow, R. M., and Wakabayashi, N. (1979). Phys. Rev. B (in press).
Fisher, E. S. and McSkimin, H. G. (1961). Phys. Rev. 124, 67.
Fisher, E. S. and Dever, D. (1968). Phys. Rev. 170, 607.
Fisher, E. S. and Dever, D. (1970). Solid State Commun. 8, 649.
Fisher, E. S., Geballe, T. H., and Schreyer, J. M. (1968). J. Appl. Phys. 39, 4478.
Geballe, T. H., Matthias, B. T., Andres, K., Fisher, E. S., Smith, T. F., and Zachariasen, W. H. (1966). Science 152, 755.
Horovitz, B., Murray, J. L., and Krumhansl, J. A. (1978). Phys. Rev. B 18, 3549.
Lander, G. H. and Mueller, M. H. (1970). Acta Cryst. B 26, 129.
Lin, W., Spalt, H., and Batterman, B. W. (1976). Phys. Rev. B 13, 5158.
Maglic, R. C., Lander, G. H., Mueller, M. H., and Kleb, R. (1978). Phys. Rev. B 17, 308.
McMillan, W. L. (1976). Phys. Rev. B 14, 1496.
Pynn, R. (1978). J. Phys. F 8, 1.
Smith, H. G., Wakabayashi, N., Crummett, W. P., Nicklow, R. M., Lander, G. H., and Fisher, E. S. (1979). In "Modulated Structures, 1979 (Kailua Kona, Hawaii)." Editor, J. M. Cowley. AIP Conf. Proc. Series, American Institute of Physics, New York.
Smith, T. F. (1972). J. Low Temp Phys. 6, 171.
Smith, T. F. and Fisher, E. S. (1973). J. Low Temp. Phys. 12, 631.
Wittig, J. (1975). Z. Physik B 22, 139.

THE CURRENT STATUS OF EXPERIMENTAL TUNNELING
RESULTS ON NIOBIUM

*J. Bostock**
*M. L. A. MacVicar**

Department of Physics
Massachusetts Institute of Technology
Cambridge, Massachusetts

Gerald B. Arnold[†]

Department of Physics
University of Notre Dame
Notre Dame, Indiana

John Zasadzinski[††]
E. L. Wolf[††]

Ames Laboratory-USDOE and Department of Physics
Iowa State University
Ames, Iowa

There has been a high level of interest in the electron-
coupled-phonon spectrum ($\alpha^2 F$) of pure Nb especially since the
discovery [Bostock *et al.*, 1976a] that nearly structurally
perfect, single crystal Nb-substrate junctions with niobium
oxide barriers formed either by *in situ* thermal oxidation or
by acid etching techniques, yield anomalous value of λ and μ^*;

*Research supported in part by the MIT Sloan Basic
Research Fund.*
[†]*Supported in part by NSF Grant DMR 77-10549 at Indiana
University.*
[††]*This work was supported by the U.S. Department of Energy,
contract No. W-7405-Eng-82, Division of Materials
Sciences, budget code AK-01-02-02.*

i.e., λ is small ($\sim$.6) and μ^* is negative ($\sim$-.05). The recent calculation of Arnold *et al.*, 1978 raises the question of the applicability of this proximity layer model to single crystal Nb-substrate tunneling data. This paper describes one such proximity analysis.

The method of fabrication of the single crystal junctions and the experimental procedures are well documented in the literature [Bostock *et al.*, 1976a] so that here we present and discuss only the data for a particular Nb-O-In junction. The first derivative scan is shown in Fig. 1 with the higher bias dI/dV-V data presented in the insert. (The measured gap value of 1.56 meV is the highest value reported to-date for niobium.) Calibrated conductance values for both the normal and superconducting states of the junction were obtained for bias energies of $\sim$5 to 31 meV beyond the sum gap edge. The solid line in Fig. 2 is ($\sigma/\sigma_{BCS}-1$), the reduced density of states (RDOS) for the single crystal Nb junction; and the solid line in Fig. 3 is the α^2F determined using the standard inversion routine. The shapes of the calculated and measured RDOS for this α^2F agree to within the precision of the measurement, however the calculated RDOS is displaced (shifted upwards) from the measured by $\sim$5x10^{-3}. Although the peaks in α^2F are located (to within the data spacing of .16 meV) at the same

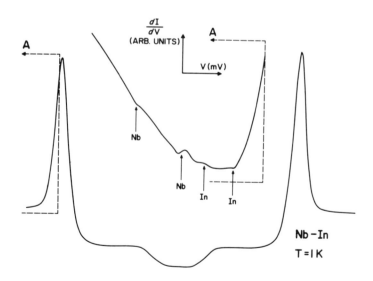

FIGURE 1. *First derivative scan of the Nb single crystal substrate junction measured at 1K. The indium gap is .55 meV, the niobium gap is 1.56 meV. The insert contains higher bias values of the conductance.*

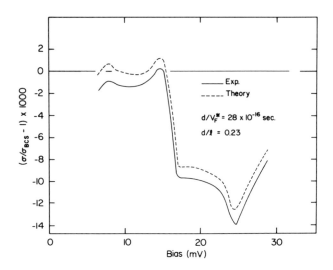

FIGURE 2. The reduced density of states determined from the measured junction data (——) and that calculated using Arnold's proximity model (———) with an RMS deviation of 1.1×10^{-3}. (The conventional deconvolution produces an offset of 5×10^{-3} from the measured RDOS.)

energies as those in the phonon spectrum (Sharp, 1969), the $\alpha^2 F$ cutoff energy is 26.6 meV, compared to the phonon spectrum cutoff of 27.2 meV. When this $\alpha^2 F$ is used in the strong coupling superconductivity equations a calculated T_C of 9.82K and values of λ and μ^* of 0.43 and $-.08$, respectively, are generated. (Acute sensitivity of these values to low bias conductance data has already been discussed by Bostock *et al.*,1976b.)

The proximity junction model incorporating diffuse scattering in the normal layer used in this analysis has been described previously [Arnold, 1978; Wolf *et al.*, 1979]. Here it suffices to point out that (1) whereas intentionally fabricated proximity junctions are characterized by definable parameters d_N, the thickness of the normal layer in the junction; v_F^* , the renormalized Fermi velocity of the N-layer; and ℓ, the effective mean free path for quasiparticles; for junctions which are suspected of having proximity layers, d_N/ℓ and d_N/v_F^* become fit parameters in the calculation and that, (2) such a junction will have an intrinsic bulk gap somewhat larger than the measured gap, because of its contact with the N-layer.

The single crystal Nb data can be fit to better than 5 x 10^4 by a range of (d_N/ℓ, d_N/v_F^*) values. For example, a d_N/ℓ of 0.25 and a d_N/v_F^* of 45 x 10^{-16} sec produces a fit to 3 x 10^{-4}(the precision of the measurement) with an rms-deviation (DEVN) of 1.4 x 10^{-4}. The associated $\alpha^2 F$ has λ, μ^* values of

1.45 and .16 respectively, and a predicted value of 1.63 meV
for the bulk energy gap. However, the T_c equation of Allen
and Dynes, 1975 predicts a T_C of 14.6K which is well outside
the measured T_c.

On the other hand, the available data extends only to 28.5
meV which is the cutoff of the phonon spectrum in this calculation so that the quality of the fit for energies beyond this
region cannot be determined. Recent experience [Wolf *et al.*,
submitted] with the proximity model indicates that an excellent fit to RDOS values over only the phonon spectrum does not
necessarily give the best fit for wider energy ranges, and
that, in fact, a minimization of DEVN over the *wider* range is
more appropriate if agreement of measured and calculated T_C-
values, for example, are to be obtained. Thus, the following
restrictions were placed on the parameter sets (d_N/ℓ, d_N/v_F^*)
chosen: (1) the measured and calculated RDOS agree to 1 part
in 10^3 (since the standard deconvolution has an offset of $\sim$5
parts in 10^3); (2) the calculated T_C agrees reasonably with
the measured value; and (3) the high energy cutoff of α^2F is in
reasonable agreement with that of the phonon spectrum.

The results of the calculation which fulfill most closely
these criteria are given in Figs. 2 and 3. For this fit d_N/ℓ
is .23, d_N/v_F^* is 28×10^{-16}sec and DEVN is 1.1×10^{-3}. The values
of λ and μ^* are 1.22 and .18, respectively, and the calculated
T_C is 10.1K which is within 10% of the measured value. (The
uncertainty in the λ and μ^* values are estimated as $\pm.2$ and
$\pm.03$, respectively.) The predicted bulk value for the energy
gap of Nb is 1.61 meV; i.e., the measured gap, 1.56 meV, appears depressed approximately 3% from this implied bulk value.
Assuming a Nb-like value for v_F^* ($\sim3\times10^7$cm/sec.), this gap suppression would correspond to a normal layer $\lesssim$ 10Å.

A precise and unique fit using this model for this single
crystal Nb data set cannot be determined since highly accurate
conductance data is not available for either the very low or
very high bias regions. Further investigation on single crystal Nb data spanning the low and high bias regions is required
before definitive evaluation of this proximity analysis for
the present junction can be made.

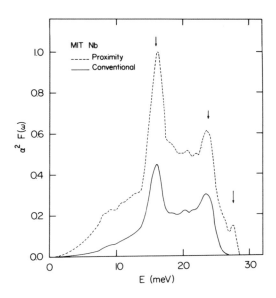

FIGURE 3.   The electron-coupled-phonon spectrum consistent
with a conventional deconvolution (———) and that determined
using Arnold's proximity with d_N/ℓ = .23, d_N/v_F^* = 28x10^{-16} sec
and superconducting parameters: λ = 1.22, μ^* = 0.18, T_c = 10.1K.
The arrows locate the phonon peaks and cutoff obtained from the
neutron determined phonon spectrum (Sharp, 1969).

REFERENCES

ALLEN, P. B., and DYNES, R. C., (1975). Phys. Rev. B 12, 905.
ARNOLD, G. B., (1978). Phys. Rev. B 18, 1076.
ARNOLD, G. B., ZASADZINSKI, J., and WOLF, E. L., (1978).
    Physics Lett. 69A, 136 and references therein.
BOSTOCK, J., LO, K. H., CHEUNG, W. N., DIADIUK, V., ROSE, R.M.,
    and MACVICAR, M. L. A., (1976a). Phys. Rev. Lett. 36,
    603 and references therein.
BOSTOCK, J., LO, K. H., CHEUNG, W. N., DIADIUK, V., and
    MACVICAR, M. L. A., (1976b). Superconductivity in d- and
    f-Band Metals, ed. D. H. Douglass (Plenum Press, New York)
    pp. 367-380, 403-412.
BOSTOCK, J., and MACVICAR, M. L. A., (1979). Physics Lett.
    71A, 373 and references therein.
GARTNER, K., and HAHN, A., (1976). Z. Naturforsch 31a, 361.
SHARP, R. I., (1969). J. of Physics C 2, 421.

WOLF, E. L., ZASADZINSKI, J., OSMUN, J. N., and ARNOLD, G. B.,
    (1979). *Solid State Commun. 31*, 321; *Phys. Rev.*
    (submitted).
WOLF, E. L. ZASADZINSKI, J., ARNOLD, G. B., MOORE, D. F.,
    BEASLEY, M. R., and ROWELL, J. M., (submitted).

A PROXIMITY ELECTRON TUNNELING STUDY OF V AND V$_3$Ga

J. Zasadzinski
W. K. Schubert
E. L. Wolf

Ames Laboratory–USDOE and Department of Physics
Iowa State University
Ames, Iowa

G. B. Arnold

Department of Physics
University of Notre Dame
Notre Dame, Indiana

Using a new method of Proximity Electron Tunneling Spectroscopy (PETS) we have determined the Eliashberg function $\alpha^2 F(\omega)$ for V and its important A-15 compound V$_3$Ga. The results are compared with neutron scattering data.

I. INTRODUCTION

The technique of electron tunneling spectroscopy has been difficult to apply in a definitive fashion to metals and alloys of transition elements because of the inability to obtain highly insulating oxide barriers on these materials. Specifically, attempts to form tunnel junctions on thin films of Vanadium by either glow discharge or thermal oxidation of the V surface have led to high leakage currents and large zero-bias conductance peaks in the tunneling characteristics (Noer, 1975). Moreover, the anomalous tunneling results on single-crystal Nb (Bostock *et al.*, 1976) may be attributable to properties of the thermally-grown oxide (Bostock *et al*, 1979) even though the tunneling characteristics were of high quality.

The importance of these materials, including the A-15 compounds, as superconductors has stimulated our development of a new Proximity Electron Tunneling Spectroscopy (PETS). The technique avoids any possible problems with transition metal oxides and in this study was incorporated to analyze the superconducting properties of V and V_3Ga. The new method has been demonstrated on Nb, and described in experimental (Wolf and Zasadzinski, 1977; Wolf *et al.*, 1979) and theoretical (Arnold, 1978; Arnold *et al.*, 1979) detail. This method requires the deposition of an ultra-thin layer of Al(N) onto the clean surface of the superconductor of interest (S). The Al is exposed to laboratory air and a self-limiting, highly insulating oxide is formed. The tunnel junction is completed with a convenient counterelectrode and the non-oxidized Al protects the superconductor from any surface contamination. Since one is now tunneling into an N-S proximity sandwich, the inversion of the data to obtain $\alpha^2F(\omega)$, $\Delta(\omega)$ and μ^* for the S layer becomes considerably more complex but is nevertheless possible in the limit of ultra-thin N layers.

II. EXPERIMENTAL TECHNIQUE

An electropolished foil of MRC Marz grade V (.005" thick) is resistively heated to near its melting point ($\sim$1800°C) in a vacuum of low 10^{-9} Torr. Impurity gases are driven from the bulk and quickly pumped by titanium coated walls maintained at 77°K. The result is a highly pure specimen with a clean, smooth surface and noticeable single-crystal grains ($\sim$.1 mm^2). The V foil is quickly cooled in the ultra-high vacuum to below room temperature and a thin layer of Al (30 Å-100 Å as determined by quartz crystal microbalance) is deposited leaving a strongly-coupled N-S proximity sandwich. The sample is then exposed to laboratory air, masked with collodion and an In counterelectrode is vapor deposited in a vacuum of about 5x10^{-6} Torr. The junctions were then studied by conventional tunneling methods at 1.4°K. To fabricate the V_3Ga junctions, the cleaned V foil was first coated with 200 Å of Ga. Annealing, started at 700°C and terminated by 20 minutes at 900°C, allowed the formation of a surface layer of V_3Ga (Tachikawa *et al.*, 1972). This surface growth process was monitored *in situ* by Auger spectroscopy demonstrating the absence of initial oxygen on the V surface before Ga deposition and verifying the final surface composition which was within the A-15 stability range. The exact Ga concentration is not known, however the measurement of gap values much larger than that of bulk V necessarily indicates that the A-15 phase is realized (Van Vucht *et al.*, 1964).

III. DATA ANALYSIS

     To invert the data and obtain the quantities of interest
i.e. $\alpha^2 F(\omega)$, $\mu*$ and $\Delta(\omega)$ for the S layer, we use a theory
developed by G. B. Arnold which calculates the Green's func-
tion for an N-S sandwich in perfect contact. For N layers of
thickness, d, much less than the coherence length of the N
layer, the tunneling density of states is, for E>>Δ_S, Δ_N

$$N_T(E) = 1 + \frac{1}{2} \text{ Re } \frac{\Delta_N^2(E)}{E^2} + \frac{1}{2} e^{-2d/\ell} \text{Re } \frac{[\Delta_S(E) - \Delta_N(E)]^2}{E^2} \exp(2i\Delta_K^N d)$$

$$+ \frac{1}{2} e^{-d/\ell} \text{Re } \Delta_N(E) \frac{[\Delta_S(E) - \Delta_N(E)]}{E^2} \exp(i\Delta_K^N d) \quad , \tag{1}$$

where

$$\Delta_K^N = \frac{2Z_N(E)E}{\hbar v_{FN}} \quad .$$

Here $\Delta_N(E)$ is the induced pair potential, $Z_N(E)$ is the renormalization function and v_{FN} is the Fermi velocity in the N-
layer. Scattering events in the N layer due to the N-S interface or to bulk impurities are characterized by the mean-free
path parameter, ℓ.
     Clearly one needs to know all of these quantities to
invert the data but in the limit as d→0 all the terms containing $\Delta_N(E)$ cancel. For ultra-thin but non-zero d values an
approximate expression for the tunneling density of states is

$$N_T(E) = 1 + \frac{1}{2} e^{-2d/\ell} \text{Re } \frac{\Delta_S^2(E)}{E^2} \exp(\frac{i4dZ_N E}{\hbar v_{FN}}) \quad . \tag{2}$$

for E>>Δ_S. Since v_{FN} and $Z_N \approx 1 + \lambda_N$ can be obtained from the
literature the only additional unknown parameter is ℓ, which
can be adjusted to so that the theoretical density of states
most closely fits the measured data. Using (2) to invert the
data produces reasonable results in cases where the Al influence is extremely small. A more precise treatment is to use
(1) which requires $\Delta_N(E)$. A detailed description of the techniques to obtain $\Delta_N(E)$ are described by Arnold et $al.$, 1979.

IV. RESULTS

     The V sample V-Al-12-7 had an Al thickness of 20 Å±10 Å

and a measured Δ_0=.80 mV±.005 mV. This Δ_0 is in excellent
agreement with values obtained by ultrasonic techniques
(Brewster *et al.*, 1963). Fig. 1 shows the second derivative
spectrum d^2V/dI^2 vs. $(V-\Delta_0)$ for V-Al-12-7. The arrows locate
the positions of peaks in the phonon spectrum for V and Al.
The influence of the Al is quite small especially near the V
peaks indicating that the thin d limit is realized. Inversion
of the V conductance data via the use of (1) and ℓ=1000 Å has
resulted in the $\alpha^2F(\omega)$ shown in Fig. 2a. The peaks line up
well with the peaks in $G(\omega)$ as determined by inelastic neutron
scattering (Schweiss, 1974), although the longitudinal peak at
28 meV is reduced by about 30%. The values of λ and μ^* are
.80 and .17 respectively, in fair agreement with phenomenological predictions (Allen and Dynes, 1975).

The Al thickness on the V_3Ga-2-2 sample was not measured
due to thickness monitor malfunction, however the second derivative spectrum indicated no Al structure. Moreover, the
gap region was broad indicating the junction area covered
regions of varying Ga concentrations and therefore varying Δ_0.
The measured Δ_0 of 1.35 meV±.05 meV is depressed from what one
would expect from V_3Ga of 15°K T_c. Nevertheless, the sample

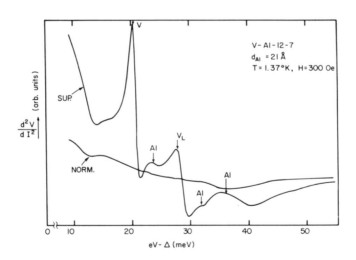

*FIGURE 1. Second derivative spectrum d^2V/dI^2 for
V-Al-12-7 at 1.4°K in 300 Oe parallel magnetic field (to remove complicating In phonon structure). Positive peaks in
d^2V/dI^2 occur at points of maximum phonon density. Both V
and Al phonon are visible.*

was definitely in the A-15 phase and exhibited strong phonon
structure. Inversion of the data using (2) resulted in the
$\alpha^2 F(\omega)$ shown in Fig. 2b. The arrows indicate peaks in $G(\omega)$
(Schweiss $et\ al$, 1976) measured at 77°K by inelastic neutron
scattering. That the $\alpha^2 F(\omega)$ peaks (determined at 1.4°K) are
lower in energy follows the trend that peaks in $G(\omega)$ shift to
lower energies from 297°K to 77°K. This phonon softening is
accompanied by a sharper resolution of the peak at 9.7 meV
which is represented as a shoulder in $G(\omega)$. The values for λ
and μ^* of 1.12 and .21 respectively are somewhat uncertain and
reduced from expected bulk values but encouragingly do not
exhibit anomalous behavior.

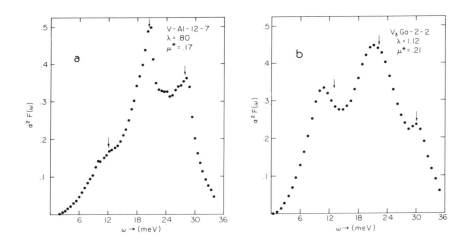

FIGURE 2.   (a) Eliashberg function $\alpha^2 F(\omega)$ for V using
Eq. (1).   The arrows are placed at peaks in the neutron phonon
spectrum (Schweiss, 1974).   (b) The $\alpha^2 F(\omega)$ spectrum for V₃Ga
obtained using Eq. (2).   The low energy peak is more clearly
resolved and lowered in energy from that observed in the
neutron result (Schweiss et al., 1976) taken at 77°K.   The
scale of this function is reduced below that expected for
the highest T_c material, but the shape is believed to be
characteristic of V₃Ga.

ACKNOWLEDGMENTS

   This work was supported by the U.S. Department of Energy,
contract No. W-7405-Eng-82, Division of Materials Sciences
budget code AK-01-02-02-2. We also thank Bell Laboratories
for donation of a vacuum chamber and Research Corporation for
a Cottrell award.

REFERENCES

ALLEN, P. B. and DYNES, R. C., (1975). *Phys. Rev. B 12*, 905.
ARNOLD, G. B., (1978). *Phys. Rev. B 18*, 1076.
ARNOLD, G. B., ZASADZINSKI, J., OSMUN, J. W., and WOLF, E. L.,
     Phys. Rev. (submitted).
BOSTOCK, J., LO, K. H. CHEUNG, W. N. DIADIUK, V., and
     MACVICAR, M. L. A., (1976). *Superconductivity in d- and
     f-Band Metals*, ed. D. H. Douglass (Plenum, New York) pp.
     367-380, 403-412.
BOSTOCK, J., MACVICAR, M. L. A., ARNOLD, G. B., ZASADZINSKI,
     J., and WOLF, E. L., (1979). Article of this volume.
BREWSTER, J. L. LEVY, J., and RUDNICK, I., (1963). *Phys. Rev.
     132*, 1062.
NOER, R., J., (1975). *Phys. Rev. B 12*, 4882.
SCHWEISS, B. P., (1974). *Karlsruhe Research Report KFR 2054*,
     p. 11.
SCHWEISS, B. P. RENKER, B., SCHNEIDER, E., and REICHARDT, W.,
     (1976). *Superconductivity in d- and f-Band Metals*, ed.,
     D. H. Douglass (Plenum, New York) p. 189.
TACHIKAWA, K., YOSHIDA, Y., and RINDERER, L., (1972). *J.
     Metal. Sci. 7*, 1154.
VAN VUCHT, J. H. N., BRUNING, H. A. C., DONKERSLOOT, H. C.,
     and GOMES DE MESQUITA, A. H., (1964). Philips Research
     Reports *19*, 407.
WOLF, E. L. and ZASADZINSKI, J., (1977). *Physics Lett. 62A*,
     165.
WOLF, E. L., ZASADZINSKI, J., OSMUN, J. W., and ARNOLD, G. B.,
     (1979). *Solid State Commun. 31*, 321; *Phys. Rev.*
     (submitted).

AN ELASTIC NEUTRON DIFFRACTION STUDY OF TRANSFORMING AND NON-TRANSFORMING SINGLE CRYSTAL ZrV2[†]

J. Bostock
M. Wong
M. L. A. MacVicar
M. Levinson[††]

Superconducting Materials Group
Massachusetts Institute of Technology
Cambridge, Massachusetts

The mosaic spread of single crystal ZrV2 is unusually narrow, ∿ 1˚ from room temperature to 130K. For non-transforming perfect single crystal the mosaic gradually increases to ∿ 1.86˚ at 4.2K; for transforming, twinned single crystal the room temperature mosaic is maintained to 110K, then increases to 2.76˚ at 94K when the crystal transforms to a mixed cubic (30%) and rhombohedral state (70%). The onset of the electronic instability (∿ 100K) is accompanied by an increase in diffuse scattering background which, for the twinned crystal, peaks at the structural transformation. The electronic instability coupled to the localized lattice stress appears to be the driving mechanism for the transformation.

Previous research [Levinson *et al.*, 1978] on the normal and superconducting state properties of the cubic Laves phase compound ZrV2, using four well characterized sample modifications (perfect and twinned single crystals, multigrained

[†]*Supported in part by the U.S. Department of Energy and the National Science Foundation.*
[††]*Present address: Universite᷄ de Genève; Genève, Switzerland.*

165

material, and 1% by volume Zr-included polycrystal) has shown
the dominant role that microstructure plays in determining the
observed properties of this "high-T_c" material. Although all
samples exhibit an electronic instability at $\sim$ 100K, only the
existence of local stress in the lattice seems to cause an actual structural transformation to occur; i.e., the perfect single crystal does not transform (down to 10K) while all other
sample types do transform (at approximately the same temperature as the onset of the electronic instability) to a rhombohedral structure. Thus, the electronic instability by coupling to localized lattice stress, drives the structural
transition.

A number of microscopic driving mechanisms are consistent
with such a coupling [Thomas *et al.*, 1978; Monceau *et al.*,
1976; Clapp, 1973; and Pan *et al.*, 1978]. The most obvious,
despite the fact that the structural transformation is first
order, is a softening phonon mode since, as Moncton [1973]
showed in earlier neutron studies of arc cast ZrV_2, the rhombohedral structure can be described as a simple contraction
along the body diagonal of the Zr diamond sublattice with a
concomitant expansion of the 2 <110> directions perpendicular
to the body diagonal. That is, a simple group theoretical
analysis [Levinson *et al.*, 1978] of the C15 structure [space
group $Fd3m(O_h{}^7)$], the inherent rhombohedral lattice strain,
and the final low temperature crystal symmetry [consistent
with space group $R\bar{3}m(D_{3d}^5)$] indicates that the transition might
be driven by a softening zone center optic mode which con-
denses into the lattice in the rhombohedral phase. An attempt
to verify the existence of such a softening mode by determining the sublattice distortion of the Zr atoms along the body
diagonal of the cube as the structural transformation proceeds
has been carried out together with a structural analysis of
both the perfect and twinned single crystals [Levinson *et al.*,
1978] at Brookhaven National Laboratory (λ = 2.55Å).

The cubic to rhombohedral transformation of the twinned
single crystal was monitored by mapping the intensity of the
(111) cubic reflection as a function of temperature. The
transformation begins at $\sim$ 99K (Fig. 1) when the cubic (111)
peak intensity starts to grow as domain formation appears[1],
increasing the crystal's mosaic spread and thus, diminishing
extinction effects. As the transition proceeds, the (111)
peak splits into a hexagonal (00.3) and 3 hexagonal (10.1) reflections of the rhombohedral phase. The intensity of the
(111) reflection continues to decrease with temperature until
at $\sim$ 87K the rhombohedral peaks have shifted enough to yield

[1]*All four possible domain types corresponding to distortions along each of the four body diagonals of the cube,
were observed in this transformation.*

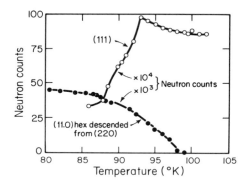

FIGURE 1. Peak height of the (111) cubic reflection as a function of temperature for the twinned single crystal. Also shown is the peak height of the (11.0) hexagonal reflection of the rhombohedral phase descended from the cubic (220) reflection.

no contribution at the (111) reciprocal space position and a residual (111) reflection remains. The residual peak height indicates that approximately 30% of the twinned single crystal volume remains untransformed.[2]

To calculate the magnitude of the sublattice distortion, neutron intensities were measured for the (1$\bar{1}$.2) reflection descended from the forbidden cubic (002) reflection and compared to the intensity of the (222) cubic reflection (of known structure factor) in the high temperature phase. The sublattice distortion is less than or equal to only a quarter of the measured lattice strain and thus, it is unlikely that a softening phonon mode is the driving mechanism for this structural transformation. [Only an upper limit on the magnitude of the sublattice distortion could be obtained because of the attendant double Bragg scattering background occurring near the (1$\bar{1}$.2) reciprocal lattice point (see Fig. 2) even though the neutron intensity was measured as close as possible to the background scattering minimum at zero degrees.] Of course, only a mapping of the phonon dispersion curves can rule out this mechanism entirely.

To fully characterize the coherent elastic scattering properties and to verify the physical perfection of the transforming and non-transforming single crystals, the mosaic spreads as functions of temperature were measured [Wong, 1979] at the MIT reactor (λ = 1.05Å). For alloys, both samples have unusually small mosaics at room temperature[3]: for the transforming and non-transforming crystals, $\sim$ .85ʹ ± .06ʹ and 1.14ʹ ± .12ʹ, respectively. Even more surprising than the narrower mosaic for the *transforming* crystal, is the fact that the

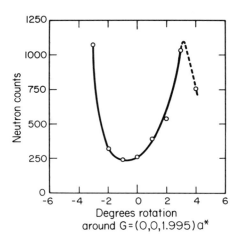

*FIGURE 2. Double Bragg scattering intensity at the re-
ciprocal space position $\vec{G}$ as a funciton of rotation about
the scattering vector measured at ∼ 80K. $\vec{G}$ corresponds to the
two hexagonal $(1\bar{1}.2)$ reflections above and below the (hhk)-
scattering plane. (a^* equals $2\pi/a_O$ with a_O the cubic lattice
constant.)*

scattering per unit volume of this crystal is only 75% that
of the non-transforming single crystal.

For the non-transforming single crystal, although both
the mosaic and integrated intensity (*II*) of the (331) rocking
curve begin to change at ∼ 134K, (see Fig. 3) there is a pre-
cipitous drop in *II* at precisely the temperature of the onset of the electronic instability (dropping to 57% of the
room temperature *II* near 77K).   From this temperature the
diffuse scattering background is seen to grow continuously
down to 4K. Simultaneously the mosaic increases gradually to
∼ 1.86´ ± .12´ at 4K. The mosaic of the twinned single crystal, on the other hand, begins to increase from its room
temperature value at 110K to ∼ 2.76´ ± .06´ at 97K where the
crystal begins to transform to the rhombohedral state; below
the transformation the mosaic (of the 30% non-transforming
material) increases beyond a meaningful value. Similarly, the
II of the (331) rocking curve begins to decrease at the onset

[2]*For hexagonal lattice vectors the c-axis is along the
cubic <111> direction and the a-axes are along the two
<110> directions normal to <111>.*
[3]*A nearly perfect Ge monochromator (.25´) with a (311)
set of planes of spacing 1.7020Å was used to monitor
the (331) set of ZrV$_2$ planes with spacing 1.7082Å.*

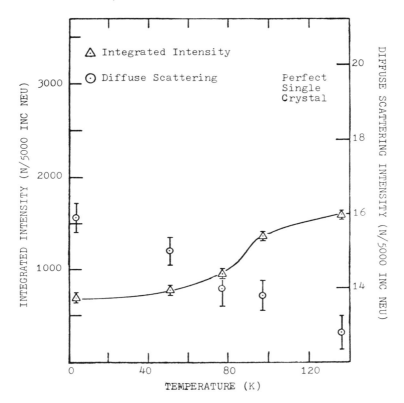

FIGURE 3. Integrated intensity of the cubic (331) rock-ing curve for the perfect ZrV₂ single crystal as a function of temperature. Also shown is the diffuse scattering back-ground. No evidence of a crystallographic transformation was seen for this specimen.

of the electronic instability and then drops precipitously at 97K and falls to only 5% of its value above the transformation (Figure 4) after the rhombohedral structure is established. Surprisingly, however, the diffuse scattering background which rises at the onset of the transformation only peaks during the rise of the rhombohedral reflections, and does not return to the pretransformation level. Thus, it would seem that the diffuse scattering background is intimately associated with the electronic instability in these crystals.

Ongoing research into the precise nature of the electronic instability and its interrelationship with the observed struc-tural transformations in less than structurally perfect ZrV₂ samples is expected to discriminate between the various non-soft-mode possibilities [Thomas *et al.*, 1978; Clapp, 1973] for

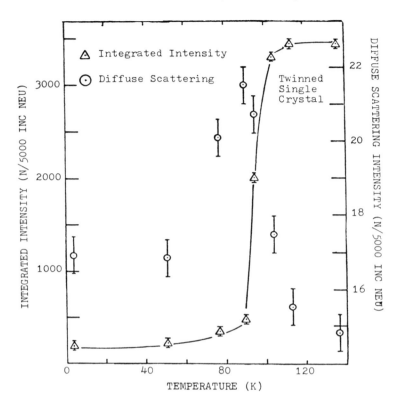

FIGURE 4.   Integrated intensity of the cubic (331) rock-ing curve for the twinned ZrV₂ single crystal as a function of temperature.   Also shown is the diffuse scattering back-ground.   Approximately 70% of the crystal transforms to a rhomohedral phase by 80K.

a microscopic basis for the anomalous transport properties of ZrV$_2$.   The evidence in hand, however, taken together with recent resistivity and susceptibility data of arc-cast ZrV$_2$ by Pan *et al.*, [1978] predispose us to believe that the elec-tronic instability is primarily of charge density wave charac-ter.   We note in passing that the C15 structure of ZrV$_2$ con-sists of six chains of vanadium atoms having V–V atomic spac-ing equal to that of the metallic solid.   The directionality implied by these chains, despite the probable interchain interaction, can be expected to result in nesting of at least some sections of the Fermi surface which, in turn, might pro-mote a charge density wave instability.

ACKNOWLEDGMENTS

The special contributions of G. Shirane, S. Collins, and C. Shull to this work are gratefully acknowledged by the authors. We would also like to thank B. Patton, D. Moncton, and P. Clapp for discussions on C-15 compounds and microscopic driving mechanisms of lattice transformations.

REFERENCES

CLAPP, P. C. (1973). *Phys. Stat. Sol. b57,* 561 and (1977). *J. Ferroel. 16,* 89.

LEVINSON, M., ZAHRADNIK, C., BERGH, R., MACVICAR, M. L. A., and BOSTOCK, J. (1978). *Phys. Rev. Lett. 41,* 899; and LEVINSON, M. (1978). Sc.D. Thesis, Massachusetts Institute of Technology (unpublished) and references therein.

MONCEAU, P., ONG, N. P., PORTIS, A. M., MEERSCHAUT, A., and ROUXEL, J. (1976). *Phys. Rev. Lett. 37,* 602.

MONCTON, D. E. (1973). *Solid State Commun. 13,* 1779, and MONCTON, D. E. (1973). S.M. Thesis, Massachusetts Institute of Technology (unpublished).

PAN, V. M., BULAKH, I. E., KASATKIN, A. L., and SHEVCHENKO, A. D. (1978). *J. Less-Common Met. 62,* 157.

THOMAS, G. A., MONCTON, D. E., WUDL, F., KAPLAN, M. and LEE, P. A. (1978). *Phys. Rev. Lett. 41,* 486.

WONG, M. W. (1979). S.B. Thesis, Massachusetts Institute of Technology (unpublished).

FROZEN PHONON CALCULATIONS FOR THE L(2/3,2/3,2/3)
PHONON OF NIOBIUM AND MOLYBDENUM

B. N. Harmon
K. M. Ho

Ames Laboratory–USDOE and Department of Physics
Iowa State University
Ames, Iowa

Materials with high superconducting transition temperatures frequently exhibit anomalous features in their phonon dispersion curves. The origin of these anomalies and their significance for superconductivity has been the subject of much debate. Two seemingly different models have been proposed to explain these anomalies in terms of the electron-phonon interaction. The first model focuses on real space charge-density distortions (CDD's), while the second emphasizes the states in reciprocal space near the Fermi energy. To test the assumptions in these models and to better understand the details of the electronic response to lattice vibrations in transition metals we have performed self-consistent frozen phonon calculations for the $\vec{q}=(2/3,2/3,2/3)$ phonon in Nb and Mo. In this paper we report our initial results for the CDD's associated with this phonon and discuss the differences between Nb and Mo.

Figure 1 illustrates the motivation for choosing to study the longitudinal (2/3,2/3,2/3) phonon in bcc transition metals. Not only is there the well known difference between Mo and Nb for this particular phonon branch, but also this branch for the high temperature bcc phase of Zr exhibits a pronounced dip which suggests a tendency for the (2/3,2/3,2/3) phonon to go soft and create the ω-phase (Stassis *et al.*, 1978). Models based on microscopic theory which have emphasized electronic screening by localized orbitals indicate the dip arises from monopolar charge fluctuations (Sinha and Harmon, 1975, 1976; Hanke, Hafner and Bilz, 1976). Phenomenological models which have been generalized to include CDD's also suggest monopolar fluctuations are responsible for the dip (Wakabayashi, 1977, and Allen, 1977). These changes in charge density involve all

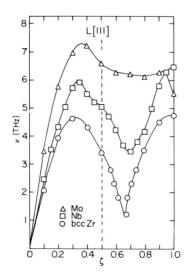

FIGURE 1.   The [111] *longitudinal* (Λ_1) *phonon dispersion curves for Mo, Nb and bcc Zr showing the large variation in the dip near* $\vec{q}$ = (2/3, 2/3, 2/3).

occupied valence states. On the other hand the very successful work of Varma and Weber (1977, 1979) suggests only the
band structure near E_F is crucial in explaining the dip.
These two models are not necessarily mutually exclusive, and a
possible reconciliation was a prime motivation for the present
work.

The calculations were performed self-consistently using
the KKR method and the muffin-tin approximation. The symmetry
of the L(2/3,2/3,2/3) phonon corresponds to three atoms per
unit cell with displacements from equilibrium of $\vec{\delta}$,- 1/2 $\vec{\delta}$,
and - 1/2 $\vec{\delta}$ with $\vec{\delta}$=(ξ,ξ,ξ)a and ξ chosen to be 0.02. The
details of the calculation will be published in a more exten-
sive paper. The band structures for bulk Nb and Mo were also
obtained using the same calculational procedures and the
results agreed with other published results. By performing
all calculations in the same manner the differences between
the results should be more meaningful with the systematic
errors minimized.

Figure 2 shows the band structure for the Nb frozen phonon
(folded out into the standard bcc symmetry directions). The
most noticeable change in the one electron energies occurs at
the zone boundaries of the new Brillouin zone. Here the
energy shifts can be quite large (±0.03 Ry) considering the
rather small nuclear displacements. The zone boundary effects

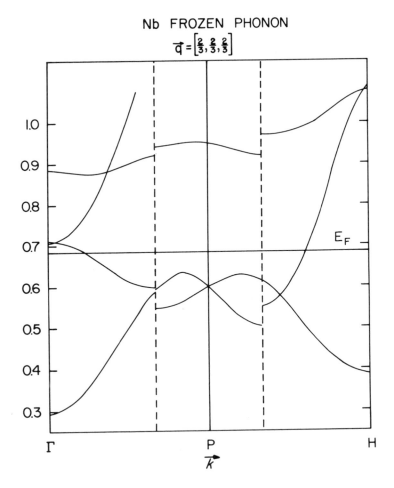

Nb FROZEN PHONON

$$\vec{q} = \left[\tfrac{2}{3}, \tfrac{2}{3}, \tfrac{2}{3}\right]$$

*FIGURE 2.   The band structure of Nb perturbed by the
L(2/3, 2/3, 2/3) frozen phonon.   The bands have been folded
into the standard symmetry lines for the bcc lattice to show
the effects of the new zone boundaries.   Energy is in Rydbergs.*

shown are however of little consequence since pairs of states
(those pushed up and those pushed down by the perturbation)
remain occupied.   There are a few less commonly shown regions
of the zone where such zone boundary effects occur near E_F,
but the splittings are smaller.   Another effect which is
related to the mechanism favored by Varma and Weber (1977) for
causing the dips is illustrated in Fig. 3.   Here, the unper-
turbed bulk Nb bands cross each other near E_F when folded into

the new frozen phonon Brillouin zone. Under the perturbation
the bands interact, with the occupied states moving down in
energy and the unoccupied moving up. If enough volume in k-
space is involved this mechanism can lower the net energy of
the system enough to cause an anomaly in the phonon dispersion
curves. However, if such mechanisms are the sole cause of the
anomalies in Nb it is surprising that they are so little
affected by temperature. This is especially true for the dip
in the L[00ξ] branch (Powell *et al.*, 1972). A strong temperature dependent anomaly is observed in hcp Zr and Ti (Stassis
et al., 1979) which can be related to the above mechanism but
which cannot involve monopolar charge fluctuations because of
symmetry. We intend to pursue these questions in future work.
Our initial calculations did not include enough sampling
points (72 pts in the irreducible 1/12th zone) to perform a
detailed analysis of small changes in band energies close to
E_F. However, the charge densities are well converged (based on
comparisons with 16, 32 and 72 pt sampling meshes), and the
results we discuss below are not expected to change significantly by increasing the number of sampling points.
    The CDD for Nb is shown in Fig. 4 for the [1$\bar{1}$0] plane.
Although the muffin-tin approximation is not to be trusted for

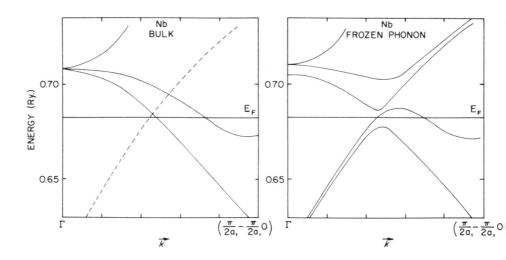

FIGURE 3. *The left figure shows the bands for bulk Nb
folded into the new frozen phonon Brillouin zone. The
effect of the phonon displacement is to split the bands. If
the splitting occurs at the Fermi level as shown on the right
the net electronic energy of the crystal is lowered.*

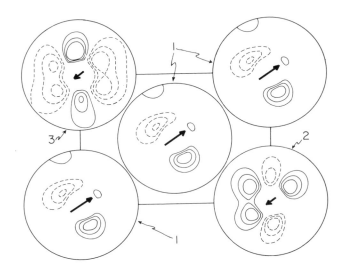

FIGURE 4.   Charge density distortions inside the Nb muffin-tin shperes for the 110 plane.   The directions and relative magnitudes of the atomic displacements are indicated by the arrows at the center of each sphere.   The solid contours at 0.001, 0.002, 0.003 el/au denote a charge increase for the frozen phonon and the dashed contours at -0.001, -0.002 and -0.003 el/au denote a charge depletion.

precise calculations of anisotropic charge densities, the *differences* betweeen the bulk and frozen phonon densities should give a fairly accurate indication of the CDD.   There is very little dipolar fluctuation parallel to the atomic displacements evident in Fig. 4, but quadrupolar fluctuations are present particularly in spheres 2 and 3.   In general the anisotropic distortions correspond to a rotation and strengthening of the bonding between spheres 1 and 2 which move closer together, and a simultaneous weaking of the bonding between spheres 1 and 3 which move apart.   These CDD's support Allen's argument (Allen, 1977) for dropping dipolar fluctuations from his model and adding quadrupolar interactions associated with nearest neighbor bond rotations.   The corresponding CDD graph for Mo indicated roughly the same behavior although the strength of the quadrupolar fluctuation was smaller and somewhat distorted.

Also evident in Fig. 4 is the monopolar fluctuation with sphere 2 picking up charge and sphere 3 losing charge.   The size of these fluctuations for the valence bands is given in Table I.   (The core charge density did not change substantially.)

TABLE I.   Valence Charge within Muffin-Tin Spheres

Calculation	Sphere		
	#1	#2	#3
Nb, bulk	3.613	3.613	3.613
Nb, F.P.	3.612	3.635	3.594
Nb, F.P. (RMT^a)	3.597	3.647	3.594
Mo, bulk	4.381	4.381	4.381
Mo, F.P.	4.390	4.403	4.373
Mo, F.P. (RMT)	4.388	4.374	4.397
Nb, F.P. (Mo^b)	4.396	4.371	4.389
Nb, F.P. (Zr^b)	2.862	2.878	2.803

^{a}RMT = *rigid muffin-tin.*
^{b}*Rigid band approximation.*

Relative to sphere 1, the fluctuation for Nb is +0.023 and
−0.020 el for spheres 2 and 3, respectively.   This should
be contrasted with the fluctuation obtained from overlapping
neutral atoms which give +0.0099 and −0.0126 el.   The neutral
atom fluctuations for Mo are larger (+0.0107 and −0.0134 el.),
but the self-consistent calculation yields only +0.013 and
−0.017 el.   This fluctuation for the self-consistent frozen
phonon is smaller than for Nb and the difference cannot be
accounted for by states changing occupation near the Fermi
level as might be expected because of the higher density of
states near E_F for Nb.   Perhaps the most interesting results
shown in Table I are the calculations for the frozen phonon
charge density using the bulk potential on each sphere (literally a rigid muffin-tin calculation).   For Mo the fluctuation
is opposite the self-consistent results, although the same
648 states were occupied in both calculations!   Using the
rigid band approximation with Nb and including states above E_F
to obtain "Mo" the fluctuations are also opposite.   This is a
general result which we have confirmed with other potentials.
A preliminary analysis suggests the states with energy between
the Nb and Mo E_F's want to fluctuate the wrong way (∿anti-
bonding), but as soon as they are allowed to respond the
states with much lower energies respond even more strongly to

maintain the "correct" charge imbalance.   There appears to be a strong coupling between the charge fluctuation behavior of states near E_F and those at much lower energies.   Much more work needs to be done before claiming that this coupling can resolve the conflict between the two theoretical models, but it suggests that such a resolution may be possible.

ACKNOWLEDGMENTS

This work was supported by the U.S. Department of Energy, contract No. W-7405 Eng-82, Division of Materials Sciences budget code AK-01-02-02. We wish to thank Drs. S. K. Sinha and D. D. Koelling for helpful discussions.

REFERENCES

ALLEN, P. B. (1977). *Phys. Rev. B 16*, 5139.
HANKE, W., HAFNER, J., and BILZ, H. (1976). *Phys. Rev. Letters 37*, 1560.
POWELL, B. M., WOODS, A. D. B., and MORTEL, P. (1972) in *Neutron Inelastic Scattering*, p. 43, IAEA, Vienna.
SINHA, S. K., and HARMON, B. N. (1975). *Phys. Rev. Letters 35*, 1515.
SINHA, S. K., and HARMON, B. N. (1976). In the Second Rochester Conference on *Superconductivity in d- and f-Band Metals* (Douglass, D. H., ed.) p. 269, Plenum, New York.
STASSIS, C., ZARESTKY, J., and WAKABAYASHI, N. (1978). *Phys. Rev. Letters 41*, 1726.
STASSIS, C., ARCH, D., HARMON, B. N., and WAKABAYASHI, N. (1979). *Phys. Rev. B 19*, 181.
VARMA, C. M., and WEBER, W. (1977). *Phys. Rev. Letters 39*, 1094.
VARMA, C. M., and WEBER, W. (1979). *Phys. Rev.* (in press).
WAKABAYASHI, N. (1977). *Solid State Comm. 23*, 737.

INTERRELATION BETWEEN WEAK ITINERANT MAGNETISM AND
SUPERCONDUCTIVITY IN SOME d–BAND METAL COMPOUNDS

Charles P. Enz

Département de Physique Théorique
University of Geneva
Geneva, Switzerland

After reviewing the properties of the weak itinerant ferromagnets Sc_3In and $ZrZn_2$ an analysis is made of the remarkable effects ascribable to the electron-phonon interaction in $ZrZn_2$ which recently led to the discovery of weak itinerant antiferromagnetism in $TiBe_2$. A recent theoretical attempt to explain the magnetism of $ZrZn_2$ and $TiBe_2$ in terms of Cooper-pairing ideas is described.

I. INTRODUCTION : WHAT IS WEAK ITINERANT MAGNETISM ?

Four decades after Stoner's first paper on band ferro-magnetism (Stoner, 1938) this problem remains a subject of controversy as can be measured at the two conflicting defini-tions of weak itinerant ferromagnets by Wohlfarth and by Matthias (Methfessel, 1979). Wohlfarth's criterium is that the moment obtained from the Curie-Weiss law, q_c , compared to the saturation moment q_s , satisfies the approximate relation (Wohlfarth, 1968; Edwards and Wohlfarth, 1968)

$$q_c/q_s \cong const \ T_m^{-1} > 1 \qquad (1)$$

where T_m is the magnetic transition (Curie-) temperature; $q_c/q_s = 1$ is the sure sign of a local moment. This distinction

181

is illustrated in Figure 1 of Rhodes and Wohlfarth (1963). Relation (1) has recently also been derived from formal spin
fluctuation theory in a paper by Moriya and Takahashi (1978)
which contains an analogous figure.

For Matthias, on the other hand, "there can be no doubt
that, in metallic ferromagnetic compounds in which the individual elements do not show any magnetic ordering of their own,
the moment must be due to itinerant electrons"(Enz and Matthias,
1978); in the extreme form his criterium even requires superconducting elements (Methfessel, 1979). This criterium leaves
only two weak itinerant magnets, $ZrZn_2$ (Matthias and Bozorth,
1958) and $Sc_{1-x}In_x$, $x \cong 0.24$ (Matthias *et al.*, 1961).

The magnetism of these substances is exceptional in almost
every respect : The Sc-In system is ferromagnetic only in the
extremely narrow range $0.238 < x < 0.242$ with a maximum $T_m \cong 7K$.
La_3In, on the other hand, is superconducting. Similarly, the
ferromagnetism of $ZrZn_2$ is easily destroyed by impurities, and
T_m varies between 4 and 35K according to sample preparation
(Ogawa and Sakamoto, 1967; Foner *et al.*, 1967; Knapp *et al.*,
1971). Furthermore, of the metals M forming ZrM_2 in the cubic
C15 structure all but Zn are transition elements, and the resulting compounds are superconductors.

In the Stoner-Wohlfarth theory (Wohlfarth, 1968) the ferromagnetic coupling constant I connects the difference between
the up- and down-spin chemical potentials $\mu_\pm$ and that between
the band electron numbers $n_\pm$ according to

$$\frac{1}{2} (\mu_+ - \mu_-) = \frac{1}{4} I(n_+ - n_-) + \mu_B H \qquad (2)$$

where H is the magnetic field and μ_B the Bohr magneton. In the
weak limit, defined by $(n_+ - n_-)/n_+ + n_-) \ll 1$, the inverse
susceptibility χ^{-1} obeys a parabolic law $2\chi_0/\chi = (T/T_m)^2 - 1$,
not too far above T_m (Wohlfarth, 1968). Here $\chi_0 = \chi(T = 0, H = 0)$.
This behaviour contrasts markedly with the linear Curie-Weiss
law observed both in Sc_3In (Matthias *et al.*, 1961) and in $ZrZn_2$
(Ogawa and Sakamoto, 1967; Knapp *et al.*, 1971), and has given
rise to theoretical tentatives to explain this linearity
(Murata and Doniach, 1972; Moriya and Kawabata, 1973 a,b). Thus
the emerging picture is that weak itinerant magnetism is as
subtle a mechanism as superconductivity, but its understanding
lags far behind.

II. EVIDENCE FOR ELECTRON–PHONON INTERACTION IN $ZrZn_2$ AND THE ANTIFERROMAGNETISM OF $TiBe_2$

The extreme sensitivity of the magnetism of $ZrZn_2$ to foreign atoms may be understood in terms of highly directional d-orbitals with small overlap. This view is supported by in-elastic polarized neutron scattering data (Shirane *et al.*, 1964; Pickart *et al.*, 1964) which show not only the absence of appreciable concentration of spin density on the atoms, as ex-pected for a weak itinerant magnet but, in addition, strong anisotropy and a maximum between Zr-atoms, reminiscent of co-valent bonds. Small overlap means a narrow d-band peak at the Fermi energy. Indeed, band calculations (Koelling *et al.*, 1971; Mueller, 1979), as well as numerical fits to the Stoner-Wohl-farth theory (Knapp *et al.*, 1971) lead to a width of the order of 0.1 eV.

Direct evidence for the crucial effect that the small overlap of the d-orbitals has on the magnetism comes from high-pressure experiments which show indeed a negative slope of the Curie temperature with pressure,

$$dT_m/dp = (-1.8 \pm 0.04) \times 10^{-3} K/bar \tag{3}$$

and a disappearance of ferromagnetism at a critical pressure $p_c = 8.5$ k bar (Smith *et al.*, 1971). This small overlap of the d-orbitals may be understood as a screening effect by the s-p orbitals of Zinc, which at the same time leads to lattice softening (Enz and Matthias, 1978). Evidence for this crucial role of Zinc comes from the substitution of Zn_2 by $Cu_{2-x}Al_x$ which leads to a broadening of the d-band peak and causes ferromagnetism to disappear (Knapp *et al.*, 1971).

The mentioned pressure dependence of the magnetism of $ZrZn_2$ immediately suggests a similar effect due to phonons. The possibility of phonon effects in magnetism had been investiga-ted in general terms by Herring (1966). He estimated that the second-order change in zero-point energy with magnetization should contribute a fraction of order ω_D/ε_F to the spin stiff-ness χ^{-1}, where ω_D is the Debye frequency and ε_F the Fermi energy. Hopfield (1968) took up this idea to calculate an iso-tope effect for weak itinerant ferromagnets. His result was

$$\alpha \equiv d \log T_m/d \log M = -\bar{I}_{e\ell-ph}/4(\bar{I}-1) \tag{4}$$

where M is the atomic mass, and

$$\bar{I} = \bar{I}_{coul} + \bar{I}_{e\ell\text{-}ph} \tag{5}$$

is the exchange coupling constant I, multiplied by the density of states at the Fermi level $N(\varepsilon_F)$, and consists of the usual Coulomb part and of a contribution due to the electron-phonon interaction. For $ZrZn_2$ Hopfield estimated α = -6 (note his numerical error of a factor 10!). An attempt to check this value experimentally, unfortunately, was inconclusive (Knapp *et al.*, 1970).

There is other evidence, however, for the conjecture that the electron-phonon interaction should have an effect in the compound $ZrZn_2$ similar to the pairing responsible for the superconductivity of the elements Zr and Zn. Indeed, there is a striking parallelity between the magnetic and superconducting transition temperatures, T_m and T_s, of the mixed Zn_2-compounds and the corresponding elements. A compilation made by Enz and Matthias (1979) is reproduced in Figure 1.

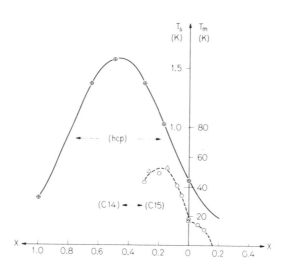

FIGURE 1. *Superconducting (full line) and magnetic (broken line) transition temperatures T_s and T_m, for $Zr_{1-x}Ti_x$ and $Zr_{1-x}Ti_xZn_2$ (left scale) and for $Zr_{1-x}Hf_x$ and $Zr_{1-x}Hf_xZn_2$ (right scale), respectively. From Enz and Matthias (1979).*

These experimental facts led to the so-called "p-state pairing" idea (Enz and Matthias, 1978). It was in this termino-logy that the idea grew out of discussions between the authors, but the expression should not be taken too literally. The idea is that formation of the compound inhibits the superconductivi-ty of the elements by somehow (see Section III) reversing the sign of the s-state pairing potential. This has two consequen-ces; the first is a Hubbard type, i.e. strongly localized but phonon-mediated, repulsion which, in the electron-hole channel, contributes to the spin fluctuation and, via eq. (5) drives the Stoner factor $(1-\bar{I})^{-1}$ through infinity, hence producing ferro-magnetism. The second, more speculative consequence is a resi-dual, less localized, i.e. p- or higher state, pairing poten-tial which might manifest itself in situations where there is no magnetization, e.g. above the critical pressure p_c.

It was of course desirable to find more substances on which to test these ideas. As noticed already by Enz and Matthias (1978) the only compound which is isostructural and isoelectronic with $ZrZn_2$ is $TiBe_2$; it has since been found to be antiferromagnetic (Matthias *et al.*, 1978). Although this discovery has been a considerable encouragement for the "p-state pairing" idea, it has also opened the even more difficult question of weak itinerant antiferromagnetism. Deferring the latter question to the following section it should be remarked here that the negative intercept in the Curie-Weiss plot of Matthias *et al.* (1978) is not a unique proof of antiferromagne-tism. Indeed, such curves may also be obtained from nearly ferromagnetic metals, as shown by Figs. 6(a) and (b) of Moriya and Kawabata (1973 a). However, Matthias *et al.* (1979) have since made extensive studies of the Ti-Cu-Be system. In the compound $TiBe_{2-x}Cu_x$ the Copper goes to Beryllium sites and the structure stays almost single phase C15 up to $x \cong 0.2$ where softening and a new phase set in. At $x \cong 0.07$, T_m switches sign to weak ferromagnetism and continues to increase up to the instability. From the continuity of $T_m(x)$ one may infer continuity of magnetic order, the sign change of T_m below $x \cong 0.07$ indicating the transition to antiferromagnetic order.

There is also experimental evidence for electron-phonon interaction in $TiBe_2$. Indeed, Matthias *et al.* (1979) have mea-sured a negative pressure dependence of T_m of about half the value given in eq. (3), and an isotope effect of roughly $T_m \propto M^{-3}$ where M is the titanium mass. It is obvious from these discoveries that with $TiBe_2$ weak itinerant magnetism has become an even more urgent theoretical problem.

III. TOWARDS A UNIFIED THEORY OF WEAK ITINERANT MAGNETISM AND
      SUPERCONDUCTIVITY ?

The basic quantity of superconductivity theory is the
Cooper-pair potential

$$V_{pp'\lambda} = |g_{pp'\lambda}|^2 D_\lambda(p-p', \varepsilon_p - \varepsilon_{p'}) \qquad (6)$$

mediated by the phonon propagator

$$D_\lambda(q,\omega) = 2\omega_{q\lambda}/[\omega^2 - (\omega_{q\lambda} - i\delta)^2] \qquad (7)$$

where $g_{pp'\lambda}$ is the electron-phonon coupling constant and $\omega_{q\lambda}$
the frequency of a phonon of wavevector q and branch index λ.
The repulsive part of $V_{pp'\lambda}$ which occurs for $\varepsilon_p - \varepsilon_{p'} > \omega_{p-p'\lambda} \cong$
$\cong \omega_D$ is inoperative due to the argument of Anderson and Morel
(1961) which gives rise to the pseudopotential μ^*. This has
been shown to be true quite generally by Bergmann and Rainer
(1973), even for soft phonons[1]. This might appear surprising
since, for soft phonons, the average pair potential per unit
energy written in standard notation (McMillan, 1968).

$$\lambda(\varepsilon) = \int d\omega \ \alpha^2 F(\omega) \ \frac{2\omega}{(\omega-i\delta)^2-\varepsilon^2} \qquad (8)$$

which occurs in the gap equation at $T \to 0$ (Scalapino, 1969),
does have negative contributions corresponding to repulsion
(Enz and Matthias, 1979). The reason for Bergmann and Rainer's
result is that a finite-temperature form of the gap equation
contains $\lambda(i\varepsilon)$ which is always positive.

In magnetism, however, the Bergmann-Rainer result is irre-
levant, and a negative contribution to $\lambda(\varepsilon)$ is indeed of interest. Therefore, attention must focus on soft phonons, in particular those with highest density of states, i.e. quasi-one
dimensional ones (Enz and Matthias, 1978, 1979). The problem
is now to calculate the contribution of the pair potential (6)
to the spin fluctuation (Doniach and Engelsberg, 1966) for a
phonon density of states $F(\omega)$ corresponding to quasi-one dimensional soft modes or, in other words, to determine the contribution to the iteration step in the electron-hole t-matrix
(Berk and Schrieffer, 1966). This gives rise to the celebrated

[1]*The general proof given by Bergmann and Rainer (1973) is
valid only for $\mu^*=0$. However, numerical tests never produced
repulsion for any parameter values (Rainer, 1979).*

Migdal vertex $\Gamma_1(p,q)$ (Migdal, 1958) which has been extensively discussed in the literature (Scalapino, 1969; Hertz *et al.*, 1976; Fay and Appel, 1979). According to this discussion Γ_1 contributes a factor ω_D/ε_F to the bare electron-phonon vertex and is thus negligible in the "isothermal" limit $q = (q_0,\vec{q}) \to 0$, $q_0/|\vec{q}| \to 0$. However, in this result, which is Migdal's theorem, a normal phonon spectrum and an isotropic Fermi surface are assumed.

The question therefore is whether an anisotropic Fermi surface and a quasi-one dimensional soft phonon spectrum are able to beat Migdal's theorem. At this point it is important to realize that for a soft mode the imaginary part of the self-energy becomes important, in fact the mode may be overdamped. This means that the infinitesimal δ in the phonon propagator (7) must be considered of comparable magnitude as $\omega_{q\lambda}$ and carry the same lables. Then evaluation of Γ_1 for $q \to 0$ and $p_0 \to 0$ leads to

$$\bar{I}_{e\ell-ph} = \sum_\lambda \int_{S_F} \frac{d^2p'}{(2\pi)^3 v_F} \; |g_{pp'\lambda}|^2 < \int_0^{\varepsilon_F} d\varepsilon (\sqrt{1+\varepsilon/\varepsilon_F} + \sqrt{1-\varepsilon/\varepsilon_F}) \times$$

$$x \; (\varepsilon + \omega_{p-p'\lambda} - i\delta_{p-p'\lambda})^{-2} - 2(\omega_{p-p'\lambda} - i\delta_{p-p'\lambda})^{-1}>_{S_F} \qquad (9)$$

where $<>_{S_F}$ is an average over $\vec{p}$ on the Fermi surface S_F. Expression (9) is still general; it vanishes in the limit $\varepsilon_F \to \infty$ which is Migdal's theorem. However, the soft mode damping $\delta_{p-p'\lambda}$ being more important in the second denominator of eq.(9) than in the first, strongly anisotropic averages over $\vec{p}'$ and $\vec{p}$ will reduce this term more significantly so that, as an order of magnitude, it may be neglected as compared to the first term. Taking again the limit $\varepsilon_F \to \infty$ one then finds (Enz and Matthias, 1978, 1979)

$$\bar{I}_{e\ell-ph} \cong \lambda(0) \equiv \lambda , \qquad (10)$$

which is the effective mass enhancement factor (McMillan, 1968; Scalapino, 1969). Of course, this result is rather speculative and requires closer investigation along the line of Fay and Appel (1979) but taking anisotropy and mode damping more serious.

According to what had been said in Section II about the d-orbitals of $ZrZn_2$ and the softening role of Zinc both, anisotropy of the Fermi surface and quasi-one dimensional soft modes

are not unlikely features of this substance. As to $\bar{I}_{e\ell-ph}$,
eq. (5) and the assumption that $\bar{I}_{coul} < 1$ yield a lower bound
of the order of 10^{-4} to 10^{-3} (Enz and Matthias, 1979). More
experimental and band structure data are obviously needed.

Of particular interest is the isotope effect which here
is due to the mass-dependence of λ (McMillan, 1968). There are
two contributions to α, defined in eq. (4), one coming from
eq. (10) and the other from the λ-dependence of $N(\varepsilon_F)$ in $\bar{I}_{coul}$;
$N(\varepsilon_F)$ acquires a renormalization factor $1+\lambda$ (see, e.g. Enz,
1968). The result is

$$\alpha \cong - \bar{I}_{e\ell-ph}(\bar{I}_{coul} + 1)/2(\bar{I}-1) \tag{11}$$

which is roughly a factor of 4 larger than Hopfield's value (4).

In view of a "unified theory" the disappearance of weak
itinerant magnetism as function of an appropriate parameter
(e.g. pressure) is of particular interest. In Section II it
was speculated (Enz and Matthias, 1978, 1979) that in this situation the residual p- or higher state pairing could induce
superconductivity and that, in a weak itinerant antiferromag-
net, it could even coexist. However, as Fay and Appel (1979)
point out, the p-state pairing potential is expected to be
very weak, if attractive at all. On the other hand, strong
spin fluctuations probably still suppress s-state pairing
(Fay and Appel, 1979).

IV. THE WEAK ITINERANT ANTIFERROMAGNETISM OF $TiBe_2$ - MORE
    SPECULATION

In looking for a distinction between $ZrZn_2$ and $TiBe_2$ the
only feature, apart from atomic radii, seems to be valency.
Since local moments are manifestly absent, Hund's rule implies
that half-filled shells are unfavourable. Looking for filled
(or empty) shells one notices that Be^{2+}, Be^{6-} and Zn^{6-} all
have stable noble gas configuration while Zn^{2+}, like Cu^+, has
not (Enz and Matthias, 1979). Thus Beryllium might like to
fluctuate between the valences 6- and 2+ while Zinc cannot.
The only way a valence configuration Ti^{4+} Be^{2+} Be^{6-} can be
realized in a regular way in the C15 structure is by having
alternating pairs on the Beryllium strings (Enz and Matthias,
1979). The interest of this configuration is that it alternates
between neighboring unit cells as shown in Figure 2. This leads

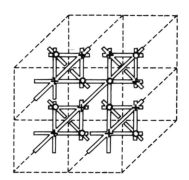

FIGURE 2. Projection on (001)-plane of the Beryllium strings drawn over 4 unit cells (broken lines) of the C15 structure of $TiBe_2$. Crosses: Be^{2+}, circles: Be^{6-}.

to an antiferro-distortive feature which in Fourier space is located at the L-center, $\vec{q}_L = (1,1,1)\pi/a$, of the Brillouin zone. If now $\bar{I}_{coul}$ has also a value close to one at $\vec{q}_L$ then an L-center soft mode may induce antiferromagnetism in the same way as ferromagnetism was derived for $\vec{q}=0$ in Section III. The point here is that an appropriate valence fluctuation frequency could easily lead to an L-center soft mode.

REFERENCES

Anderson, P.W., and Morel, P. (1961). *Phys. Rev. 123*, 1911.

Bergmann, G., and Rainer, D. (1973). *Z. Physik 263*, 59.

Berk, N.F., and Schrieffer, J.R. (1966). *Phys. Rev. Letters 17*, 433.

Doniach, S., and Engelsberg, S. (1966). *Phys. Rev. Letters 17*, 750.

Edwards, D.M., and Wohlfarth, E.P. (1968). *Proc. Roy. Soc. (London) A 303*, 127.

Enz, C.P. (1968). *In* "Theory of Condensed Matter", p. 729. Internat. Atomic Energy Agency, Vienna.

Enz, C.P., and Matthias, B.T. (1978). *Science 201*, 828.

Enz, C.P., and Matthias, B.T. (1979). *Z. Physik B33*, 129.

Fay, D., and Appel, J. (1979). *Phys. Rev. B.* To be published.

Foner, S., McNiff, E.J. Jr., and Sadagopan, V. (1967). *Phys. Rev. Letters 19*, 1233.

Herring, C. (1966). *In* "Magnetism" (G.T. Rado and H. Suhl, eds.), p. 290-297. Vol. 4. Academic, New York.

Hertz, J.A., Levin, K., and Beal-Monod, M.T. (1976). *Solid State Commun. 18*, 803.

Hopfield, J.J. (1968). *Physics Letters 27A*, 397.

Knapp, G.S., Corenzwit, E., and Chu, C.W. (1970). *Solid State Commun. 8*, 639.

Knapp, G.S., Fradin, F.Y. and Culbert, H.V. (1971). *J. Appl. Phys. 42*, 1341.

Koelling, D.D., Johnson, D.L., Kirkpatrick, S., and Mueller,F.M. (1971). *Solid State Commun. 9*, 2039.

Matthias, B.T., and Bozorth, R.M. (1958). *Phys. Rev. 109*, 604.

Matthias, B.T., Clogston, A.M., Williams, H.J., Corenzwit, E., and Sherwood, R.C. (1961). *Phys. Rev. Letters 7*, 7.

Matthias, B.T., Giorgi, A.L., Struebing, V.O., and Smith, J.L. (1978). *Physics Letters 69A*, 221.

Matthias, B.T. *et al.* (1979). To be published.

McMillan, W.L., (1968). *Phys. Rev. 167*, 331.

Methfessel, S. (1979). "Unusual Conditions of Superconductivity and Itinerant Magnetism in d-Materials", DFG-Honnefer Rundgespräch, organized by S. Methfessel, Bad Honnef, Germany, May 21 to 23, 1979. Unpublished.

Migdal, A. (1958). *J. Exptl. Theoret. Phys. (U.S.S.R.) 34*, 1438 [engl. transl. *Soviet Physics JETP 7*, 996 (1958)].

Moriya, T., and Kawabata, A. (1973 a). *J. Phys. Soc. Japan 34*, 639.

Moriya, T., and Kawabata, A. (1973 b). *J. Phys. Soc. Japan 35*, 669.

Moriya, T., and Takahashi, Y. (1978). *J. Phys. Soc. Japan 45*, 397.

Mueller, F.M. (1979). *Reported in* Methfessel (1979). To be published.

Murata, K.K., and Doniach, S. (1972). *Phys. Rev. Letters 29*, 285.

Ogawa, S., and Sakamoto, N. (1967). *J. Phys. Soc. Japan 22*, 1214.

Pickart, S.J., Alperin, H.A., Shirane, G., and Nathans, R. (1964). *Phys. Rev. Letters 12*, 444.

Rainer, D. (1979). Private communication (Methfessel, 1979).

Rhodes, P., and Wohlfarth, E.P. (1963). *Proc. Roy Soc.(London) A273*, 247.

Scalapino, D.J. (1969). *In* "Superconductivity" (R.D. Parks,ed.), p. 449. Vol. 1. Dekker, New York.

Shirane, G., Nathans, R., Pickart, S.J., and Alperin, H.A. (1964). *Proc. Internat. Conf. on Magnetism*, Nottingham, p. 223. The Inst. of Phys. and the Phys. Soc., London.

Smith, T.F., Mydosh, J.A., and Wohlfarth, E.P. (1971). *Phys. Rev. Letters 27*, 1732.

Stoner, E.C. (1938). *Proc. Roy. Soc. (London) A165*, 372.

Wohlfarth, E.P. (1968). *J. Appl. Phys. 39*, 1061.

THE CURRENT EXPERIMENTAL STATUS OF CuCl-RESEARCH

C. W. Chu

Department of Physics and Energy Laboratory
University of Houston
Houston, Texas

The CuCl-experiments are reviewed. Various possibilities giving rise to a diamagnetic ac susceptibility shift are examined. The current experimental situation on the CuCl research is summarized and evaluated.

I. INTRODUCTION

Cuprous chloride (CuCl) is an unusual compound. In contrast to predictions, it possesses very small elastic moduli[1] and undergoes some intermediate phase transitions[2] under pressure before collapsing into a highly insulating NaCl-structure. Increases of the electrical conductivity (σ)[2] and the optical opacity[3] were also reported in some of these intermediate phases, suggesting the possible existence of a metallic phase in CuCl at a moderate pressure, e.g. < 60 kb. In view of the predominantly d-characteristic upper valence band[4] and the large inherent instabilities[1] of CuCl, to look for high temperature superconductivity was the underlying motivation[5] behind the extensive studies which have been carried out on this otherwise unobtrusive compound, assuming that the suggested metallic state could be stabilized. Attempts[6] to search for a thermally equilibrium metallic phase induced by pressure were made but failed. However, more anomalies in CuCl have since been observed. The most unusual observation[6,7] of all is the ac diamagnetic susceptibility (χ) shift at high temperature in some samples under proper conditions. Therefore, in the present paper, attention will be focused only on this peculiar aspect of CuCl, i.e., the diamagnetic χ-anomaly. The poor reproducibility of the χ-anomaly has kept the CuCl-research still in in a stage of confusion. A clarification and assessment of the present experimental situation of the CuCl-problem is in order. Consequently, we shall review the CuCl-experiments with emphasis on the Russian experiments[7] only briefly described previously and examine the different possible explanations of the χ-anomaly before a brief summary is given.

II. χ-EXPERIMENTS ON CuCl

Signs of transitions to a diamagnetic state were detected a few years ago in the polycrystalline CuCl samples freshly prepared by direct fusion of pure CuCl powder, during both an isobaric run[8] at 22 kb and ∿85 K and the isothermal runs[9] at 4.2 K and ∿20 kb. Further isobaric measurements[8] on the same sample indicated a decreased diamagnetic signal with a decreased pressure while leaving the anomaly-temperature unshifted. Unfortunately, subsequent effort[10] to yield isobarically any such χ-anomaly in the similarly prepared CuCl samples down to 1.2 K and up to 32 kb failed. A self-clamp technique using the 1:1 n-pentane and isoamyl-alcohol pressure medium was used in the isobaric experiments and a cryogenic press with a piston-cylinder arrangement using a Grafoil pressure medium was employed in the isothermal experiments. A standard ac inductance bridge technique with both the primary and secondary coils outside the high pressure chamber was used to measure the χ. The operating frequency was 27 Hz.

In April 1977,[11] an ac χ-anomaly with a width of 10-20°K in a diamagnetic sense was detected in a sample under pressures up to 32 kb on rapid warming at a rate of ∿5°K/min but not on cooling (<5°K/min). As shown in Fig. 1, the χ-anomaly was accompanied by a σ-increase and was subtended by two extrema in the differential thermal analysis (DTA) signal. Similar anomalies were found in all 10 fresh CuCl ingots investigated, although not in all samples cut from them. The size of the χ-anomaly depended on the sample, the warming rate of the sample and the pressure applied to the sample. However, the temperature ranging from 90 to 250 K where the χ-anomaly occurs depended only on the sample. The maximum χ-anomaly observed was ∿7% of a perfect diamagnet, i.e., ∿7% of $-(1/4\pi)$ and the corresponding σ-increase was by a factor of ∿40. No χ-shift larger than $10^{-3} \times (1/4\pi)$ was induced by pressures up to 32 kbar at 300 K. Erratic and irreproducible χ-oscillations were occasionally seen. The samples were all prepared by condensation from the vapor of purified CuCl. Their resistivity showed a sharp drop at ∿40 kb in a Bridgman anvil rig. A high pressure Be-Cu clamp (∿56 mm O.D. × 70 mm) with the 1:1 n-pentane and

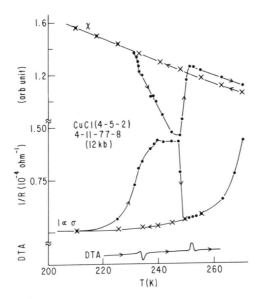

FIGURE 1.   χ, σ, and DTA of CuCl under 12 kb on rapid
warming (•) and cooling (×).   (After Ref. 6.)

isoamyl-alcohol pressure medium was used to generate the high
pressure environment. The rapid warming was achieved by direct-
ing a heat-gun against the clamp after being removed from the
He-cryostat. An ac inductance bridge technique was employed to
determine the χ. The two secondary bucking coils (∿0.6 mm dia.
× ∿ 3 mm each) were oppositely wound on the sample and a superconducting Pb-manometer (and/or reference) inside the high pressure region. The primary coil (37 mm dia.× 50 mm) was mounted
inside the clamp but outside the high pressure chamber. The
driving magnetic field of the primary was about 3 G. The
bridge was operated at 400 Hz, and was also tested at 10 Hz.
The relative size of χ-anomaly was not found to depend on the
frequency. The sign and size of the χ-anomaly were obtained by
comparing them with those of the superconducting Pb-manometer
of the same geometry and volume. The transient nature of the
appearance of the χ-anomaly makes any definitive diagnosis of
the anomaly extremely difficult if not impossible.
      Later in July 1977,[12] a similar but more dramatic χ-
anomaly was observed. On a second rapid cooling at a rate
>20°K/min, an initially sluggish transition at ∿120 K of a
polycrystalline CuCl sample to a weakly diamagnetic state developed into a sharp transition to a strongly diamagnetic state,
as shown in Fig. 2. The magnitude of the diamagnetic signal

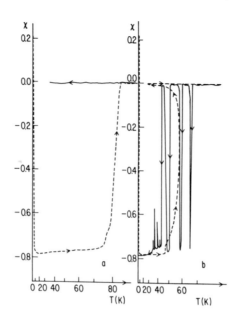

FIGURE 2.   Temperature dependence of χ *of CuCl: a) heating (dashed curve) and slow cooling (solid curve, and b) rapid cooling from* ∿*300 K (solid curve), heating and halt at the transition, followed by cooling and heating (dashed curve). (After Ref. 7.)*

and the transition temperature were found to depend on the cooling rate. No transition was detected on slow cooling. The maximum diamagnetic signal obtained was ∿75% that of a perfect diamagnet, and the highest transition temperature observed was 120 K. Occasional χ-oscillations between the weakly and strongly diamagnetic states preceded immediately the aforementioned transition. The same large diamagnetic signal was unambiguously observed in a total of seven consecutive runs on the same sample over a period of ∿15 hours by varying the temperature between 4.2 and 350 K. An instability check was also made on this sample. It was found, as shown in Fig. 2, that after warming the sample to the temperature when χ started returning to normal, subsequent cooling could not regenerate the transition to a diamagnetic state, suggesting the transition was metastable near the transition temperature. However, rapid cooling reproduced the χ-anomaly following the warming of the sample to above 300 K. Unfortunately, the χ-anomaly disappeared after the sample was accidentally warmed up to above 77 K overnight. Prior to the observation of the above transition, a χ-anomaly of 3-5% of a perfect diamagnet with a 30°K width

appeared in a different sample cut from the same ingot at
∿150 K during warming. Then in December 1977,[12] following the
similar experimental procedure, a transition to a diamagnetic
state was observed at 5 kb in another sample from a different
ingot. The maximum diamagnetic signal was ∿10% of a perfect
diamagnet and the maximum transition temperature was ∿165 K.
The transition to a state with ∿10% of a perfect diamagnet
signal was reproduced ∿15 experimental runs on the same sample
between 4.2 and 300 K over a period of four days. The transition disappeared after the sample was accidentally warmed up
to >300 K for about two days. However, the transition re-
appeared but with a smaller diamagnetic signal of ∿2% after the
pressure was reduced to 1 kb. The signal grew to ∿4% when the
pressure was increased to 3 kb. The failure in an attempt to
determine the magnetic field effect on the transition led to
an unexpected disruption of the experiment and the inevitable
warming-up of the sample. Consequently, the transition in this
particular sample could not again be recovered. A similar diamagnetic transition but with a 1-3% signal has also been observed in several other samples from different CuCl ingots
since December 1977, according to Rusakov.[13] The CuCl samples
which displayed a large χ-anomaly (75 or 10%)[7] were supposedly
prepared differently from samples used in other investigations.[6]

The χ-measuring technique used here was similar to that
previously used,[6] except for the following few details. The
dimensions of the Be-Cu clamp were ∿18 mm O.D. × ∿75 mm. Rapid
change of temperature was accomplished by introducing a controlled amount of exchange-gas into a vacuum chamber where the
small high pressure clamp was located. The pressure medium
was an oil and kerosene mixture. The two secondary bucking
coils (∿0.6 mm dia. × ∿3 mm each) were oppositely wound on a
plastic coil-form, in different parts of which the sample and
a superconducting manometer were respectively situated. The
primary coil had the dimensions of 11 cm I.D. × 50 cm and was
immersed in a liquid nitrogen bath. The measuring magnetic
field generated by the primary was ∿30 G. The operating frequency was 20 Hz. The fact that the reported diamagnetic state
could stay at low temperature for hours made this Russian experiment very attractive for a definitive diagnosis of the
χ-transition.

Following the announcements of the χ-anomalies,[6,7] various other experiments have been carried out on CuCl. Only those
measuring the χ under pressures will be briefly described.
Lefkowitz and co-workers[14] observed a diamagnetic χ-anomaly at
17 Hz under 5 kb between 120 and 180 K, similar to that in
Ref. 6, using a BN anvil set housed inside a stainless steel
cylinder. No pressure medium was used. The color of the
samples investigated varied from colorless transparent to gray

with blue or black flecks. Shelton et al.[15] using an ungask-
eted high pressure diamond cell, found some kind of diamagnetic
χ-shift (but of much smaller magnitude than those in Refs. 6
and 7) in purified CuCl powder under pressure at 300 K as the
optical opacity of the sample increased. Fast cooling of the
sample in a Be-Cu clamp did not produce any χ-anomaly. The
CuCl sample underwent a structural transformation at 60 kbar.
Guertin and co-workers[16] employing a dc technique by combining
a mini BeCu high pressure clamp with a vibrating magnetometer,
failed to prove or disprove unambiguously the existence of the
Meissner effect in CuCl. Dietrich,[17] using a Be-Cu self-clamp
technique, observed a few percent diamagnetic χ anomaly[6] in
some of the samples, but suggested the freezing of the pressure medium being the possible cause. With the same technique,
Smith et al.[18] did not detect any χ-anomaly in their samples.
Chu and co-workers[19] had increased the temperature variation
rate to ∿50°K/min, by reducing the size of their Be-Cu high
pressure clamp. They observed a ∿2% χ-anomaly similar to
Ref. 6, both during rapid warming and cooling, but did not
succeed in stabilizing at low temperature the diamagnetic state
as described in Ref. 7.

III. POSSIBILITIES FOR THE χ-ANOMALY

From the above CuCl-experiments, there clearly exist two
important questions, namely: i) why is the χ-anomaly so poorly
reproducible? and ii) is the χ-anomaly really unusual? To
provide some insights, the following possibilities[20] leading
to a χ-anomaly are examined to see if the χ-anomaly is due to
some experimental artifacts and if the χ-anomaly can be explained in terms of some simple conventional models:

1. Interference of equipment--A sudden change in the sample
current due to a change of σ associated with a phase transition
can result in a drastic change in χ when both the χ and σ are
measured simultaneously. However, the χ-anomaly remained when
the σ-probe was deactivated and only χ was determined.
2. Volume change of the sample--A sudden change in volume
associated with a phase transition may give rise to a relative
movement of the whole sample-secondary and/or a partial slip-
page of the sample-secondary, if the secondary is directly
wound on the sample. This in turn can induce a step in χ.
However, the χ-anomaly stayed when a free secondary coil was
used. In addition, the estimated χ-anomaly obtained by dis-
placing the secondary by 10% of the sample dimension along the
coil axis inside the primary described in Ref. 6, should be

only $\sim 10^{-3}$ of that observed. This estimate is consistent with the failure to detect any χ-anomaly by replacing the sample with either a piece of plastic or ceramic, the two kinds of material with drastically different thermal expansion coefficients.

3. Freezing of the pressure medium—When the freezing of a fluid pressure medium is first order, a volume change of the medium is expected. This may induce a coil movement of the kinds described in 2) and thus a χ-anomaly. Such a suggestion cannot be reconciled with the two DTA extrema observed to subtend the χ-anomaly, not to mention the large spread of the anomaly-temperature over $\sim 150^{\circ}$K even for the same fluid used, and the large size of the χ-anomaly. It should be noted that Lefkowitz et al. observed the χ-anomaly when no fluid pressure medium was used.

4. Eddy current effect—A high σ arising from a new phase can generate an apparent diamagnetic ac χ-shift, due to an eddy current shielding. However, to obtain a signal size of 7% of a perfect shielding[6] would have required the sample to have a bulk σ about 10 times that of pure copper at 200 K for an operating frequency of 400 Hz, assuming a penetration depth of 0.5 mm or 0.6 times the radius of the sample used and an optimal, although unlikely, pattern of current flow. This is not consistent with the small overall σ observed. While a closely connected shielding network of fine filamentary conductor could lead to the same apparent diamagnetic signal, the corresponding required reduction in the penetration depth (< the dimension of the filaments) would impose a much larger lower bound on σ to generate the same signal.

5. Thermal excitation of trapped charges—Detrapping charge carriers by thermal excitation in a semiconductor can result in an increase in the electrical current through the sample under the measuring electric field for σ. This can, in turn, induce a jump in χ. However, the χ-anomaly existed in the absence of an electric field which was required to trap charge carriers on cooling. In addition, the size, the polarity and the shape of the χ-anomaly did not agree with that expected from such a suggestion. In fact, the estimated size was 10^{-3} times too small[6] even if one assumed an optimal flow pattern for the current of the detrapped charge carriers.

6. Acoustic resonance—Near a ferroelectric transition, the mobility of the charge carriers increases and resonates at a *fixed* acoustic frequency. An ac σ-increase and thus a χ-jump may be apparent at such a frequency. However, the χ-anomaly was reported to occur in different frequencies used, e.g., 10, 17, 20, and 400 Hz. Furthermore, the χ-anomaly was diamagnetic and the σ-increase was dc for the case of Ref. 6.

7. Antiferromagnetic transition--If CuCl under pressure first undergoes a nonmagnetic to weakly-magnetic transition (Dzialoshinsky-Moriya type) and the strain due to any rapid temperature excursion subsequently induces a weakly-ferromagnetic to antiferromagnetic transition, a sudden diamagnetic χ-shift can result. However, no χ-shift, either paramagnetic or diamagnetic, larger than $10^{-3} \times (1/4\pi)$, which was 10-10^3 times smaller than the χ-anomaly reported,[6] was detected at 300 K up to 32 kb. In addition, an ESR study[6] indicated that at 1 bar, there existed no magnetic state in the CuCl samples investigated.

8. Piezoelectric and magnetostrictive effect--A strain is generated in a magnetic compound in the presence of a strong magnetic field. If the compound is also piezoelectric, an electric current will be induced and so will a χ-anomaly. However, CuCl is nonmagnetic and the driving magnetic field of the primary coil was only as low as ~ 3 G, although CuCl is piezoelectric.

9. Martensitic transformation--The sensitive dependence of the appearance of the χ-anomaly on thermal cycling reminds one of a martensitic type of transition. The effect of such kind of phase transition on χ is through the volume change. Such an effect was already examined in 2).

10. Superconductivity--The observed diamagnetic χ-anomaly and the σ-increase can easily be understood in terms of super-conductivity. With regard to this suggestion, one should keep in mind that aside from our lack of knowledge of the mechanism giving rise to a high transition temperature such as $\sim 10^2$ K, the existence of superconductivity in CuCl has yet to be proved.

11. Superdiamagnetism--A new phenomenon is proposed to account for the observed large diamagnetism and small σ, if the finite and small σ is characteristic of the bulk sample. The mechanism leading to such an effect lies beyond the realm of the present-day physics.

IV.   SUMMARY AND COMMENTS

    Based on the above analysis and all data available, the current experimental status of CuCl can be summarized as:

1. The diamagnetic ac χ-anomalies of CuCl observed both in the US and USSR may be of similar nature and cannot be attributed to any possible experimental artifacts just described.

2. The appearance of the χ- and σ-anomalies of CuCl depends sensitively on the sample and experimental conditions, and may be associated with the metastable phase or phases of CuCl.

3. The so called "IIa" phase between $\sim$40 and 60 kb although quite conducting is not metallic.

4. There exists no consistent evidence for an indirect band gap of $\sim$0.3 eV in the thermally equilibrated phases of CuCl.

5. There exists no convincing evidence for an overall large σ-increase (e.g. 10^4-10^6) associated with the χ-anomaly, other than an increase of σ by a factor of $\sim$40.

6. The χ-oscillations preceding the χ-anomaly on rapid cooling may be due to an effect of no real consequence.

A recent optical study[21] showed that the behavior of CuCl under pressure depended sensitively on the sample age, light exposure, pressure inhomogeneity, and the sample preparation technique. The chemically and physically unstable nature of CuCl and the transient characteristic of the appearance of the χ-anomaly in CuCl have further complicated the study of CuCl. In addition to the conflicting reports on the χ-results of CuCl, already mentioned in Section II, inconsistent data also exist in the lattice parameter, the low pressure phase diagram, electrical conductivity, etc. Therefore, the most effective way to tackle the CuCl-problem will be to characterize systematically the samples and to determine the most favorable conditions under which the χ- and other anomalies occur, so the nature of the state giving rise to a diamagnetic ac χ-anomaly can be definitively diagnosed. To find consistency within inconsistencies reported on CuCl may prove to be particularly fruitful in improving the reproducibility of the χ-anomaly in CuCl.

In short, we believe that CuCl is a very unusual compound[22] in which new physics may yet be discovered. However, the unusual observations on CuCl, although suggestive enough, are not sufficient to prove the existence of superconductivity in this compound. In fact, neither of the two necessary and sufficient conditions for superconductivity, namely, the Meissner effect and infinite conductivity, has been unambiguously met by CuCl. More studies are needed for the clarification of the anomalous behavior of CuCl.

V.   ACKNOWLEDGMENTS

I would like to thank A. P. Rusakov, N. B. Brandt, S. V. Kuvschimnikov, M. V. Semyonov, Yu. A. Lisovskii, and their colleagues for useful conversations concerning some of the details about the CuCl experiments carried out in USSR. Valuable discussions with T. H. Geballe, A. A. Abrikosov and my colleagues in the US are very much appreciated. The work was supported in part by NSF Grant Nos. DMR77-23204 and 79-08486 and the Energy Laboratory of the University of Houston.

REFERENCES

1.   R. M. Martin, *Phys. Rev.* *B1*, 4005 (1970).
2.   R. S. Bradley, D. C. Munro, and P. N. Spencer, *Trans. Faraday Soc.* *65*, 1912 (1969).
3.   A. Van Valkenburg, *J. Res. Natl. Bur. Std. 68A*,97 (1964).
4.   M. Cardona, *Phys. Rev. 129*, 69 (1963).
5.   See for example, A. P. Rusakov, V. Int. High Pressure Conference at Moscow, May 1975.
6.   C. W. Chu, A. P. Rusakov, S. Huang, S. Early, T. H. Geballe, and C. Y. Huang, *Phys. Rev. B18*, 2116 (1978) and references therein.
7.   N. B. Brandt, B. V. Kuvshinnikov, A. P. Rusakov, and M. V. Semyonov, *Pisma JETP 27*, 37 (1978).
8.   C. W. Chu, S. Early, T. H. Geballe, A. P. Rusakov, and R. E. Schwall, *J. Phys. C8*, L241 (1975).
9.   A. P. Rusakov, A. V. Omelchenko, V. N. Laukhin, and S. G. Grigoryan, *Fig. Trerd. Tela (Leningrad) 19*, 1167 (1977) [*Sov. Phys.-Solid State 19*, 680 (1977)].
10.   C. W. Chu, unpublished.
11.   Ref. 6 and unpublished results.
12.   Ref. 7 and private communications with A. P. Rusakov, B. V. Kuyshinnikov, and N. B. Brandt, April (1979).
13.   A. P. Rusakov, private communication, April (1979).
14.   I. Lefkowitz, J. S. Manning, and P. E. Bloomfield, private communication at the March APS Meeting (1979).
15.   R. N. Shelton, NRL Symposium on "Superconducting Materials-A Forecast," September (1978); R. N. Shelton, G. W. Webb, F. J. Rachford, P. C. Taylor, F. L. Carter, C. T. Ewing, P. Brant, I. L. Spain, and S. C. Yu, *APS Bull. 24*, 498 (1979).
16.   R. P. Guertin, S. Foner, G. W. Hull, Jr., T. H. Geballe, and C. W. Chu, *APS Bull. 24* , 498 (1979).
17.   M. Dietrich, private communication, November (1978).
18.   T. F. Smith et al., private communication,December (1978).
19.   C. W. Chu, V. Diatschenko, D. Harrison, and S. Z. Huang, unpublished.
20.   Some of these possibilities have been discussed in Ref.6.
21.   C.W. Chu and H. K. Mao, to appear in *Phys. Rev.*
22.   For a review, see J. A. Wilson, *Phil. Mag. B38*,427 (1978).

THE PAIRING INTERACTION IN SURFACE SUPERCONDUCTIVITY

Werner Hanke
Alejandro Muramatsu

Max-Planck-Institut
für Festkörperforschung
Stuttgart, W.-Germany

ABSTRACT

The microscopic expressions for the superconducting parameters are derived for a surface system. Our formulation fully
takes surface and periodicity induced nonlocality in the electron-phonon interaction and in the Coulomb electron-electron
interaction into account. Thus, in contrast to the commonly
used jellium models, it is applicable to transition-metal as
well as semiconductor surfaces. As an example the exciton mechanism postulated for superconductivity is discussed. It is
shown that surface nonlocality, in contrast to local bulk treatments, allows for electron-exciton-electron coupling and thus
can induce attractive interaction.

I. INTRODUCTION

There has long been an interest in the mechanism of superconductivity at surfaces and interfaces. A typical example of
this is furnished by the flurry of interest in what might be
called the superconducting Schottky barrier (Allender, Bray
and Bardeen, 1973). Following an initial suggestion of Ginzburg (1970) efforts have been made to estimate what effect the
presence of a semiconductor upon the metal surface would have
on the superconducting transition temperature of that metal
(thin film). The metal electrons interact by way of virtual

excitons in the semiconductor so that an extra attractive interaction is obtained (Bardeen, 1976). More generally, the basic idea is to replace the normal phonon exchange interaction
by same other boson exchange. Further work (Inkson and Anderson, 1973; Cohen and Louie, 1976) seemed to indicate that the
attractive nature of the exciton effect was offset by the repulsive residual Coulomb interaction of the semiconductor so
that the net effect would rather be to depress the transition
temperature. However, two questions were left completely unre-
solved: (a) the exchange via phonons was simply considered in
both metal and semiconductor to be of the same value, and approximated by a free-electron bulk metal result, (b) the effects of non-locality, as manifested in the image potential and
collective excitations at the interface, were neglected and a
(local) bulk screening assumed.

We present a description of the pairing interaction at surfaces
and interfaces which treats both Coulomb interaction and phonon exchange on a consistent microscopic footing, and express
the superconductivity parameters in a specific (LCAO, muffin-
tin) electronic basis, immediately accessible to numerical com-
putations. In particular our nonlocal surface formulation shows
that, in contrast to local bulk-type formulations (Inkson and
Anderson, 1973), the elementary excitation exciton manifests
itself as a *pole* in the screening function ε^{-1} thus couples to
the electrons and thereby can at least in principle give rise
to an attractive character of the screened electron-electron
interaction around excitonic (electron-hole) frequencies.

II. THE PAIRING INTERACTION AT A SURFACE

    We use the following notation: position vectors $\vec{r} = (\underline{r}, z)$,
with $\underline{r}$ being parallel to the surface; $\vec{K} = (\underline{k}, k_z)$ with $\underline{k}$ being
a good quantum number and k_z a label of the eigenstates. Our
starting point is the Eliashberg interaction - I $(\vec{r}\ t,\ \vec{r}'t')$
which represents the change in energy of an electron at $(\vec{r}\ t)$
due to the presence of an electron at $(\vec{r}'\ t')$.

Like in the bulk (Allen, 1979) the calculation of the phonon
contribution to I can be split into three parts:

    (i) The electron at $(\vec{r}'t')$ creates an impulsive force
$F_{\alpha'}(\underline{\ell}', \ell_z'; \tau)$ acting on atom $\vec{\ell}' = (\underline{\ell}', \ell_z)$ in the α' direction

$$F_{\alpha'}(\underline{\ell}', \ell_z', \tau) = -\nabla_{\underline{\ell}', \ell_z'; \alpha'}\ W(\vec{r}' - \vec{\ell}')\delta\ (\tau - t'),\qquad (1)$$

where W denotes the screened ionic potential.

(ii) The impulse causes a lattice displacement u_α $(\underline{\ell}, \ell_z; t)$
at subsequent time t which is determined by the displacement-
displacement correlation function $D_{\alpha\alpha'}(\underline{\ell}, \ell_z, \underline{\ell}', \ell'_z; t-t')$ of the
surface system

$$u_\alpha(\underline{\ell}, \ell_z; t) = \sum_{\underline{\ell}', \ell'_z} D_{\alpha\alpha'}(\underline{\ell}, \ell'_z; t-t') \, F_{\alpha'}(\underline{\ell}', \ell'_z; t), \qquad (2)$$

where

$$D_{\alpha\alpha'}(\underline{\ell}, \ell_z; \underline{\ell}', \ell'_z; t-t') =$$

$$= -\Theta\ (t-t') \sum_{\underline{q}, j} \frac{\vec{e}(\ell_z; q, j) \vec{e}(\ell'_z; q \cdot j)}{M\omega(\underline{q}, j)^{-1}} \sin\ (\omega(q, j)(t-t')) e^{iq(\underline{\ell}-\underline{\ell}')}$$

$$(3)$$

M is the ion mass and we assume for simplicity of notation only
one atom per (bulk) unit cell. The dynamical matrix is given
by $D_{\alpha\beta}(\ell_z \ell'_z; q)$ determining the phonon frequencies $\omega(q, j)$ via
the eigenvalue equation

$$\sum_{\ell'_z, \beta} D_{\alpha\beta}(\ell_z \ell'_z; q) e_\beta(\ell'_z; q, j) = \omega^2(q, j) e_\alpha(\ell_z; q, j) \qquad (4)$$

$j = 1 \ldots 3N$ labels the $\omega(q, j)$, N being the number of atomic
layers.

(iii) Finally, the energy shift of an electron $(\vec{r}, t)$ is

$$V_{El,Phon}(\vec{r}, t) = - \sum_{\underline{\ell}, \ell_z} \nabla_{\underline{\ell}, \ell_z} W(\vec{r}-\ell) < \vec{u}(\underline{\ell}, \ell_z; t) > \qquad (5)$$

Putting the previous equations together, and transforming the
interaction to wavevector and frequency space by taking the matrix elements corresponding to scattering a Cooper pair

$$\left(\psi_{n,\underline{k}\uparrow} \, k_z (\vec{r}) \psi_{n,\underline{k}\downarrow k_z}(\vec{r}') \right) \text{ to} \left(\psi_{n',\underline{k}'\uparrow, k'_z}(\vec{r}) \, \psi_{n',-\underline{k}'\downarrow, k'_z}(\vec{r}') \right), \text{ we}$$

arrive at the Eliashberg interaction

$$I^{Phon}(n, \underline{k}, k_z; n', \underline{k}', k'_z; \omega) =$$

$$= (N_{\shortparallel})^2 \sum_j |<n',\underline{k}+q,k_z'| \sum_{\substack{\vec{r},\ell_z \\ \ell z}} \vec{\nabla}_{\underline{r},\ell_z} W(\vec{r}-(o,\ell_z))| n,\underline{k},k_z> \vec{e}(\ell_z;q,j)|^2.$$

$$\cdot D_j(q,\omega) \tag{6}$$

where $D_j(q;\omega) = \left(M(\omega^2-\omega(q,j)^2)\right)^{-1}$, $q=\underline{k}-\underline{k}'$, (7)

and $N_{\shortparallel}$ is the number of $\underline{\ell}$'s in the surface plane.

The scattering of the Cooper pairs contains of course also the Coulomb part determined by the matrix elements of the dynamical-ly screened Coulomb interaction $v(\vec{r}-\vec{r}')$ (2D-Fourier transform $v(\underline{q};|z-z'|)$)

$$V_{sc}(\underline{q}+\underline{G},\underline{q}+\underline{G}';z,z';\omega) = \int v(|\underline{q}+\underline{G}|;|z-z''|)\ \varepsilon^{-1}(\underline{q}+\underline{G},\underline{q}+\underline{G}';z'',z';\omega)dz''$$

(8)

where $\varepsilon^{-1}(\underline{q}+\underline{G},\underline{q}+\underline{G};z,z';\omega)$ is the non-local surface screening function, with $\underline{G}$ and $\underline{G}'$ denoting 2D reciprocal lattice vectors.

After having cast the pairing interaction into a form appropri-ate for a surface system, we can gain detailed insight into its microscopic nature by introducing an explicit expression which we have derived for the surface response function ε^{-1} (Wu and Hanke, 1977).

We expand the wave function of the surface system

$$\psi_{n\vec{k}}(r) = \frac{1}{(N_{\shortparallel}N)^{1/2}} \sum_{iR} c_{im}(n\vec{k})e^{i\underline{k}\cdot\underline{R}} a_i(\vec{r}-\vec{R}), \quad m = R_z, \tag{9}$$

where a_i denotes a Wannier, LCAO or muffin-tin orbital. In this local basis the density response integral equation $\varepsilon^{-1} = 1+v\chi\varepsilon^{-1}$ is separable, with the kernel $\chi = ANA^+$. Therefore

$$\varepsilon^{-1}(\underline{q}+\underline{G},\underline{q}+\underline{G}';z,z';\omega) = \underset{\underline{G},\underline{G}'}{\delta}\ \delta(z-z')+\int v(\underline{q}+\underline{G};|z-z''|)\cdot$$

$$\sum_{ss'} A_s(\underline{q}+\underline{G};z'') \left(N^{-1}(q;\omega)- V_{xc}(q)\right)^{-1}A_{s'}^+(\underline{q}+\underline{G}';z')dz'' \tag{10}$$

Here s and s' are short for band indices and locations of the surface orbitals a entering the density wave A_s. N^{-1} is the in-verse one-particle polarization and V_{xc} the density-density interaction corrected for exchange (x) and correlation (c). With this we can directly obtain the following results:

(i) The electronic contribution to the dynamical matrix,

$$D_{\alpha\beta}^{E}\ (\ell_z,\ell_z';q) = \sum_{ss'} F_{\alpha}^{s}(q;\ell_z)\ \left(N^{-1}(q)+V(q)\right)^{-1} F_{\beta}^{s'}(q,\ell_z') \qquad (11)$$

F is the force experienced by an ion at $\vec{\ell}= (\ell,\ell_z)$ due to inter-action with the density wave A_s in direction α. D^E plus the ionic contribution allows for a calculation of the surface pho-non energies.

(ii) Using the expansion (9) and ε^{-1} from eq. (10) the ma-trix element of the screened electron-ion interaction is found to be

$$<n',\underline{k+q},k_z'|\ \sum_{\ell z} \vec{\nabla}_{\underline{r},\ell_z}\ W(\vec{r}-(o\ell_z))\ |\ n,\underline{k},k_z> \vec{e}(\ell_z;q,j) =$$

$$= \frac{N_{\shortparallel}}{N}^{1/2} \sum_{im'} c_{im'}^{*}(n',\underline{k+q},k_z')c_{jm_z}(n,\underline{k},k_z)e^{-i\,(\underline{k+q})\,Rs}\ e_{\alpha}(\ell_z;q,\lambda) \cdot$$

$$\left(F_{\alpha}^{s}(\underline{q};\ m_z-\ell_z) + \sum_{s'} V_{ss'-\ell_z}(q)\left(N^{-1}V_{xc}\right)^{-1}F_{\alpha}^{s'\,*}(q,m_z)\right) \qquad (12)$$

(iii) Finally, the dynamically screened Coulomb interaction V_{sc} in eq. (8) can similarly be expressed in terms of the scree-ning matrix $\left(N^{-1}(q;\omega)-V_{xc}(q)\right)_{ss'}^{-1}$ using eq. (10).

III. COMMENTS ON THE "EXCITON MECHANISM" OF SUPERCONDUCTIVITY

Allender, Bray and Bardeen (1973) have explored the possi-bility to induce Cooper pairing in an interface system, where the metal electrons at the Fermi surface tunnel into the semicon-ductor gap, where they interact exchanging "virtual excitons". After that, Inkson and Anderson (1973), using a (bulk) model dielectric function, estimated the pairing interaction which was found repulsive for energies of the order of interband transi-tions. In their local screening treatment the essential point was that the exciton corresponds to a *pole* in ε, thus to a zero in ε^{-1} and therefore does not couple to the electrons. Cohen and Louie (1976) also demonstrated this in terms of a bulk semicon-ductor (Ge) dielectric function based on a pseudopotenial band calculation. However, they proposed, like Allender et al., that if an attractive interaction is possible via exciton exchange, umklapp processes (for the surface $\underline{G} \neq \underline{G}'$ terms in ε^{-1}) corresponding to a non-local screening are necessary. Now, the main source of non-locality in a surface system is simply the surface boundary itself (z, z' dependence in ε^{-1}). Solving then for the self-consistent surface response we have the screening matrix $\left(N^{-1} - V_{xc} \right)^{-1} = N \left(1 - V_{xc} N \right)^{-1}$ in eq. (10). It corresponds to the infinite summation over "ladder" (c),(d) and RPA diagrams (a),(b) displayed in fig. 1. In this non-local microscopic formulation the zeros of the determinant of $\left(N^{-1} - V_{xc} \right)$ determine the (longitudinal) excitons. This becomes evident (Hanke, 1979) by realizing that N^- corresponds to $\Delta E_{Gap} + \omega$ and V_{xc} to the screened electron-hole $1/\varepsilon r$ interaction. Another well-studied example is furnished by the response function ε^{-1} of a metal-oxyde-semiconductor (MOS) system which, in addition to the acoustic plasmon poles, has poles at the interband transitions (Dahl and Sham, 1978), We have carried out preliminary calculations of the response function of a Si(111) surface system (slab) involving band structure effects and from ε^{-1} extracted the Fermi-surface averaged electron-electron interaction $V_{sc}(\omega)$. V_{sc} gives attractive contributions for ω around the dominant electron-hole transitions (Muramatsu and Hanke, 1979).

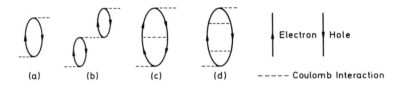

FIGURE 1. Polarization diagrams.

REFERENCES

Allender D., Bray J., and Bardeen J. (1973) Phys. Rev. B7,1020, and Phys. Rev. B8,4433.
Allen P.B. (1979). "Dynamical Properties of Solids". Vol. III, North-Holland.
Bardeen J. (1976). "Superconductivity in d-and f-Band Metals" Plenum, N.Y., 1.
Cohen M.L., and Louie S.G. (1976) "Superconductivity in d-and f-Band Metals" Plenum, N.Y., 7.
Dahl D.A., and Sham L.J. (1977). Phys. Rev. B16, 651.
Ginzburg V.L. (1970). Sov. Phys.-Usp. 13, 335.
Hanke W. (1979) "Festkörperprobleme". Adv. in Sol. St. Phys. XIX Vieweg.
Inkson J.C., and Anderson P.W. (1973). Phys. Rev. B8, 4429.
Muramatsu A., and Hanke W. (1979). To be published.
Wu C.H. and Hanke W. (1977). Sol. State Comm. 23, 829.

SUPERCONDUCTIVITY IN 2D
INHOMOGENEOUS NbN FILMS

S.A. Wolf
D.U. Gubser
J.L. Feldman

Naval Research Lab
Washington, D.C.

Y. Imry

Department of Physics and Astronomy
Tel Aviv University
Ramat Aviv, Israel

I. INTRODUCTION

There has been much recent interest in phase transitions in lower dimensional systems. Of particular interest both theoretically and experimentally is the two dimensional superconducting phase transition and whether in fact it occurs at finite temperature (Beasley et al., 1979, Doniach et al., 1979 and Kosterlitz et al., 1973). Recently, universal current scaling was reported in the critical region above a 2D superconducting transition in a granular NbN film (Wolf et al., 1979). This strongly suggested that indeed a phase transition occurred at finite temperature (T_{cj}) and also provided critical exponents and the experimental scaling function for the transition.

In this paper the universal scaling behavior is shown to extend below T_{cj} with the constants and exponents that were determined from the data above T_{cj} . The form of the scaling function in this low temperature regime predicts a temperature dependence of the critical current in the critical region which agrees remarkably well with the experimental values

ISBN 0-12-676150-7

for $T_{cj}-T \equiv \Delta T < 0.5$ K. Critical currents were also measured in the temperature regime below the critical region. A sharp rise in the critical current was observed at a characteristic temperature $T*$ ($\approx.6\ T_{cj}$) which may be associated with yet another transition in the sample. Data on three samples show that this behavior is quite universal.

The samples are composed of small NbN grains whose dimensions are slightly smaller than the superconducting coherence distance ξ_0. These grains are embedded in a non-superconductive matrix produced by anodization of the rf reactively sputtered NbN film An oxidation procedure produces isolated islands of 200Å NbN which are fairly uniformly distributed on the substrate. A schematic illustration of a cylindrical sample is shown in Fig. 1. The cylindrical geometry was chosen to minimize edge effects

II. RESULTS

At room temperature the sheet resistances $R_\square$ of the granular regions are between 6-10 x $10^4\ \Omega/\square$. As the temperature is lowered, $R_\square$ increases by almost a factor of 2 before zero dimensional superconducting fluctuations begin to reduce the resistance at about 2 T_{CG} ($\approx$25K) where T_{CG} is the mean field critical temperature for the isolated grains (Wolf et al., 1977). Near T_{CG} the resistance drops rapidly as the grains become strongly superconducting and begin to couple to one another by Josephson tunnelling. Below T_{CG}, a long resistive tail is observed until at a characteristic temperature T_{cj} all experimentally detectable resistance vanishes.

In a previous paper (Wolf et al., 1979), current (I) voltage (V) characteristics at and above T_{cj} were analyzed in analogy to a magnetic phase transition and were found to scale to an equation of the following form:

$$V = bI^x \times (\frac{d(T - T_{cj})}{I^\lambda})$$ (1)

with the following limits:

$$\text{as } I \rightarrow 0 \qquad \frac{V}{I} = a\Delta T^\mu$$

At $T = T_c \qquad V = bI^x$

where $\qquad x - \lambda\mu = 1$

and $\qquad d^\mu = a/b.$

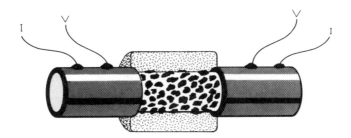

FIGURE 1. A schematic diagram of a cylindrical sample. The NbN film is deposited on a 1mm diameter quartz rod. The central portion is anodized which leaves the NbN grains embedded in an oxide matrix. Current and voltage leads are attached as indicated.

Here x, λ and μ are critical exponents, a, b and d are experimentally determined constants and χ is a scaling function. Data on three samples all scaled remarkably well to Eq. (1) with consistent exponents. The excellent agreement with this equation was strong confirmation that we indeed had observed the critical region above a superconducting phase transition. This transition at T_{cj} is believed to be associated with an intergranular ordering of the phases associated with the individual grains along a continuous path between the two NbN electrodes (see Fig. 1) giving rise to a phase coherent state.

In this paper, the characteristics of the samples in the temperature region surrounding T_{cj} are examined more carefully, especially the region near and just below T_{cj}. Data were obtained from V versus T plots at various current values and a typical family of curves is illustrated in Fig. 2. Note that T_{cj} is indicated by the vertical dotted line. From the curve at the lowest current one can extract a, μ, and T_{cj} ($V_{I \to 0} = aI\Delta T^{\mu}$). From the points at T_{cj} (intersections of dotted line with the family of curves) one can find b and x ($V = bI^{x}$). Thus all exponents and parameters are determined. By plotting V/bI^{x} versus $\left(\dfrac{d(T - T_{cj})}{I^{\lambda}}\right) \equiv z$ one can generate the scaling function from any of the curves shown in Fig. 2 if it is indeed a universal function. A result of such a plot is shown in Fig. 3 for all the curves in Fig. 2. Note that the scaling extends to negative values of z corresponding to the temperature region below T_{cj}. Furthermore the inset in Fig. 3 shows just how well the scaling behavior is obeyed below T_{cj}. The data shown extends to temperatures about 0.5 K below T_{cj} and within experimental uncertainties fits on the same curve.

This is further confirmation of our interpretation of having observed the critical region of a superconducting phase transition. Results for three samples are summarized in Table I.

Table I

Sample	$R_\square$ $\Omega/\square$	$I_c(0)$ µA	T_{cj} K	µ	λ	x
1	6600	80	6.70 ⎫			
2	9090	41	5.50 ⎬	3.7 ± 0.1	$.55\pm.05$	$3.0\pm.2$
3	7500	14	3.58 ⎭			

These data can be used to predict the temperature dependence of the critical current I_c if we define the critical current as the current necessary for the onset of a voltage. The z intercept of the scaling function $\chi(z)$ defines the point at which V/bI^x goes to zero at finite currents, i.e., is the critical current. To find the intercept, all the data on the scaling function between $z = -0.9$ and $z = +3$ were fit to an equation of the form $\chi(z) = (A + z)^\gamma$ where A and γ were

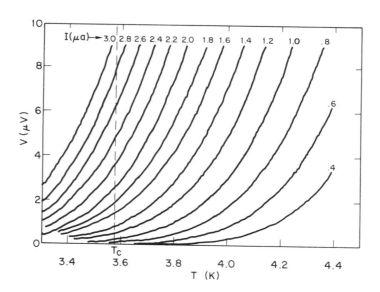

FIGURE 2. Voltage versus temperature at various current values for sample 3. T_c is indicated by a vertical dotted line.

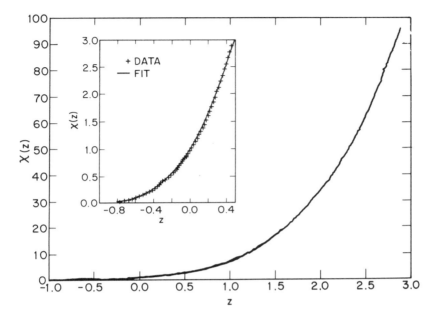

FIGURE 3. The scaling function χ(z) plotted as a function of z. The inset is an expansion of the low χ region of the curve.

computer adjusted to provide a best fit. The result is shown plotted in the inset of Fig. 3 where $A = 1.0$ and $\Upsilon = 2.764$ $(\Upsilon \approx \mu{-}1)$. This fit has a z intercept of -1; hence, I_c is given by:

$$z_c = \left(\frac{d(T - T_{cj})}{I_c^\lambda}\right) = -1$$

or

$$I_c = d(T_{cj} - T)^{1/\lambda}. \tag{2}$$

The temperature dependence of the critical current in the region just below T_{cj} is seen to be a power law in ΔT with a critical exponent determined from the scaling function.

Figure 4 shows a plot of Eq. (2) for one sample using d and λ determined from the high temperature data along with experimental I_c's determined from voltage onset points (0.05 μV voltage criterion) using $I - V$ characteristics at constant temperature. Notice the excellent agreement down to about 3.0 K.

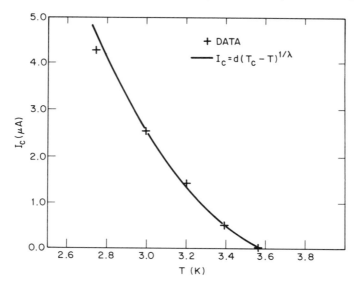

FIGURE 4.   Critical current versus T for sample 3.   The solid line is the expression based on the data above T_c.   The (+)'s are the experimental points.

Below 3.0 K for this sample, the critical current reverts first to a mean field temperature dependence and then at 2 K exhibits a sharp rise coincident with a change in the nature of the $I - V$ characteristics. All of the measured samples show the same normalized behavior although $I_c(0)$ and T_{cj} are quite different (see Table I).   These data suggest that a further transition is occurring at some lower temperature T^* ($\approx 0.6\ T_{cj}$) which may represent the stabilization of phase coherence throughout the entire 2D granular region and may be related to some very recent theories on 2D superconductive phase transitions. The low temperature critical current data will be reported on in detail in a subsequent publication.

III.   CONCLUSION

Universal scaling has been shown to extend below T_{cj} with the exponents and constants determined above T_{cj}.   The extrapolated z intercept of the scaling function predicts the temperature dependence of I_c in the critical region and the measured I_c's are in excellent agreement with the prediction. A sharp rise in the critical current at T^* may be associated with yet another transition in the samples.

REFERENCES

Doniach, S., and Huberman, B.A., (1979) *Phys. Rev. Lett.*
42, 1169.
Nelson, D. and Kosterlitz, J.M. (1977) *Phys. Rev. Lett.*
39, 1201.
Wolf, S.A. and Lowrey, W.H., (1977) *Phys. Rev. Lett.*
39 (1038).
Beasley, M., Mooij, J., and Orlando, T. (1979) *Phys. Rev. Lett.*
42, 1165.
Wolf, S.A., Gubser, D.U., and Imry, Y. (1979) *Phys. Rev. Lett.*
42, 324.

EXOTIC PAIRING IN METALS:
FORMALISM AND APPLICATION TO Nb AND Pd

F. J. Pinski[*]
P. B. Allen[*]

Department of Physics
State University of New York
Stony Brook, New York 11794, U.S.A.

W. H. Butler[†]

Metals and Ceramics Division
Oak Ridge National Laboratory
Oak Ridge, Tennessee 37830

Previous treatments of $\ell > 0$ pairing assume spherical symmetry, which we find give large errors in metal like Pd. A general formalism is developed, including the influence of impurities. Calculations of the electron-phonon Γ_{15} ("p-wave") coupling constants for Nb and Pd suggest that exotic pairing will be hard to observe in metals.

I.   FORMALISM

Foulkes and Gyorffy (1977) and Appel and Fay (1977) have used strong-coupling theory to discuss the possibility of BCS pairing in states with angular momentum $\ell > 0$ in metals

[*]Supported in part by NSF Grant no. DMR79-00837.
[†]Research sponsored by the Materials Science Division, U.S. Department of Energy under contract W-7405-eng-26 with Union Carbide Corporation.

like Pd. Both papers derive the equation

$$T_c(\ell) = \omega_o \exp[-(1 + \lambda_o)/\lambda_\ell] \tag{1}$$

where $\lambda_\ell = \lambda_\ell^{ph} + \lambda_\ell^{sf} + \ldots$ is the sum of the ℓ-wave coupling constants due to phonons, spin fluctuations, and so
forth. We believe that eq. (1) is correct in spherical
symmetry but very inaccurate in metals like Pd with non-
spherical Fermi surfaces. In this section we describe the
generalization of eq. (1) to include anisotropic metals
(Butler and Allen, 1976, referred to henceforth as BA) and
impurity effects (Allen, to be published, referred to as A)
The basic equation for T_c is (A)

$$\Delta_J(i\omega_n) = \pi T \sum_{J',i\omega_{n'}} [\lambda_{JJ'}(\omega_n - \omega_{n'}) - \mu_{JJ'}^*] \Delta_{J'}(i\omega_{n'})/|\omega_{n'}|$$

$$-\pi T \sum_{J',i\omega_{n'}} \Lambda_{JJ'}(\omega_n - \omega_{n'}) \operatorname{sign} \omega_{n'} \Delta_{J'}(i\omega_{n'})/\omega_n$$

$$+\sum_{J'} (\gamma_{JJ'} - \Gamma_{JJ'}) \Delta_{J'}(i\omega_n)/|\omega_n| \tag{2}$$

The index J runs over Fermi surface harmonics $F_J(k)$ (Allen,
1976), a complete set of "angular" functions (generalizations
of spherical harmonics except orthogonal on the actual Fermi
surface instead of a sphere.) The anisotropic gap is given
by

$$\Delta(k,i\omega_n) = \sum_J \Delta_J(i\omega_n) F_J(k) , \tag{3}$$

and $i\omega_n$ is a "Matsubara frequency" $2\pi i(n+\tfrac{1}{2})T_c$. In the
special case where only the constant function $F_o(k) = 1$ is
important, Γ_{oo} cancels γ_{oo}, Λ_{oo} equals λ_{oo}, $\lambda_{oo}(\omega_m)$ and
μ_{oo}^* are the usual strong coupling parameters, and eq. (2)
becomes eq. (1) of Bergmann and Rainer (1973). Some further
definitions are

$$\lambda_{JJ'}(\omega_m) \equiv \int_0^\infty d\Omega \; \alpha_{JJ'}^2 F(\Omega) \; [2\Omega/(\Omega^2 + \omega_m^2)] \tag{4}$$

$$\alpha_{JJ'}^2 F(\Omega) \equiv N(o)^{-1} \sum_{kk'} |M_{kk'}|^2 F_J(k) F_{J'}(k')$$

$$\times \; \delta(\Omega - \omega_{k-k'}) \; \delta(\varepsilon_k) \; \delta(\varepsilon_{k'}) \tag{5}$$

$$\gamma_{JJ'} = \pi N(o)^{-1} n_{imp} \sum_{kk'} |V_{kk'}|^2 F_J(k) F_{j'}(k') \delta(\varepsilon_k) \delta(\varepsilon_{k'}) \quad (6)$$

where $M_{kk'}$ and $V_{kk'}$ are the electron-phonon and electron-impurity matrix elements, ω_Q is the phonon frequency, and sums run over all electron states at the Fermi surface. The average electron lifetime $1/\tau$ for impurity scattering is $2\gamma_{oo}$. The only difference between the upper case quantities $(\Lambda_{JJ'}^{oo}(\omega_m), \Gamma_{JJ'})$ and the corresponding lower case quantities $(\lambda_{JJ'}(\omega_m), \gamma_{JJ'})$ is that $F_{j'}(k')$ in eqs. (5,6) is replaced by $F_{j'}(k)$. Thus it follows for all J (including J=0), that $\lambda_{oJ} = \Lambda_{oJ}$ and $\gamma_{oJ} = \Gamma_{oJ}$. These equations are all exact in the sense that they include anisotropy effects and make no additional approximations beyond Migdal's approximation and the related foundations of strong-coupling theory. Spin fluctuations have not been included explicitly, but to some extent are implicitly included in $\mu_{JJ'}^*$.

Equation (2) is a matrix equation which can be block-diagonalized by choosing the functions $F_J(k)$ to transform according to irreducible representations of the crystal point group. In spherical symmetry, F_J becomes $Y_{\ell m}$ and the matrices λ, μ^*, Λ, γ, and Γ are diagonal. Equation (1) then follows by application of McMillan's (1968) two-square-well procedures, and yields the unadjusted version of McMillan's equation in the $\ell = 0$ (s-wave) case. For cubic symmetry, the matrices are block-diagonal. The highest symmetry (Γ_1) block gives a theory for T_c with an anisotropic gap Δ_k which has the full symmetry of the crystal and the Fermi surface. In practice it is quite accurate in the Γ_1 block to forget that λ, μ^*, etc. are matrices, and keep only the $J = J' = 0$ parts. This neglects the anisotropy of Δ_k which is commonly found to occur in clean samples at a level of 10-20%. However, because of a variational principle (BA), T_c is affected only to second order in the anisotropy, i.e. 1-4%. Impurities occur only in the $J \neq 0$, $J' \neq 0$ parts and affect T_c only by washing out the (weak) anisotropy enhancement.

There are a total of 10 irreducible representations of the cubic group, and in principle, superconductivity could occur with any of these 10 symmetries. After Γ_1 ("s-wave"), the most likely channel is Γ_{15} ("p-wave"). Unfortunately, the algebra is no longer as simple as in the Γ_1 case. *There is no a priori way to make a good zeroth order variational estimate of the shape of Δ_k except in the Γ_1 channel where $\Delta_k \propto F_o = 1$.* To illustrate this important difficulty, consider the two simplest Γ_{15} ("p-wave") choices, (a) $F_X^a \propto k_x$ and (b) $F_X^b \propto v_{kx}$. These very different

choices are both x-components of a vector (Γ_{15} symmetry)
and both reduce to Y_{10} in spherical symmetry. If Δ_k were
approximately proportional to either F_x^a or F_x^b, then the
Γ_{15} block of eq. (2) could be replaced by a single part
involving only the matrices λ_{xx}, μ_{xx}^*, and so forth.
Applying McMillan's procedures, an accurate variational
lower bound for T_c would be

$$T_c(\Gamma_{15}) = 1.13\omega_D \exp[(1 + \Lambda_{xx})/(\lambda_{xx} - \mu_{xx}^*)] . \qquad (7)$$

$$\frac{1 + \Lambda_{xx}}{\lambda_{xx} - \mu_{xx}^*} = \psi \left[\frac{\omega_D}{2\pi T_c} + \frac{(\Gamma_{xx} - \gamma_{xx})/(1 + \Lambda_{xx})}{2\pi T_c} + 1 \right]$$

(pure)

$$- \psi \left[\frac{(\Gamma_{xx} - \gamma_{xx})/(1 + \Lambda_{xx})}{2\pi T_c} + 1/2 \right] \qquad (8)$$

(dirty)

where ψ is the digamma function. However, since we have no
a priori way of knowing the shape of Δ_k, eqs. (7,8) are not
expected to be accurate, only to give lower bounds.

The McMillan two-square-well procedures can be applied
to the otherwise exact matrix eq. (2), yielding an eigenvalue equation which generalizes eqs. (7,8):

$$\underset{\sim}{\delta} = \underset{\approx}{\eta} \left[\psi((\omega_D/2\pi T_c + 1) \underset{\approx}{1} + \underset{\approx}{\alpha}) - \psi (1/2 \underset{\approx}{1} + \underset{\approx}{\alpha})\right] \underset{\sim}{\delta} \qquad (9)$$

where $\underset{\sim}{\delta}$ is a vector in J-space related to Δ_J and

$$\underset{\approx}{\eta} = (\underset{\approx}{1} + \underset{\approx}{\Lambda})^{-\frac{1}{2}} (\underset{\approx}{\lambda} - \underset{\approx}{\mu}^*) (\underset{\approx}{1} + \underset{\approx}{\Lambda})^{-\frac{1}{2}} \qquad (10)$$

$$\underset{\approx}{\alpha} = (\underset{\approx}{1} + \underset{\approx}{\Lambda})^{-\frac{1}{2}} [(\underset{\approx}{\Gamma} - \underset{\approx}{\gamma})/2\pi T_c] (\underset{\approx}{1} + \underset{\approx}{\Lambda})^{-\frac{1}{2}} \qquad (11)$$

Equations (9-11) reduce to (8) when only the xx component is
kept of the various matrices. For small values of α the
Taylor series of the digamma function of eq. (9) cañ be used,
yielding a more manageable eigenvalue equation valid only
when $kT_c \gg \hbar/\tau$,

$$[\underset{\approx}{\eta}^{-1} + (\pi^2/2) \underset{\approx}{\alpha}] \underset{\sim}{\delta} = \log (1.13\omega_D/T_c) \underset{\sim}{\delta} \qquad (12)$$

Thus T_c is determined by the minimum eigenvalue of
$\underset{\approx}{\eta}^{-1} + (\pi^2/2) \underset{\approx}{\alpha}$. For a pure metal, this is the minimum

eigenvalue of $\underset{\sim}{\eta}^{-1}$ or the maximum eigenvalue of $\underset{\sim}{\eta}$. Let us denote the corresponding eigenvector $\underset{\sim}{\delta}_{jo}$, and transition temperature T_{jo}, where j denotes the Γ_j irreducible representation. Then we can write

$$T_{jo} = 1.13\omega_D \exp\left[-1/\lambda_{eff}(\Gamma_j)\right] \tag{13}$$

$$\lambda_{eff}(\Gamma_j) = \text{max eigenv. of } \left[(1 + \underset{\sim}{\Lambda})^{-\frac{1}{2}}(\underset{\sim}{\lambda}-\underset{\sim}{\mu}^*)(1 + \underset{\sim}{\Lambda})^{-\frac{1}{2}}\right]_{\Gamma_j} \tag{14}$$

$$\log (T_j/T_{jo}) \underset{\sim}{\sim} \delta T_j/T_{jo} \underset{\sim}{\sim} -\pi^2/2 \, (\underset{\sim}{\delta}_o \cdot \underset{\sim}{\alpha} \cdot \underset{\sim}{\delta}_o)/(\underset{\sim}{\delta}_o \cdot \underset{\sim}{\delta}_o) \tag{15}$$

where $\delta T_j = T_j - T_{jo}$ is the change in T_c to first order caused by impurities, and the right hand side of (15) is standard first-order matrix perturbation theory. Defining a new vector $\underset{\sim}{\Delta} = (1 + \underset{\sim}{\Lambda})^{-\frac{1}{2}} \underset{\sim}{\delta}$, eq. (15) becomes with the use of (10), (1ĩ),

$$\delta T_j = -(\pi/4) \left[\underset{\sim}{\Delta}_o \cdot (\underset{\sim}{\Gamma}-\underset{\sim}{\gamma}) \cdot \underset{\sim}{\Delta}_o\right]/\left[\underset{\sim}{\Delta}_o \cdot (1 + \underset{\sim}{\Lambda}) \cdot \underset{\sim}{\Delta}_o\right] \tag{16}$$

Because of the structure of eqs. (5,6), the matrices $\underset{\sim}{\lambda}$, $\underset{\sim}{\gamma}$ are small when $J = J' = 0$, but the upper case matrices $\underset{\sim}{\Lambda}$, $\underset{\sim}{\Gamma}$ are equal to $\lambda 1$, $(1/2\tau) 1$ respectively, plus small corrections, where $\lambda = \lambda_{oo}$ and $1/2\tau = \gamma_{oo}$ are the isotropic coupling constant and impurity scattering rate. Thus (16) becomes in first approximation

$$\delta T_j = - \pi\hbar/8\tau \, (1 + \lambda) \, k_B \tag{17}$$

a relation valid when $T_{jo} > 1/\tau$. The residual resistance ratio (rrr) can be written as

$$\rho(300)/\rho(o) = (rrr) = 2\pi \lambda_{tr} k_B(300K) \, \tau_{tr}/\hbar \tag{18}$$

and λ_{tr}, τ_{tr} can be replaced to good approximation by λ, τ. Thus

$$\delta T_j = - \pi^2\lambda \, (300K)/4(1 + \lambda) \, (rrr) \tag{19}$$

This relation is valid for all channels Γ_j except Γ_1. It is remarkable that there is such a simple relation for the impurity reduction of T_{jo} in spite of the absence of any simpler formula than eq. (13) for T_{jo}.

II.   CALCULATIONS

We have previously described in detail calculations of $\alpha^2 F$ and λ for Nb (Butler, Pinski, and Allen, 1979) and Pd (Pinski and Butler, 1979). These calculations begin by calculating γ_Q, the decay rate of a phonon into electron-hole pairs. The experimental verification of our predictions seems to us adequate reason for considerable confidence in our procedures. These calculations use approximately 1000 points on the irreducible $1/48^{th}$ of an accurate (KKR) Fermi surface, experimental phonon dispersion with Born-von Karman interpolation, and complete KKR wavefunctions. The largest uncertainty attaches to the rigid-muffin-tin procedure for the matrix elements, but this seems experimentally confirmed.

The extension of our calculations to the Γ_{15} channel is straightforward. We have not attempted to compute the maximum eigenvalue (eqs. 13,14) to convergence, but have used one and three-parameter variational procedures. The one parameter procedure is the trial solution $\Delta_k \sim v_{kx}$. This choice seems preferable to k_x on the grounds that the latter introduces an unphysical discontinuity on several pieces of Fermi surface which intersect the Brillouin zone boundary. The three parameter procedure is the trial solution $\Delta_k \sim c_i v_{kx}$ where c_i is allowed to take different constant values on the different sheets of Fermi surface. The dc electrical resistivity has also been computed in the same approximation and agrees with experiment to better than 10% for both Pd and Nb in the range 20K < T < 300K. The results are shown in table 1. The results for Pd were previously published in preliminary form (Pinski, Allen, and Butler, 1978), where $\alpha^2_{XX} F(\Omega)$ and $\rho(T)$ can be seen. The results for λ_{eff} (Γ_{15}) are very small and suggest that Γ_{15} superconductivity will not be seen unless a significant attraction comes from spin fluctuations or some other mechanism. Similarly small coupling constants have been found by Appel and Fay (1978) for Pd. The spin-fluctuation-induced coupling was estimated to be small by Fay and Appel (1977).

On the optimistic side, the impurity-reduction of T_{jo} (eq. 19) is less severe by a factor 8 $(1+\lambda)$ than was previously estimated by Foulkes and Gyorffy (1977). The factor $(1+\lambda)$ is a strong-coupling correction omitted from their work, while the factor of 8 appears to be an algebraic error. Webb et al. (1978) have found no evidence for superconductivity in Pd down to 1.7 mK. Their best sample had a resistance ratio 38,300. From eq. (19) we estimate

TABLE I. *Calculated Coupling Constants*

Material	Pd	Nb
λ_{xx}	-0.019	0.54
Λ_{xx}	0.44	1.17
λ_{xx}	0.41	1.12
λ_{tr}	0.46	1.07
λ_{eff}(1 parameter)	-0.013	.025
λ_{eff}(3 parameter)	0.002	.030
c_{eff} (Γ-centered)	1.0	1.0
c(open surface)	-0.67	.36
c(ellipsoids)	7.2	.39

that if pure Pd had a non-s-wave transition temperature
greater than 7.3 mK, a transition would have been observed
above 1.7 mK. This assumes that eq. (19) can be extrapolated
down to $\delta T_j = - T_{jo}$, far beyond the nominal range of
validity. We suspect that this extrapolation is fairly
reliable. The experiment by Webb *et al.* thus seems a
stringent test. A further measurement with a resistance
ratio increased by a factor of four or more would be
interesting. However, our calculations indicate that if
superconductivity should be discovered in Pd, the Γ_1 channel
is a likelier candidate than Γ_{15}.

REFERENCES

Allen, P.B. (1976). *Phys. Rev.* B13, 1416.
Appel, J. and Fay, D. (1978). *Sol. State Commun.* 28, 157.
Bergmann, G. and Kainer, D. (1973). *Z. Phys.* 263, 445.
Butler, W.H., and Allen, P.B. (1976). *In* "Superconductivity
    in d- and f-Band Metals" (D.H. Douglass, ed.), p. 73.
    Plenum Press, New York.
Butler, W.H., Pinski, F.J., and Allen, P.B. (1979). *Phys.
    Rev.* B19, April 1.
Fay, D. and Appel, J. (1977). *Phys. Rev.* B16, 2325.
Foulkes, I.F., and Gyorffy, B.L. (1977). *Phys. Rev.* B15, 1395.
McMillan, W.L. (1968). *Phys. Rev.* 167, 331.
Pinski, F.J., Allen, P.B., and Butler, W.H. (1978). *Phys.
    Rev. Lett.* 41, 431.

Pinski, F.J. and Butler, W.H. (1979). *Phys. Rev. B19*,
    June 15.
Webb, R.A., Ketterson, J.B., Halperin, W.P., Vuillemin, J.J.,
    and Sandesara, N.B. (1978). *J. Low Temp. Phys. 32*, 659.

PROPERTIES OF s-p AND s-d TYPE A-15
SUPERCONDUCTORS - A COMPARISON*

A. L. Giorgi

University of California
Los Alamos Scientific Laboratory
Los Alamos, New Mexico

The A-15 crystal structure, shown in Fig. 1, was discovered
in 1931 (Hartmann, et al, 1931) when an oxide of tungsten was
mistaken for a polymorphic form of tungsten and the new phase
called beta tungsten (β-W). Later when the error was discovered, crystallographers renamed the structure A-15 or Cr_3Si-type.
However, the name beta tungsten has persisted so that at present there are four designations which are used interchangeably:
A-15, β-W, Cr_3Si and Cr_3O. Superconductivity was first observed in an A-15 phase in 1953 (Hardy and Hulm, 1953) when it was
found that V_3Si was superconducting with a transition temperature, T_c, of 17.1 K, a record high T_c at the time. Since then
the record high T_c superconductors have consistently been A-15
phases and include Nb_3Sn at 18.4 K (Matthias et al, 1954),
$Nb_3Al_{.8}Ge_{.2}$ at 20.4 K (Matthias et al, 1967), Nb_3Ga at 20.3 K
(Webb et al, 1971) and Nb_3Ge at 23.2 K (Gavaler et al, 1974;
Testardi et al, 1974). At present there are over 70 intermetallic compounds with an A-15 structure and approximately 50
of these are superconducting.

The idealized composition for the A-15 phase is A_3B with
the A atoms occurring in pairs on each of the faces of a cubic
cell and forming three orthogonal lines or "chains of atoms as
illustrated in Fig. 1. The B-atoms are arranged in a body-
centered-cubic configuration. The interatomic distance between
the A-atoms in the chains is much closer than in the pure A
metal crystal and the superior superconducting properties of
this structure is believed associated with these chains.

The conditions considered favorable for the formation of
the A-15 phase are:

*
*Work performed under the auspices of the U.S. Department of
Energy.*

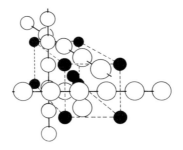

FIGURE 1.   A-15 structure showing orthogonal A chains.

1. As shown in Fig. 2, the A-atoms are limited to the
transition elements in groups 4A, 5A and 6A while the B-atoms
can be either the transition elements in groups 7A through 10A
or the nontransition elements in groups 1B through 5B.
2. The ratio of the Goldschmidt CN12 radii of the A and B
atoms (r_A/r_B) should be within the limits of 0.84 and 1.12 with
a value close to unity being the most favorable.
3. The valence electron per atom ratio (e/a) most favorable for the occurrence of the superconducting phase are 4.75
and 6.5 (Roberts, 1969) as shown in Fig. 3. These values are
remarkably close to the values proposed by Matthias (1955) for
the occurrence of high T_c in compounds. They also correspond
to the maxima in the density of states at the Fermi level, N(0),
versus e/a curve as determined by Morin and Maita (1963).

It was originally believed that the A-15 phases were nicely
ordered compounds with narrow composition ranges which occurred
close to the ideal A_3B composition and always on the A rich
side. However, it was soon learned that this was not always
true. Work by Waterstat and van Reuth (1966) on the Cr-Pt and
Cr-Os systems showed not only were some binary A-15 phases

Transition elements							Nontransition elements				
4A	5A	6A	7A	8A	9A	10A	1B	2B	3B	4B	5B
A							**B**				
									Al	Si	P
Ti	V	Cr	Mn	Fe	Co	Ni	Cu	Zn	Ga	Ge	As
Zr	Nb	Mo	Tc	Ru	Rh	Pd	Ag	Cd	In	Sn	Sb
Hf	Ta	W	Re	Os	Ir	Pt	Au	Hg	Tl	Pb	Bi

FIGURE 2.   Table of elements of groups 4A through 5B.

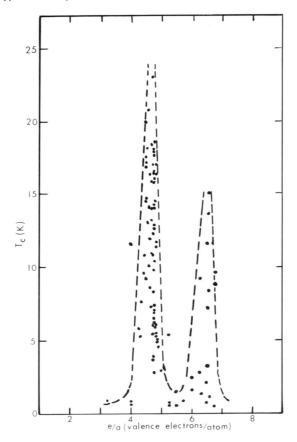

FIGURE 3. T_c versus e/a ratio for A-15 compounds.

stable over an extended composition range but often the composition range did not even include the A3B composition.   Other
investigators (Blaugher et al, 1969; Hartsough, 1974; and Flukiger et al, 1974) later pointed out that there are actually two
different types of A-15 compounds whose superconducting properties depended on whether the B atoms were transition elements
or nontransition elements.   It has been suggested (Flukiger et
al, 1974) that the two classes be designated as typical and
atypical A-15 compounds to differentiate between the two types.
A more reasonable classification would be one based on the electron characteristics of the B atoms.   For the purpose of the
present paper the two types of A-15 compounds will be distin-
guished as "s-p type" for the compounds with nontransition elements as B atoms, and "s-d type" for the compounds in which the
B atom is a transition element.   The various s-d type of A-15

compounds are shown in Fig. 4. A total of 26 binary compounds are represented including the recently discovered A-15 phase in the V-Re system (Giorgi et al, 1978). Superconductivity has been observed in 23 of the 26 s-d type A-15 compounds.

The superconducting properties in which the s-d type of A-15 superconductors show marked differences in behavior from the s-p type include:

  I.   Stoichiometry and range of composition.
  II. The strong influence of the $N(0)$, as shown by the e/a ratio, on the stability and T_c of the A-15 phase.
  III. The effect of composition and atomic ordering on the T_c of the A-15 phase.

I.   STOICHIOMETRY AND RANGE OF COMPOSITION

An empirical correlation between the T_c and the range of composition (solid solution range, SSR) has been proposed (Wang, 1974; Wang, 1977) which shows that in general the T_c of the superconducting A-15 compounds is proportional to the range of

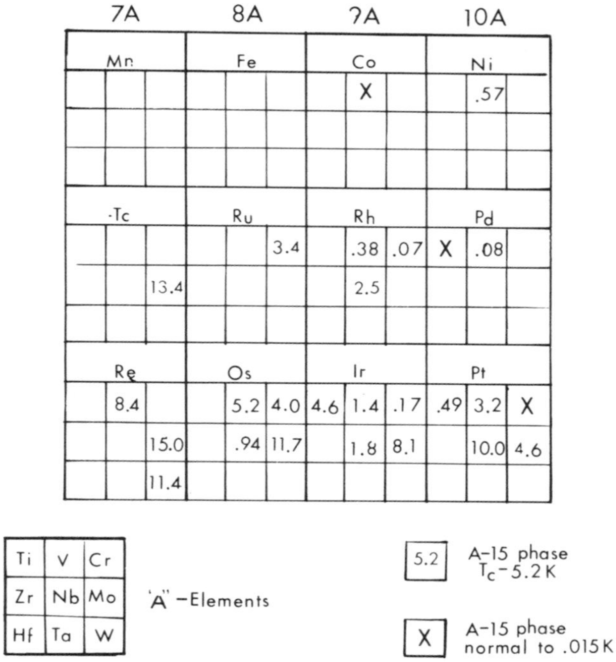

7A			8A			?A			10A		
Mn			Fe			Co			Ni		
						X			.57		
·Tc			Ru			Rh			Pd		
				3.4		.38	.07	X	.08		
	13.4					2.5					
Re			Os			Ir			Pt		
8.4			5.2	4.0	4.6	1.4	.17	.49	3.2	X	
	15.0		.94	11.7		1.8	8.1		10.0	4.6	
	11.4										

Ti	V	Cr
Zr	Nb	Mo
Hf	Ta	W

'A' – Elements

5.2	A-15 phase T_c ~ 5.2 K

X	A-15 phase normal to .015 K

FIGURE 4.   Occurrence of binary s-d type A-15 compounds.

composition of the phase. However, when this correlation is
applied to all superconducting A-15 materials, large variations
are found for a number of compounds, e.g. V-Pt, V-Rh, Nb-Ir,
etc., and these variations occur predominately for the s-d
type of A-15 compounds. If only the s-p type of A-15s are considered a much closer correlation between T_c and the range of
composition results, as shown in Fig. 5. Further, for the s-p
type, the range of composition occurs close to the ideal composition, A_3B, and extends in the direction of higher A-atom
content. The chains of A atoms appear to be continuous with
either formation of vacancies on the B sublattice, or the excess A atoms occupy positions on the B atom sites. The T_c increases towards a maximum value as the phase approaches the
stoichiometric A_3B composition.

The s-d type A-15 compounds, on the other hand, show wide
variations in composition which can extend in either the A rich
or the B rich direction. The range of composition has little
relation to the T_c, and for phases with a composition range
extending on both sides of the stoichiometric A_3B composition,
the maximum T_c does not occur at the A_3B composition. In fact
some of the s-d type of A-15 phases with fairly high T_c form
at composition which show extreme variation from the ideal
$A_{.75}B_{.25}$ atomic composition, e.g. $V_{.50}Os_{.5}$ with a $T_c=5.15$ K
(Blaugher et al, 1969); $Mo_{.4}Tc_{.6}$ - $T_c = 13.4$ K (Darby and Zegler, 1962); and $V_{.3}Re_{.7}$ - $T_c= 8.4$ K (Giorgi et al, 1978).

II.   EFFECT OF e/a ON STABILITY AND T_c

According to Laves (1956) for the A-15 structure to form,
the Goldschmidt CN12 radius ratio of the constituent atoms must
be close to unity. After an extensive examination of the various stable A-15 phases, it was later concluded (Hartsough, 1974)
that this radius ratio must lie between 0.84 and 1.12. However, the size factor alone is not a sufficient condition for
determining whether an A-15 structure will form. Compounds
such as Nb_3As and Nb_3Pd have favorable size factors but the
phases do not form. Zr and Hf have almost identical radii, yet
no A-15 compounds of Hf exist. So another factor, namely the
electron relationship of the component atoms, exerts a strong
influence on the formation of these phases. This is shown in
Fig. 3 by the large number of A-15 compounds which have e/a
ratioes close to 4.7 and 6.5. Since the s-p type A-15 compounds
form at compositions close to the A_3B stoichiometry, they form
over a wider range of e/a ratios depending on the nature of
the B atom. However, all of the compounds with high T_c values
form at the favorable e/a ratios.

In the s-d type A-15 compounds the e/a ratio exerts a major

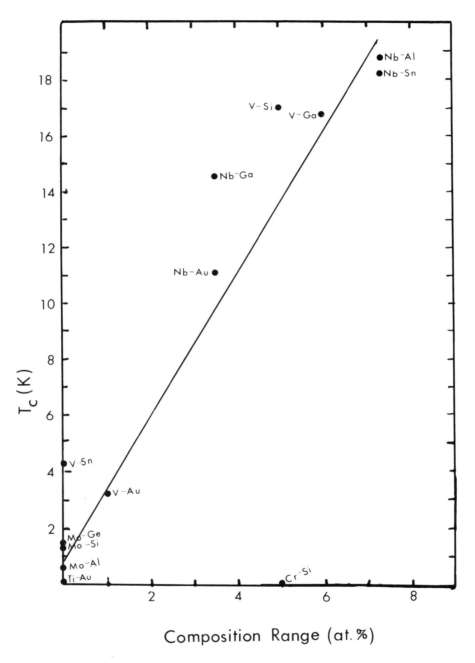

FIGURE 5. T_c versus composition range for s-p type A-15 compounds.

influence and essentially dictates the compositions where the
A-15 phase is most likely to form. Thus the s-d type A-15s
form over a narrower range of e/a ratioes and a much wider
range of compositions. The higher T_c values again occur near
the favorable e/a ratio of 6.5. This effect of the e/a ratio
was first confirmed by the changes in compositions which
occurred in the phases Cr-Pt, Cr-Ir and Cr-Os (Waterstrat and
van Reuth, 1966). A more dramatic example is shown in Fig. 6
which includes the variation of composition for the A-15 series
of compounds V-X, where X = Re, Os, Ir, Pt and Au (Giorgi et al,
1978). These compounds are found to vary from the idealized
atomic composition of $V_{.75}Au_{.25}$ and $V_{.75}Pt_{.25}$ to the extreme
atomic composition of $V_{.29}Re_{.71}$ in a continuous manner. Mean-
while the e/a ratio of all of these compounds remains essentially constant a a value of 6.4 $\pm$ 0.1.

III. EFFECT OF ATOMIC ORDERING

    The effect of atomic ordering on the T_c of A-15 phases has
been the subject of a number of investigations. The importance

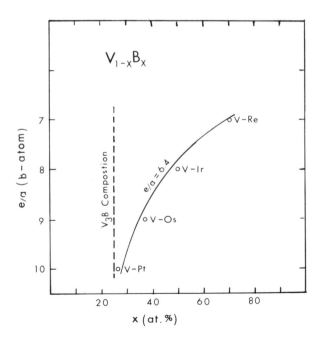

FIGURE 6.   Variation of composition for a vanadium based
series of s-d type A-15 compounds.

of long range crystallographic order (LRO) in A-15 phases was
first suggested by Hanak (Hanak et al, 1961) to explain the
variation of T_c for vapor deposited Nb_3Sn. Van Reuth and Waterstrat (1968) using x-ray diffraction methods determined the
Bragg-Williams long range order parameter, S, for a number of
A-15 compounds. Their studies show that the s-p type of A-15
phases have compositions close to the stoichiometric A_3B and are
highly ordered, while the s-d type A-15s have compositions off
stoichiometry and are disordered. Blaugher et al, (1969) found
that atomic ordering by low temperature anneals caused T_c to
increase in a number of s-p type systems. They also found that
for the s-d type A-15 phases in the Cr-Ir and V-Ir systems, the
T_c increased when the compositions departed from stoichiometry.
Similar increases in T_c by low temperature anneals have been
found in $Nb_3Al_{.8}Ge_{.2}$ (Matthias et al, 1967), Nb_3Ga (Webb et al,
1971) and V_3Ga (Koch 1973). The most interesting ordering effect was observed in the V_3Au system (Hein et al, 1971) where
the A-15 phase occurs at the stoichiometric composition and has
an extremely narrow range of homogeniety. The superconducting
behavior was found to vary from no superconductivity above
0.015 K to a T_c = 3.22 K as the order parameter was increased
by low temperature anneals.

A method that has proven very effective in producing relatively large degrees of disorder in A-15 compounds is the irradiation by high energy neutrons (E > 1 Mev) to a fluence (neu-
trons/cm^2) of approximately 10^{19}. The method permits progressive amounts of site-exchange disorder to occur uniformly
throughout the sample without any change in either the crystal
structure or the stoichiometry. Using high energy neutron irradiation, Sweedler et al, (1974,1975) studied the effect of
ordering on the T_c of the s-p type superconductors: Nb_3Sn,
Nb_3Al, Nb_3Ga, Nb_3Ge and V_3Si. Depression in T_c from 18% to
greater than 90% resulted as the fluence increased from 10^{18} to
10^{19}. A similar neutron irradiation study conducted on the s-d
type Mo_3Os (Sweedler et al, 1977) resulted in much smaller depression in T_c (3% to 16%), and a more recent study on the irradiation of the A-15 phase $Mo_4Tc_{.6}$ (Giorgi et al, 1979) resulted in an even smaller depression in T_c (0 to 3%). Of interest is the observation that similar depressions in T_c were
observed when the disorder was effected by converting the A-15
structure to an A-2 (bcc) structure at the same composition
using quick-quench techniques. Bucher et al, (1964) succeeded
in converting the A-15 form of Nb_3Au to the A-2 structure and
found the T_c decreased from 11.1 K to 1.2 K. A recent specific heat study on this system (Stewart et al, 1978) showed that
the drop in T_c was due to a corresponding drop in N(0). Willins
et al, (1966) converted Nb_3Al from the A-15 to the A-2 form
with a depression in T_c from 18.0 K to 3.1 K. Recently Giorgi
et al, (1978) succeeded in converting the highly disordered s-d

type A-15 Mo_2Tc_3 from A-15 to A-2 and observed essentially no change in T_c. These results suggest that A-15 compounds will have the highest T_c values when they have the ideal stoichiometry and the A strings form a totally undisturbed monatomic system with only B atoms in the B sublattice. For an entirely disordered A-15 lattice, the superconducting behavior for both the s-p and the s-d type A-15s will more or less correspond to that of a bcc lattice with a comparable e/a ratio.

REFERENCES

Blaugher, R. D., Hein, R. E., Cox, J. E., and Waterstrat, R. M. (1969). J. Low Temp. Phys. 1, 539.

Bucher, E., Laves, F., Muller, J., and van Philipsborn, H. (1964). Phys. Letters 8, 27.

Darby, J. B. and Zegler, S. T. (1962). J.Phys. Chem. Solids 23, 1825.

Flukiger, R., Paoli, A., and Muller, J. (1974). Solid State Commun. 14, 443.

Gavaler, J. R., Janocko, M. A., and Jones, C. K. (1974). J. Appl. Phys. 45, 3009.

Giorgi, A. L., Matthias, B. T., and Stewart, G. R. (1978). Solid State Commun. 27, 291.

Giorgi, A. L. and Matthias, B. T. (1978). Phys. Rev. B17, 2160.

Giorgi, A. L., Stewart, G. R., and Szklarz, E. G. (1979). The Electrochemical Society Extended Abstracts, Boston Meeting, May 6-11, Abstract 403.

Hanak, J. J., Cody, G. D., Aron, P. R., and Hitchcock, H. C. (1961). "High Magnetic Fields"(lax ed.) p. 592, Wiley, New York.

Hardy, G. F. and Hulm, J. K. (1953). Phys. Rev. 87, 884.

Hartmann, H., Ebert, F., and Bretschneider, O. (1931).Z.Anorg. und Allgemein. Chem. 198, 116.

Hartsough, L. D. (1974). J. Phys. Chem. Solids 35, 1691.

Hein, R. A., Cox, J. E., Blaugher, R. D., Waterstrat, R. M., and van Reuth, E. C. (1971). Physica 55, 523.

Koch, C. C. (1973). J. Phys. Chem. Solids 34, 1445.

Laves, F. (1956). Trans. A.S.M., Ser. A, 48, 124.

Matthias, B. T. (1955). Phys. Rev. 97, 74. (1954). Phys. Rev. 95, 1435.

Matthias, B. T., Geballe, T. H., Longinotti, L. D., Corenzwit, E., Hull, G. W., Jr., Willens, R. H., and Maiter, J. P. (1967). Science 156, 645.

Morin, F. J. and Maita, J. P. (1963). Phys. Rev. 129, 1115.

Roberts, B. W. (1969) "Superconductive Materials" NBS Technical Note 482.

Stewart, G. R. and Giorgi, A. L. (1978). Solid State Commun. 26, 669.

Sweedler, A. R., Schweitzer, D. G., and Webb, G. W. (1974). Phys. Rev. Letters 33, 168.

Sweedler, A. R. and Cox, D. E. (1975). Phys. Rev. B12, 197.

Sweedler, A. R., Moehlecke, S., Jones, R. H., Viswanathan, R., and Johnston, D. C. (1977). Solid State Commun. 21, 1007.

Testardi, L. R., Wernick, J. H., and Royer, W. A. (1974). Solid State Commun. 15, 1.

Van Reuth, E. C. and Waterstrat, R. M. (1968). Acta Cryst. B24, 186.

Wang, F. E. (1974). J. Phys. Chem. Solids 35, 273.

Wang, F. E. (1977). Solid State Commun. 23, 803.

Waterstat, R. M. and van Reuth, E. C. (1966). Trans. Met. A.I.M.E. 236, 1232.

Webb, G. W., Vieland, L. J., Miller, R. E., and Wicklund, A. (1971). Solid State Commun. 9, 1769.

Willens, R. H. and Buehler, E. (1966). Trans. TMS A.I.M.E. 236, 171.

A-15 COMPOUNDS AND THEIR
AMORPHOUS COUNTERPARTS

C. C. Tsuei

IBM Thomas J. Watson Research Center
Yorktown Heights, NY 10598

I. INTRODUCTION

The A-15 compounds are known to favor the occurrence of high temperature superconductivity (transition temperature $T_c > 15K$). The origin of superconductivity in these materials is a subject of much controversy and importance. A useful approach to this problem is to study comparatively the superconducting and normal-state properties of the A-15 superconductors and their amorphous counterparts. Efforts along these lines have yielded some insight into the mechanisms responsible for high temperature superconductivity.

It is interesting to note that most high-T_c A-15 compounds contain one glass-forming element such as Ge, Si or Al and are thus conducive to the formation of a non-crystalline phase. The amorphous (or higher disordered) state of the A-15 compounds can be achieved, for example, by one of the following techniques: 1) sputtering or co-evaporation onto substrates held at relatively low temperatures, (Tsuei, et al, 1977, 1978, and references therein) 2) particle irradiation, (Sweedler, et al 1974, Gurvitch, et al, 1978) and 3) ion-mixing (Tsaur, et al, 1979). It should be mentioned that one can study systematically various properties of an A-15 material as a function of the degree of disordering by varying certain experimental parameters such as the particle fluence in technique 2). The A-15 phase can also be recovered from the amorphous or highly disordered state by appropriate thermal annealing.

The transition temperature, T_c, of an A-15 superconductor has been shown to change drastically as it is made into an amorphous state. Examples of T_c values for various A-15 superconductors and their amorphous counterparts are shown in Table I.

Table I T_c *values for several A-15 compounds*
                  and their amorphous counterparts

T_c (K)

	crystalline (A-15)	amorphous or highly disordered
Nb_3Ge	22-23	3-4
Nb_3Sn	18	3
Nb_3Al	18.5	2
V_3Si	17	<1.5
V_3Ge	6.1	<1.5
Mo_3Ge	1.5	7.2
Mo_3Si	1.4	7.5

From Table I, one can see that most A-15 compounds suffer a large reduction in the T_c values when they are made into the amorphous or highly disordered state. Eminent exceptions to this general observation are amorphous Mo_3Ge and Mo_3Si, of which the T_c is enhanced as compared to the crystalline (A-15) state. It is hoped that a comparative study as described in this article will lead to a better understanding of superconductivity in the crystalline (c) and the amorphous (a) phases. In the following a brief review of recent experiments on this topic with emphasis on Nb_3Ge and V_3Si will be given.

II. ATOMIC SCALE STRUCTURE

Atomic scale structural studies of the amorphous phase of the A-15 compounds such as Nb_3Ge include x-ray (Cargill and Tsuei, 1978), electron scattering (Alessandrini, et al, 1978) and EXAFS (Brown, et al, 1977) measurements. Results of these experiments can be summarized as follows:

A. Large angle x-ray scattering measurements were made for the r.f. sputtered thin (a few 1000 Å to 10 μm) and thick (200 μm) a-Nb_3Ge samples prepared under different conditions. Only very minor differences were observed in the reduced interference functions and distribution functions for the thick and thin samples. Thus, the amorphous Nb_3Ge is structurally a well-defined phase.

It should be emphasized that the irradiation damage of energetic particles in A-15 high T_c superconductors such as Nb_3Ge and Nb_3Sn results in a highly disordered state which is characterized by a variety of structural defects, but there are *no* extensive regions of the amorphous phase in such samples. For example, in Fig. 1 of an article by D.E. Cox et al in this proceeding, the x-ray diffraction pattern for a CVD Nb_3Ge sample irradiated to a fluence of 3.4×10^{19} n/cm² shows very little, if any, contribution from the amorphous phase. The T_c of this particular sample is 4.4K, — a drastic reduction from the T_c of 20.9K before irradiated. The diffraction pattern of a Nb_3Ge sample irradiated to 6.5×10^{19} n/cm² (with $T_c = 3.4K$) does indicate a broad band that can be attributed to the amorphous phase.

B. Salient features of the total radial distribution (RDF) function for a-Nb_3Ge and a-Nb_3Si cannot be reproduced by microcrystalline models based on the structure of A-15 Nb_3Ge, Cu_3Au—type Nb_3Si or bcc Nb. First maximum in RDF (r) for r.f. sputtered a-Nb_3Ge is shown in Fig. 1 (a), along with that of A-15 microcrystallites with a lattice parameter of 5.16 Å and a size of 40 Å in Fig. 1 (b) for comparison. From these figures, one can see that the Nb-Nb peak (at r=2.6 Å in Fig. 1. (b)) corresponding to the three orthogonal Nb chains in the A-15 structure cannot be seen in the RDF for a-Nb_3Ge. The asymmetrical feature in the first peak of the RDF for a-Nb_3Ge bears no resemblance to the RDF at r ~3 Å for A-15 microcrystallites. Attempts to interpret the experimental results

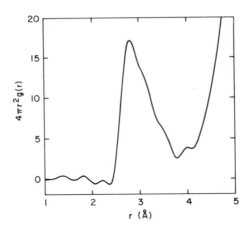

FIGURE 1a. The radial distribution function of
a-Nb₃Ge.

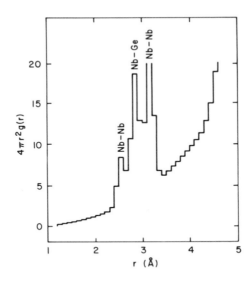

FIGURE 1b. The radial distribution function for A-15
Nb₃Ge crystallites with lattice constant ~5.16Å and size ~
40°Å

for a-Nb_3Ge with structural models of relaxed, binary dense random packing of hard spheres were also unsuccessful.

C.   Based on an assumption that amorphous alloys M_3X of group VA transition metals (M=Nb, Ta) and group IVB metalloids (X=Si, Ge) are isostructural, approximate partial distribution functions for M-M and M-X pairs have been calculated from the experimentally determined total RDF's.   Results on nearest neighbor distances (r_{ij}), distance distribution width (FWHM: full-width-at-half-maximum) and coordination number (CN_{ij}) are listed in Table II.   These parameters are also compared with those of the c-Nb_3Ge with A-15 structure:

Table II   Nearest neighbor configurations for M_3X alloys

	A-15	Nb_3Ge	amorphous $M_3X(Nb_3Ge, Ta_3Si)$		
	$r_{ij}(Å)$	CN_{ij}	$r_{ij}(Å)$	CN_{ij}	$FWHM(Å)$
MM	2.58	2	2.99	10.7	0.45
	3.16	8			
MX	2.87	4	2.66	2.1	0.10
XX	no "nearest neighbor"		information not obtainable from present data		

From Table II, it is interesting to note that the width of the MX (Nb-Ge, for instance) distance distribution is only about 25% that of the MM (Nb-Nb) distance distribution in the amorphous M_3X alloys, and the average MX distance and coordination number are significantly smaller than those of the A-15 crystal.   This result is very useful in later discussions on the degree of d-p hybridization and its relationship with the strength of electron-phonon interaction.

D. The Nb-Ge nearest neighbor distance in a-Nb_3Ge is found to be 2.64 ± 0.06 Å by EXAFS spectroscopy in excellent agreement with the result obtained from x-ray scattering measurements (see Table II).

III. SUPERCONDUCTIVE TUNNELING AND THE STRENGTH OF ELECTRON-PHONON COUPLING

Many high T_c A-15 superconductors such as Nb_3Ge and Nb_3Sn are characterized by a strong electron-phonon coupling. It is thus of interest to investigate the effect of structural disorder on the strength of electron-phonon interaction in such materials. In particular, there are theoretical predictions (Bergmann, 1976) that the electron-phonon in amorphous superconductors can be significantly enhanced due to the absence of long-range order in the atomic arrangement. This is apparently in accordance with the fact that all soft-metal non-crystalline superconductors such as amorphous Pb-Bi alloys exhibit very strong electron-phonon coupling. For A-15 strong-coupled superconductors, therefore, it is important to find out how much is the change in electron-phonon coupling strength and how it is related to T_c on going from the crystalline to the amorphous state.

In the absence of experimental data on the Eliashberg function, $\alpha^2(\omega)F(\omega)$, a simple but reliable indication of the electron-phonon coupling strength for superconductors can be obtained from the value of $2\Delta(O)/k_BT_c$, the ratio of energy gap to transition temperature. In the weak electron-phonon coupling limit, the BCS theory predicts that $2\Delta(O)/k_BT_c$ should be 3.52. To a good approximation, the amount of deviation from the BCS value (i.e. $2\Delta(O)/k_BT_c - 3.5$) is a measure of the strength of electron-phonon coupling.

As recently derived from superconductive tunneling measurements (Tsuei et al, 1977), values of $2\Delta(O)/k_BT_c$ for a number of amorphous transition-metal based superconductors (including a-Nb_3Ge) are found to be 3.5 ± 0.1 in agreement with the BCS value for the weak-coupling limit. In sharp contrast, the value of $2\Delta(O)/k_B \; T_c$ for the high T_c A-15 Nb_3Ge is reported to be 4.2 (Rowell, et al, 1976), suggesting a strong electron-phonon coupling. Therefore, it appears there is a fundamental change in the electron-phonon interaction as the A-15 Nb_3Ge is made into the amorphous state.

Recently, a systematic study of electron tunneling into the A-15 thin films of Nb_3Sn, Nb_3Ge and V_3Si as a function of composition and film-deposition condition (Moore, 1978). An interesting conclusion from this work is that the A-15 films of Nb-Sn and Nb-Ge are strong-coupled (i.e. $2\Delta(O) \approx 4.2k_BT_c$) near the stoichiometric composition and become weak-coupled $(2\Delta(O) \approx 3.5k_BT_c)$ in the Nb-rich region. On the other hand, the V-Si system exhibits a quite different trend with composition. In the single phase (A-15) region (~ 20-25 at. % Si), the value of $2\Delta(O)/k_BT_c$ varies relatively little $(2\Delta(O) \approx 3.5\text{-}3.8 \; k_BT_c)$. This result is extremely interesting in the sense that it suggests that the V-Si alloy system is a weak-coupled electron-phonon system even when the composition is at the stoichiometry and the T_c is relatively high $(T_c \sim 16$ K). The implication of this observation will be discussed later.

IV. DENSITY OF STATES AT FERMI ENERGY

The electronic density of states at Fermi level, $N(E_F)$, plays a crucial role in determining superconducting properties such as T_c as well as those anomalous behavior in the normal state for A-15 superconductors. According to various theoretical models based on tight binding approximation (Weger and Goldberg, 1974), a sharp peak in $N(E)$ near E_F is especially essential to achieve high T_c's in A-15 compounds. This is thought to be due to the one-dimensional and the localized nature of the d-electrons associated with the transition metal chains in the A-15 crystal structure. In view of this, it is expected that a study of the relationship between $N(E_F)$ and T_c in A-15 superconductors as a function of disorder should shed light on the origin of high temperature superconductivity in this class of materials. Among others, an estimate of changes in $N(E_F)$ with disorder can be obtained by measuring A) low temperature specific heat, B) the temperature coefficient of the upper critical field, dH_{c2}/dT, and the normal-state resistivity ρ_0 near T_c, or C) photoemission spectrum. In the following, a brief account of the results from these measurements will be presented.

A.   Low-Temperature Specific Heats

Results of specific heat measurements for high-T_c (21.8K) single-phase crystalline (A-15) Nb_3Ge by G. Stewart et al (1978) are shown in Fig. 2. Recently, low-temperature specific heats of amorphous Nb_3Ge films have been reported by Tsuei et al (1978). These results are presented in Fig. 3. The inset in Fig. 3 shows the resistive phase transition of the same sample used in the calorimetric measurement. As shown in Figs. 2 and 3, the specific heat data ($T>T_c$) both of the A-15 and amorphous Nb_3Ge can be approximately described by the standard equation:

$$c = \gamma T + \beta T^3 \tag{1}$$

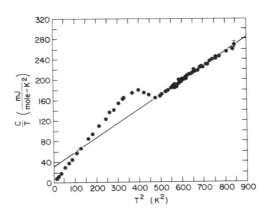

FIGURE 2.   Total specific heat, as a function of temperature, of single-phase *A-15 Nb_3Ge obtained by CVD technique. The T_c of this sample is 21.8K.*

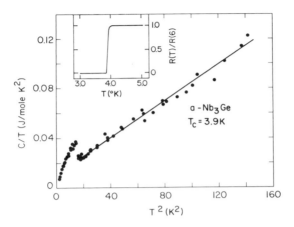

FIGURE 3. Total specific heat of a-Nb$_3$Ge as a function of temperature (1.5K < T < 16K). The inset shows the resistance R(T), normalized at T=6K, of the same specimen used in the calorimetric measurement.

A least-squares fit of the normal-state heat capacity data to Eq. 1 yields the following results:

	c $-$ Nb$_3$Ge	a $-$ Nb$_3$Ge
$T_c(K)$	21.8	3.9
$\gamma(mJ/moleK^2)$	30.3 $\pm$ 1	12 $\pm$ 2
$\beta(mJ/moleK^4)$	0.282 $\pm$ 0.002	0.71 $\pm$ 0.02
$\theta_D(K)$	304 $\pm$ 1	222 $\pm$ 3
$2\Delta(O)/k_B T_c$	4.2	3.5 $\pm$ 0.1

Table III Parameters Derived From Specific Heat Data

From the data below T_c, one can also estimate the values of $2\Delta(O)/k_B T_c$. As listed in Table III, these results are in excellent agreement with those obtained by tunneling measurements.

By taking into account of the renormalization effect arising from the electron-phonon interaction, $N(E_F)$ can be related to γ obtained from specific heat by $\gamma = 2/3\pi k_B^2(1+\lambda)N(E_F)$ where λ is the McMillan electron-phonon coupling parameter ($\lambda_a \sim 0.6$, $\lambda_c \sim 1.7$ for a— and c—Nb_3Ge). It is concluded by Tsuei et al (1978) that $N(E_F)_c/N(E_F)a = (1 + \lambda_a)\gamma_c/(1+\lambda_c)\gamma_a \approx 1.7$. This is somewhat unexpected in view of those theoretical models mentioned before. In fact, the experimental results indicate that $N(E_F)$ of Nb_3Ge changes relatively little while T_c decreases from 22K to 4K.

B.   Measurements of (dH_{c2}/dT) and ρ_o near T_c.

Wiesmann et al (1978) recently measured (dH_{c2}/dT) and ρ in order to estimate the density-of-states change with the structural disorder arising from α-particle irradiation in A-15 Nb_3Ge and Nb_3Sn. It was concluded that the density of states changes dramatically, as does T_c, in both alloy systems. The data by Wiesmann et al were re-analyzed by taking into account the change in the electron-phonon coupling strength with T_c and the effect of mass enhancement arising from the electron-phonon interaction. The results of this new analysis indicates that $N(E_F)c/N(E_F)a = 1.9$ for Nb_3Ge (Tsuei, 1978) in good agreement with the result of specific heat measurement. On the other hand, the results suggest that $N(E_F)$ in Nb_3Sn decreases by a factor of ~ 4 as T_c changes from ~ 18 to $\sim 3K$. Similar conclusions were reported by Ghosh et al (1978) in a paper where new data and a refinement of an earlier estimate of $N(E_F)$ were presented. The relatively low $N(E_F)$ of c—Nb_3Ge, however, was attributed to anomalies in the sample properties. For instance, it was argued that the usual samples of Nb_3Ge contain too much non-A-15 material. This is currently a point of much controversy. It should be emphasized that the specific heat data on high T_c A-15 Nb_3Ge used in our $N(E_F)$ analysis were obtained from a "single phase" sample (Stewart, et al, 1978). It may be added that the technique of using the dH_{c2}/dT and ρ_o data to determine $N(E_F)$ has recently gained credibility. Orlando et al (1978), for example, have demonstrated that, for Nb_3Sn and V_3Si, the γ values obtained by this technique agree well with those obtained by specific heat measurements.

C. Photoemission Spectroscopy

As far as the determination of $N(E_F)$ is concerned, photoemission data should be used only as supporting evidence due to the relatively poor energy resolution of the technique. An XPS study of the high T_c crystalline and the low T_c amorphous phases of Nb_3Ge reveals that, within the 500 meV XPS resolution, $N(E_F)$ is comparable in c– and a–Nb_3Ge (Fig. 4).

V. ORIGIN OF HIGH T_c IN A-15 COMPOUNDS

As presented in section IV, the experimental data obtained from three independent techniques, strongly suggest that $N(E_F)$ of Nb_3Ge does not change much as a result of the transition from the crystalline to the amorphous (or the highly disordered) state. A discussion of other alternative mechanisms for high T_c in this material is in order. As shown in the well-known McMillan formula:

$$T_c = \left(\frac{\theta_D}{1.45}\right)\exp\left[\frac{-1.04(1+\lambda)}{\lambda-\mu^*(1+0.62\lambda)}\right] \qquad (2)$$

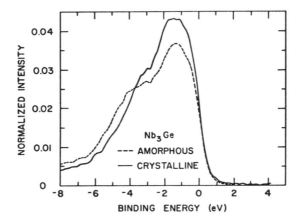

FIGURE 4.   Valence band XPS spectra for a-Nb_3Ge (T_c=3.7K) and c-Nb_3Ge (T_c=16K).

the value of T_c is essentially determined by λ ($\lambda = N(E_F)<I^2>/M<\omega^2>$, where $<\omega^2>$ is an average square phonon frequency, and $<I^2>$ the Fermi-surface average of the electron-phonon interaction strength). Given the values of λ, θ_D, and $N(E_F)$ for c– and a–Nb$_3$Ge and the assumption of $<\omega^2> \propto \theta^2_D$, one can estimate the ratio of $<I^2>_c/<I^2>_a$

$$\frac{<I^2>_c}{<I^2>_a} = \frac{N(E_F)_a<\omega^2>_c\lambda_c}{N(E_F)_c<\omega^2>_a\lambda_a} \sim 3. \tag{3}$$

This result is in good agreement with the conclusions from the tunneling and the specific heat measurements that there is a fundamental change in the electron-phonon coupling (i.e. from strong to weak-coupled) upon the crystalline (A-15) to amorphous transition. All these findings are consistent with a microscopic theory of dielectric screening and lattice dynamics (Hanke, et al, 1976) for transition-metal intermetallic compounds. According to this theory, hybridization of the relatively localized electronic states (i.e. 4d states of Nb atoms) and the more extended states (i.e. the s, p states from a non-metal element such as C, Ge, etc.) at E_F can produce a resonance-like increase in $<I^2>$. As shown in Fig. 4 and in a paper by Pollak, et al, (1977), the XPS data do show a reduction in d-p hybridization at E_F for a-Nb$_3$Ge. It should be mentioned that the fact the coordination number for Nb-Ge pairs is only 2 in a–Nb$_3$Ge as compared to 4 in c–Nb$_3$Ge (see Table II) is also in accordance with the conclusion that there is a dehybridization in Nb$_3$Ge as a result of structural disorder. It should be noted that an alternative theoretical approach to the enhancement of T_c by localized states (Cowan and Carbotte, 1978) leads to similar conclusions. In short, the high T_c superconductivity of Nb$_3$Ge does not appear to stem only from an unusually high $N(E_F)$ but rather from a relatively strong electron-phonon coupling arising probably from a large d-p hybridization at E_F.

On the other hand, $<I^2>$ does not play a dominant role in determining the T_c of A-15 superconductors such as V$_3$Si. This is due to the facts: 1. The tunneling data indicate that V$_3$Si is a BCS weak-coupled superconductor. 2. The $N(E_F)$ of V$_3$Si is about twice larger than that of A-15 Nb$_3$Ge and is, in fact, higher than any other A-15 compounds except V$_3$Ga. It should be pointed out here that V$_3$Ga (T_c=16K) is a convincing example to demonstrate that, in a given class of materials (the A-15 compounds), $N(E_F)$ does not correlate with T_c. 3. The XPS results for A-15 V_3Si (Riley et al, 1976) exhibit a very weak d-p hybridization, again suggesting a weak electron-phonon interaction.

ACKNOWLEDGMENT

The author wishes to thank G.S. Cargill III, R.A. Pollak and G.R. Stewart for helpful discussions.

REFERENCES

Alessandrini, E.I., Laibowitz, R.B., and Tsuei, C.C. (1978). *J. Vac. Sci. Technol. 15*, 377.

Bergmann, G. (1976). *Phys. Rept.* (Phys. Lett. C) *27*, 159.

Brown, G.S., Testardi, L.R., Wernick, J.H., Hallak, A.B., and Geballe, T.H. (1977). *Solid State Commun.23*, 875.

Cargill III, G.S., Tsuei, C.C. (1978) in *"Rapidly Quenched Metals III,"* (B. Cantor, ed.), p. 337, The Metal Society, London.

Cowan, W.B. and Carbotte, J.P (1978). *J. Phys. C 11*, L265.

Ghosh, A.K, Gurvitch, M., Wiesmann, H., and Stongin, Myron (1978). *Phys. Rev. B 18*, 6116.

Gurvitch, M. Ghosh, A.K., Gyorffy, B.L., Lutz, H., Kammerer, O.F., Rosener, J.S., and Stongin Myron (1978). *Phys. Rev. Lett. 41,*, 1616.

Hanke, W., Hafner, J., and Bilz, H. (1976). *Phys. Rev. Lett. 37*, 1560.

Klein, B.M., Boyer, L.L., and Papaconstantopoulos, D.A. (1978). *Phys. Rev. B 18*, 6411.

Moore, D.F. (1978). Ph.D. thesis, Stanford University.

Orlando, T.P., McNiff Jr., E.J., Foner, S., and Beasley, M.R. (1978). *Phys. Rev. B 19*, 4545.

Pollack, R.A., Tsuei, C.C., and Johnson, R.W. (1979). *Solid State Commun. 23*, 879.

Riley, J., Azonlay, J., and Ley, L. (1976). *Solid State Commun. 19*, 993.

Rowell, J.M. and Schmidt, P.H. (1976). *J. Appl. Phys. Lett. 29,* 622.

Stewart, G.R., Newkirk, L.R., and Valencia, F.A. (1978). *Solid State Commun. 26,* 417.

Sweedler, A.R., Schweitzer, D.G. and Webb, G.W. (1974). *Phys. Rev. Lett. 33,* 168.

Tsaur, B.Y., Liau, Z.L., and Mayer, J.W. (1979). *Appl. Phys. Lett. 34,* 168.

Tsuei, C.C., Johnson, W.L., Laibowitz, R.B., and Viggiano, J.M. (1977). *Solid State Commun. 24,* 615.

Tsuei, C.C., (1978). *Phys. Rev. B18,* 6385.

Tsuei, C.C., von Molnar, S., and Coey, J.M. (1978). *Phys. Rev. Letter 41,* 664.

Weger, M. and Goldberger, I.B. (1974). In *Solid State Physics* (H. Ehrenreich, F. Seitz and D. Turnbull, editors), Vol. 28, p. 1, Academic Press, New York.

Wiesmann, H., Gurvitch, M., Ghosh, A.K., Lutz, H., Kammerer, O.F., and Strongin, Myron (1978). *Phys. Rev. B 17,* 122.

X-RAY DETERMINATION
OF ANHARMONICITY
IN V_3Si*

J.-L. Staudenmann[1]

Department of Physics
and
Research Reactor Facility
University of Missouri
Columbia, Missouri

and

L.R. Testardi

Bell Laboratories
Murray Hill, New Jersey

X-ray intensities from over 1000 reflections in single crystal V_3Si at 300K, 78K, and 13.5K, previously obtained by one of the authors (J.-L.S.) have been analyzed for anharmonicity when $\sin\theta/\lambda > 0.7Å^{-1}$ where diffraction occurs from core electrons only. At 78K we find $\langle\mu^2\rangle$ and $\langle\mu^4\rangle$ moments for V which are ~50% and 500% larger than expected for a harmonic lattice, and evidence for localization of the V atoms off the lattice sites by ~0.2A. No evidence of such large anharmonicity is found for Si atoms at all three temperatures.

Work supported for J.-L.S. by Swiss N.F. (request #820.360.75) and U.S.N.S.F. through grant #PHY 76 08960 at the University of Missouri at Columbia.

[1]*Present address: Ames Laboratory, U.S.D.O.E., Ames, Iowa.*

247

II.   INTRODUCTION AND EXPERIMENTAL

Many of the physical properties of A15 compounds such as
V_3Si appear to be linked in some way with the existence of
linear chains of tightly bounded transition metal atoms. As
examples, anharmonicity studies (Shier and Taylor, 1966;
Patel and Batterman, 1966; Garcia et al., 1971; Testardi,
1973, 1975; Knapp et al, 1975; Allen et al., 1976; Smith et
al., 1976; Kimball et al., 1978) as well as structural instabilities (Shull, 1964; Batterman and Barrett, 1966; Labbé
and Friedel, 1966) are always discussed in terms of peculiar
features of these chains. In this letter we report the anharmonicity of the V atom motion in V_3Si based on accurate
X-ray intensities measured at 300, 78, and 13,5K. We find
large anharmonicity at 78K in the V atom displacements such
that $<\mu^2>$ and $<\mu^4>$ are ~50% and 500% larger than expected for
a harmonic lattice. No such unusual behavior is found for
the V atoms at 300K or for the Si atoms at any of the temperatures studied. The results indicate a tendency for the
V atoms to localize ~0.1Å-0.2Å off the lattice sites at 78K
in anticipation of the incipient structural transformation.
Among the consequences of this result are microscopic evidence of instability in the transition metals chains, high
frequency phonon softening and anharmonicity, and probable
anomalous contributions to the electrical resistivity.
Over 1000 diffracted integrated intensities were collected
for each of the three data sets from two different spherical
single crystals. The sample studied at 13.5K exhibits a
structural transformation (Staudenmann, 1978). Both spheres
came from a large single crystal and were annealed simultaneously after the grinding process. Further experimental
details will be published later (Staudenmann and Testardi, to
be published). The structure factors used in this letter are
the A15 cubic averaged integrated intensities scaled and
corrected for extinction (Becker and Coppens, 1974). These
corrections will not significantly change the conclusions
drawn here because extinction is much weaker for reflections
in the core region than those in the valence region, and
because the main results can be obtained from ratios of observed structure factors (see Fig. 1,2,3) where these corrections do not apply.

III.   METHOD AND LEAST SQUARES REFINEMENT

From x-ray measurements at wavelength λ the mean square atomic displacements $\langle \mu^2 \rangle$ perpendicular to diffracting planes (angle θ) can be obtained from the measured structure factor F at two different temperatues T_1 and T_2 using the Debye-Waller result for harmonic motion and neglecting thermal expansion (< 0.2% as referred to lattice constant as a function of temperature)

$$\lambda^2 \ln[F(T_1)/F(T_2)] / 8\pi^2 \sin^2\theta = \langle \mu^2(T_2) \rangle - \langle \mu^2(T_1) \rangle$$

Fig. 1 shows a plot of the left hand side of eq. 1 (T_1 = 78K and T_2 = 300K) versus $\sin^2\theta/\lambda^2$ for reflections from V_3Si where only V atoms contribute to F. The data does not confirm the expectation of zero slope behavior, but show reasonable grouping about line of nonzero slope above $\sin\theta/\lambda > 0.7 \overset{\circ}{A}^{-1}$ where diffraction arises from core electrons only (Fukamachi, 1971). "Core electrons only" means that the total contribution of the valence electrons is less than 0.1 electrons. For $\sin\theta/\lambda < 0.7 \overset{\circ}{A}^{-1}$ the values of these structure factor ratios are influenced by the distribution and temperature dependence of

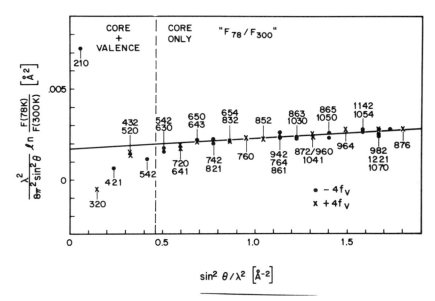

FIGURE 1.   *Normalized logarithm of the structure factor ratio for T_1 = 78 K and T_2 = 300 K as a function of $\sin\theta^2/\lambda^2$. Only 4 V atoms contribute to the structure factors. (hkl) of the diffracting planes has been indicated.*

valence charge density (see conclusion). Fig. 2 where T_1 = 13.5K and T_2 = 78K exhibits behavior similar to that in Fig.1 with points fitting equally well a straight line but having a slope of opposite sign. Fig. 3 where T_1 = 78K and T_2 = 300K clearly shows that similar features are also seen in reflections involving both V and Si atoms (here 2V + 2Si contribute to the structure factors). The systematic trend of these plots cannot be due to anisotropic $\langle \mu^2 \rangle$ alone, but strongly suggests the existence of a large contribution to $\langle \mu^4 \rangle$.

Borie (1974) first explained how to introduce a cubic term (in atomic displacements) for the thermal parameter of V atoms sitting on chain sites. This contribution is zero for Si atoms occupying the BCC positions of the Al5 structure because the cubic symmetry of those sites. We have extended Borie's procedure to fourth order in the atomic displacements (Staudenmann and Testardi, to be published). These results yield the following consequences for the data of Fig. 1,2, and 3:

i)   isotropic $\langle \mu^4 \rangle$ gives straight lines in plots vs. $\sin^2\theta/\lambda^2$; anisotropic $\langle \mu^4 \rangle$ will cause "scatter" above and below the mean slope

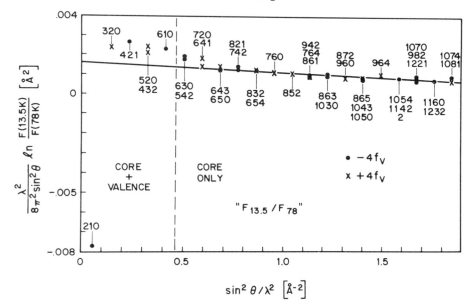

FIGURE 2.   *Normalized logarithm of the structure factor ratio for T_1 = 13.5 K and T_2 = 300 K as a function of $\sin\theta^2/\lambda^2$. Only 4 V atoms contribute to the structure factors. (hkl) of the diffracting planes has been indicated.*

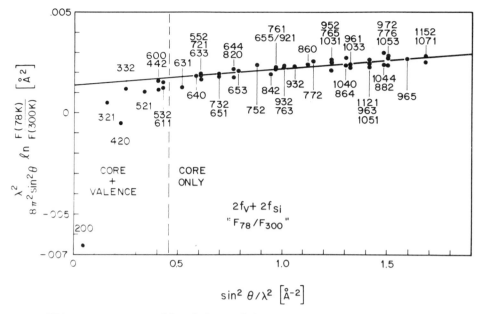

FIGURE 3. *Normalized logarithm of the structure factor ratio for T_1 = 78 K and T_2 = 300 K as a function of $sin^2\theta/\lambda^2$. Only 2V + 2Si atoms contribute to the structure factors. (hkℓ) of the diffracting planes has been indicated.*

ii) cubic anharmonicity contributes additional "scatter" above and below the mean slope

iii) anisotropic $\langle\mu^2\rangle$ displaces the points from the slope in one direction only by an amount which depends on the diffracting plane (hkℓ) as well as on the anisotropy.

Thus, in Fig. 1,2, and 3 the anomalous slope arises mainly from unusually large $\langle\mu^4\rangle$ terms.

To obtain numerical values of the displacement moments (static and dynamic displacements cannot be distinquished), a least squares fit of the observed F (for V atoms only) at each temperature was made by using atomic form factors from the International Tables for X-ray Crystallography. We found agreement to within ~0.6% (average, with maximum discrepancy ~1% for all reflections of Fig. 1, 2, and 3 with $sin^2\theta/\lambda^2 > 0.5 \overset{\circ}{A}^{-2}$ assuming two second moments, parallel $\langle\mu_{\parallel}^2\rangle_V$, and perpendicular $\langle\mu_{\perp}^2\rangle_V$ to the chains, one third moment $\langle\mu_{\perp}^2\mu_{\parallel}\rangle_V$, and one isotropic $\langle\mu^4\rangle_V$. Since the experimental error is estimated at ~1% further information on the anisotropy of $\langle\mu^4\rangle$ cannot be obtained although its macrosopic (average) value is small. The Si atom moments were obtained by analyzing a somewhat larger number of reflections involving both V and Si contributions, and using the V results from proceeding least squares fit. Data for V at 13.5K are of insufficient accuracy to obtain $<\mu^2_\perp \mu_{||}>$ and $<\mu^4_\perp>_V$. Least squares results are given in Table 1.

TABLE 1.

T	$<\mu^2_{||}>$		α	C.T.	$<\mu^4>$		A	
	V	Si	V	V	V	Si	V	Si
[K]	$[10^{-3} A^{\circ 2}]$				$[10^{-5}\overset{\circ}{A}{}^4]$			
300	4.80	5.96	0.07	+3.0	5.6	11.1	0.81	1.04
	.15	.15	.02	2.3	1.1	1.0	.2	.1
78	2.93	4.14	0.03	-2.1	5.6	5.3	2.22	1.03
	.10	.13	.02	1.1	.6	.4	.3	.1
14	1.34	3.54	0.07	<1*	<2	3.9	<3	1.04
	.07	.1	.03			.4		.1

$<\mu^2>_V \equiv (2<\mu^2_\perp>_V + <\mu^2_{||}>)/3$

$\alpha \equiv (<\mu^2_\perp> - <\mu^2_{||}>)/<\mu^2_\perp>$ *anistropic $<\mu^2>$ coefficient*

$C.T. = <\mu^2_\perp \mu_{||}>$ *expressed in $[10^{-5}\overset{\circ}{A}{}^3]$*

$A = <\mu^4>/3<\mu^2>^2$ *=1 for harmonic behavior*

second line numbers for each temperature are
standard deviations

*Refers to absolute magnitude

Table 1 also confirms that plots (not shown here) similar to Fig. 1,2, and 3 but with T_1 = 13.5K and T_2 = 300K do not exhibit non-zero slope. This is because the anharmonicity at 300K is relatively small and the one at 13.5K is not sufficiently well determined with the present data. The anisotropy in $<\mu^2>$ for V is <10% at all three temperatures. On the other hand the electron density maps at 300K as well as at 13.5K also exhibit a contour density anisotropy of   <10% around V sites (Staudenmann et al., 1976; Staudenmann, 1978). Those two anisotropies should be the same because the anisotropy in $<\mu^2>$ is deduced from structure factors which are the Fourier transform of the electron charge distribution in real space. The same kind of least squares refinement done

without anharmonic corrections shows an anisotropy in $\langle\mu^2\rangle$
greater than 15% at all three temperatures. This means that
a part of the scattering due to anharmonicity was put into
anisotropy in $\langle\mu^2\rangle$ (see iii above).   One can see that
the 10% anisotropy could also come from the static com-
pression of V atoms on a chain in V₃Si as referred to the
V-V distance in pure V metal.

    Results for the cubic anharmonicity are uncertain. Its
magnitude indicates a preferred displacement of the V atoms
~0.01Å along the chains. A positive/negative sign indicates
pairing of the V atoms toward the center/edge of the unit
cell. However, this potentially important result, which
could indicate a precursor of Peierls pairing at the structural transformation, cannot be firmly established by the
experiments except perhaps at 78K where the sign seems reliable.

IV. DISCUSSION AND "FLATTENED" POTENTIAL WELL

    By far the most significant anharmonic contribution is
that from the $\langle\mu^4\rangle$ of the V atoms. Its magnitude at 78K is
equal to that at 300K to within the uncertainties. The
simplest measure of this anharmonicity is the ratio

$$A = \langle \mu^4 \rangle / 3 \langle \mu^2 \rangle^2$$

which equals unity for harmonic behavior. Table 1 shows
that the 300K behavior is nearly harmonic. At 78K, however,
there is a pronounced anharmonicity[a] such that the V atoms
spend more time at the extremities of their travel than that
obtained in harmonic motion. The large $\langle\mu^4\rangle$ indicates a
softening for displacements off the lattice sites (probably
by shear), and is a possible signature of the "flattened"
potential well previously proposed (Testardi, 1972). Such
behavior could be a manifestation of incipient "chain buck-
ling" (e.g. $\langle\mu_\perp^4\rangle$) suggested from radiation damage and anomalous temperature properties (Testardi et al., 1977)[b] or
incipient Peierls pairing (e.g. $\langle\mu_\parallel^4\rangle$). Note from Table 1
that A≈1 for the Si atoms at all temperatures. Our x-ray

----

[a]The value of $\langle\mu^4\rangle_V$ obtained from the least squares fit is
within ~20% of that extracted directly from the slope of Fig.
1,2, and 3 assuming isotropic $\langle\mu^2\rangle_V$ neglecting $\langle\mu_\perp^2\mu_\parallel\rangle$, and
ignoring anharmonic effect at 300K. Thus, our results do not
depend strongly on the least squares fitting routine.

results do give one of the first clear microscopic confirma-
tions of the unique behavior of the transition metals chains.

The occurrence of the anharmonicity factor $A \simeq 2.22$ cannot
be readily obtained by the usual anharmonic series expansions
of a particle well potential. A sufficient condition for
$A>1$ is the existence of a secondary maximum in the position
$P(\mu)$ at $\mu > \mu_{rms}$. Weber (1978) has calculated that, for harmonic behavior of the V atoms, $<\mu^2(78K)>_{harm} \simeq <\mu^2(300K)>_{harm}/$
$2.45 \simeq 2.0 \times 10^{-3} \overset{\circ}{A}^2$ using the measured $<\mu^2>_V$ (assumed harmonic)
at 300K. The observed value $<\mu^2>_V = 2.93 \times 10^{-3} \overset{\circ}{A}^2$ indicates an
anharmonic enhancement of ~50% in $<\mu^2>$ also, and this in-
dicates important anharmonic contributions to the electron-
phonon interaction, and the electrical resistivity[b]. With
three available numbers, $<\mu^2>_{harm}$, $<\mu^2>_V$, and $<\mu^4>_V$ the
function $P(\mu)$ for V in one dimension can be modeled by a
gaussian (harmonic) with rms displacement $<\mu^2>_{harm}^{\frac{1}{2}} = 0.044\overset{\circ}{A}$,
and a δ-function located at $\mu=0.2\overset{\circ}{A}$ and having 2%-3% of the
spectral weight. We have chosen a more physically reasonably

$$P(\mu) \propto exp(- \mu^2 / 2 < \mu^2 >_{harm.})$$
$$+ D exp(- (\mu - \mu_o)^2 / 2 b^2)$$

such that the excess probability (last term) is a gaussian
with width b comparable to the harmonic part. For D=0.015
and $\mu_0=0.17\overset{\circ}{A}$ one obtains $<\mu^2>_V = 3 \times 10^{-3}\overset{\circ}{A}^{-2}$, $<\mu^4>_V = 5.8 \times 10^{-5}\overset{\circ}{A}^4$,
and thus A=2.1, both within 5% of the experimental values.
The modeled three dimensional radial distribution function
$(RDF) \propto r^2 P$ (we assume isotropy and replace μ by r) of the V
atoms about the lattice sites at 78K (lower part of Fig. 4)
shows the clear tendency for the atoms to localize at $\sim 0.2\overset{\circ}{A}$
off the proper sites. To further illustrate the anharmon-
icity at 78K we use Maxwell-Boltzmann statistics to obtain
qualitative results for the potential U from the model probability function

$$P \propto exp(- U / k T)$$

and we represent it in the upper part of Fig. 4. Note the
local minimum at $r=0.17\overset{\circ}{A}$ where U~25meV (~center of the phonon

[b]Since $<\mu^2>$ is ~50% larger than $<\mu^2>_{harm}$ at 78K one also
expects a large anomalous contribution to the electrical
resistivity ρ at this temperature which would not occur at
300K. This may be a part of the cause of the negative curvature in $\rho(T)$ (Woodard et al., 1964; Marchenko, 1973; Fisk and
Lawson, 1973; Taub and Williamson, 1974; Flükiger, 1975).

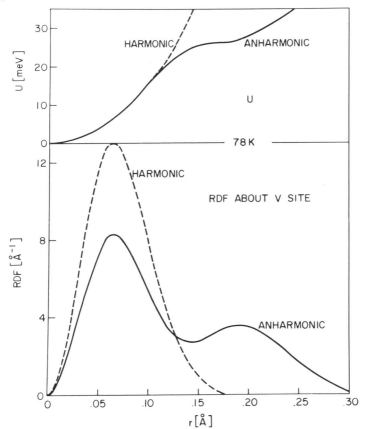

FIGURE 4.   Potential U and radial distribution function (RDF) (modeled) as a function of the distance r from a V site position.

spectrum for V_3Si (Karlsruhe group, 1974, 1975)). Furthermore the average elastic modulus is

$$< c > = < d^2 U / d\mu^2 >$$

Calculated for our model it comes to be 1.0×10^{12} dyn/cm^2 for the harmonic case, which agrees with the observed value at 300K. With the introduction of the anharmonicity at 78K one finds a calculated softening in <c>~15% compared to 20% for the long wavelength (Debye-averaged) elastic moduli, and a smaller value for the (V+Si) high frequency modes[11]. Fig. 4 graphically exhibits an incipient instability for V atoms at 78K displaced from their lattice sites by ~0.1-0.2Å. (A similar result for static defects was previously concluded

from channeling (Testardi et al., 1977). Presumably, at
lower T the minimum in U moves to smaller μ and becomes
larger in magnitude until it finally drives the structural
transformation with new coherent displacements. Our failure
with the present least squares results to extract information
on the anisotropy of $<\mu^4>_V$ means that no specification of
possible sublattice distortions can be made.

V. CONCLUSION

   We conclude with a remark concerning reflections with
$\sin\theta/\lambda < 0.7\overset{\circ}{A}^{-1}$. While reflections in the core region show a
behavior marked mainly by unusual $<\mu^4>$, reflections involving
the valence electrons exhibit a temperature dependence wholly
unexpected and unaccountable in terms of the above discussion.
It thus appears that the origin of this is a temperature dependent valence electron density in real space. While the
absolute accuracy of electron density maps (Staudenmann et
al., 1976; Staudenmann, 1978) is subject to some uncertainties
(form factor, extinction, etc.), Fig. 1,2,3 and others not
presented here clearly indicate that there appear to be
significant changes in the real space electron density with
temperature. This result supports differences seen between
electron density maps at 300K and at 13.5K (Staudenmann,
1978). Furthermore, any model based on thermal excitation
over ~100K fine structure in the density of states should
also produce the significant changes in real space electron
density indicated by the x-ray results. Density difference
maps to facilitate such a comparison are in progress.
   Finally a study of the diffuse scattering in V_3Si at 78K
would help in the distinction between static and dynamic
atomic displacements (Georgopoulos and Cohen, 1977).

ACKNOWLEDGMENTS

J.-L.S. wishes to express his thanks to Prof. S.A. Werner,
Prof. R.M. Brugger and Prof. J. Muller for their support,
and the hospitality of Bell Labs for a stay in Oct. 1978.

REFERENCES

Allen, P.B., Hui, J.C.K., Pickett, W.E., Varma, C. M. and
    Fisk, Z. (1976). *Solid State Comm. 18, 1157.*
Batterman, B.W., and Barrett, C.S., (1966). *Phys. Rev. 145,*
    296.
Becker, P.J. and Coppens, P. (1974). *Acta Cryst. A30, 129.*
Borie, B. (1974). *Acta Cryst. A30, 337.*
Fisk, A., Lawson, A.C. (1973) *Solid State Comm. 13, 277.*
Flükiger, R. (1975) *Private Communication.*
Fukamachi, T. (May 1971). *Technical Report of the Institute*
    for Solid State Physics, The University of Tokyo-Serie
    B#12.
Garcia, P.F., Barsch, G.R. and Testardi, L.R. (1971). *Phys.*
    Rev. Lett. 27, 944.
Georgopoulos, P., and Cohen, J.B. (1977). *Journal de Physique*
    C7, 191.
Geshko, E.I., Lotoskii, V.B., and Mikhal'chenko, V.P. (1976)
    Ukr. Fiz. Zh. 21, 186.
Kimball, C.W. van Landuyt, G., Barnett, C., Shenoy, G.K.,
    Dunlap, B.D., and Fradin, F.Y. (1978). *J. de Physique*
    C6, 367.
Knapp, G.S., Bader, S.D., Culber, H.N., Fradin, F.Y. and
    Klippert, J.E. (1975). *Phys. Rev. B11, 1331.*
Labbé, J., and Friedel, J. (1966). *J. de Physique 27, 15.*
Marchenko, V.A. (1973). *Soviet Phys. Solid State 15, 1261.*
Patel, J.R., and Batterman, B.W. (1966). *Phys. Rev. 148, 662.*
Schweiss, P. et al., Karlsruhe Group (1974, 1975). *Progress*
    Reports KFK 2056, KFK 2183, Gesellschaft für
    Kernforschung, M.B.H., Karlsruhe, W. Germany.
Shier, J.S. and Taylor, R.D. (1966). *Phys. Rev. 174, 346.*
Shull, C.G., (1964). *M.I.T., Annual Report-Research in Met.*
    Sci.
Smith, T.F., Finlayson, T.R., and Taft, A. (1976). *Comm. on*
    Physics 1, 167.
Staudenmann, J.-L., Coppens, P., and Muller, J. (1976). *Solid*
    State Comm. 19, 29.
Staudenmann, J.-L. (1978). *Solid State Comm. 26, 461.*
Taub, H., and Williamson, S.J. (1974). *Solid State Comm. 15,*
    181.
Testardi, L.R. (1972). *Phys. Rev. B5, 4342.*
Testardi, L.R. (1973). *Phys. Rev. Lett. 31, 37.*
Testardi, L.R. (1975). *Rev. Mod. Phys. 47, 637.*
Testardi, L.R., Poate, J.M., Werner, W., Augustyniak, and
    Barnett, J.H. (1977). *Phys. Rev. Lett. 39, 716.*
Woodard, D.W., and Cody, G.D. (1964). *Phys. Rev. 136, A166.*

PREFERRED ORIENTATION AND LATTICE REGISTRY
EFFECTS OF Nb3Sn DIFFUSION LAYERS[†]

J. DeBroux
V. Diadiuk[††]
J. Bostock
M. L. A. MacVicar

Department of Physics
Massachusetts Institute of Technology
Cambridge, Massachusetts

Nb3Sn diffusion layers grown on single crystal niobium exhibit physical properties which consistently discriminate between the bcc-(111) orientation substrate and other bcc major symmetry directions. These layers show preferred orientations in their surface region which do not extend into the layer bulk except for the A15-(100) surface texturing, occuring at the Nb-(110). The Nb3Sn-Nb interface character is distinctly different for the Nb-(111) orientation as compared to other orientations around the niobium crystal circumference.

A preliminary study showing the existence of structural affinities in Nb3Sn layers fabricated on oriented single crystal niobium substrates has been published [Diadiuk *et al.*, 1979] in which we reported detailed microstructure, compositional and superconducting characterizations of the layers as a function of underlying bcc substrate direction. The techniques used (SEM, RED, EDX, X-ray, Auger, and superconducting tunneling) all showed striking differences in properties between A15 Nb3Sn layers fabricated at the Nb-(111) substrate orientation and those fabricated at non-(111) bcc high

[†]*Supported in part by the Department of the Air Force through Lincoln Laboratory and by the Department of Energy through the Plasma Fusion Center, M.I.T.*
[††]*Present Address: Lincoln Laboratory, M.I.T.*

259

symmetry orientations: (100), (110), and (211). For example, Nb_3Sn fabricated on non-(111) bcc orientations exhibits fairly uniform grain size and shape in contrast to the irregular mixed grain size, sometimes microfractured, microstructure of layers on the bcc-(111) substrates; Auger depth profiles on bcc-(111) layers indicate excessive oxygen and unbonded carbon; and tunneling studies show no indications of superconductivity down to 1K for the bcc-(111) layers despite their 3:1 stoichiometry, while for non-(111) bcc substrate layers an energy gap of 3.5 meV $[2\Delta/kT_c = 4.5]$ is measured. Here we report on the orientational character of the diffusion layer A15 material as it relates to the underlying oriented Nb-substrate.

Examination of the exterior Nb_3Sn layer *surface*, which is the *initial growth* surface of the A15 "into" the bcc single crystal, with RED indicates a preferred orientation, or structural texturing, depending on the orientation of the niobium substrate (see Fig. 1). Hypothesizing [Diadiuk *et al.*, 1979] from the generic family of the brightest rings, this correspondence takes the form: bcc-(100)||A15-(110), bcc-(110)||A15-(100), bcc-(111)||A15-(111), and bcc-(211)||A15-(110). Mindful that one must proceed with caution in interpreting RED

sample	A15 Rings						
				RED INDEXED REFLECTIONS			
bcc (100)-disc	200	320	(330)	400	{431 510}	610	541
bcc (110)-disc	200	211	310	320	(400)	{411 330}	{431 510}
	610	541	614	{554 741}	761		
bcc (111)-disc	200	(222)	330	422	{431 510}	440	541
	534	614	752				
bcc (211)-disc	320	(330)	422	534			
RCA-X	220	321	{411 330}	{433 530}	541	613	615
SF film	200	211	310	320	(400)	422	
	440	610	534	614	615	752	{743 750}

FIGURE 1. Tabulated values of reflection electron diffraction rings for Nb_3Sn layers fabricated on niobium single crystal discs of the major bcc orientations. The brightest ring is circled for each orientation. Brackets indicate ambiguity in indices. Data for the RCA (110) single crystal of Nb_3Sn is also tabulated for comparison as is a high quality Stanford tunneling film[(Jacobsen et al., (1978)].

reflection patterns, this hypothesis is particularly attractive both because the errant bcc-(111) substrate layers of
Nb3Sn exhibit brightest-ring indices different from those exhibited by the other layers as a group and because at least
one of the Nb orientations, the bcc-(110), yields strongly
textured A15-(100) Nb3Sn, in agreement with observations by
Stanford of a natural growth direction of electron beam co-
evaporated Nb3Sn films deposited on a high-temperature substrate [Jacobson *et al.*, 1978]. Lattice registry matches at
the initial growth surface have been found both for the bcc-
(111) and the bcc-(110) relationships. The bcc-(111)||A15-
(111) registry requires a uniform expansion in all directions
of the A15-(111) surface of ∿8% aided, perhaps, by an accumu-
lation of C and O stabilizers. The lattice registry of the
bcc-(110)||A15-(100), shown in Fig. 2, requires a translation
of the A15-(100) direction and a slight rotation relative to
a standard Nb-[111] epitaxy.

Extensive x-ray investigations of the Nb3Sn layers by dif-
fractometry and direct pole figure analysis show that the preferred orientations of the surface indicated by the RED data
do not extend inward into the bulk of the Nb3Sn layers except
in the case of bcc-(110)||A15-(100). Figure 3 shows this
correspondence to be a strong one: the (200) and (400) reflections are greatly enhanced over the corresponding reflections of a powder pattern. The strength and ease of this texturing for both surfaces and bulk may well be related to the

Nb-(110)||A15-(100)

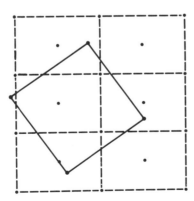

*FIGURE 2. The solid lines (——) represent sides of the
A15-(100) cell face drawn to scale, overlaid on dashed lines
(---) representing sides of niobium-(110) faces also drawn to
scale. Dots indicate niobium atom positions.*

| RELATIVE INTEGRATED INTENSITIES OF X-RAY REFLECTIONS* | | | | | | | | | |
A15 Reflection / Sample	200	210	211	222	320	321	400	420	440
powder standard	.51	1.43	1	.061	.25	.54	.18	.14	.15
(100)-disc	.52	1.36	1	.110	.41	.86	.40	.31	.38
(110)-disc	1.75	1.77	1	.087	.33	.52	1.10	.25	.17
(111)-disc	.38	1.15	1	.084	.37	.74	.30	.35	.49
(211)-disc	.38	.75	1	.046	.27	.71	.33	.24	.28

*[normalized to the A15 (211) reflection]

FIGURE 3.   Tabulated values of integrated intensities of x-ray reflections for Nb_3Sn layers fabricated on oriented niobium single crystal substrates.   Intensities of a Nb_3Sn randomly-oriented powder standard are also listed for comparison.

near-coincidence of the Nb-[111] and the Nb3Sn-[100] which are the directions in the two materials corresponding to the most dense Nb linear atomic packing. Layers fabricated at the bcc-(100) and bcc-(211) have orientational enhancements very similar to one another and appear to consist of a mix of A15-(100) and A15-(110) characters; i.e., the initial surface orientation plus the natural growth orientation of the A15. The Nb3Sn layer at the bcc-(111) exhibits no enhancement of the A15-(111) orientation within the layer bulk. Rather, the data indicate a modest preference for the A15-(110) with an admixture of A15-(100).

Examination of the Nb3Sn-Nb interface, which is the *final growth* surface of the Nb3Sn, shows that the interface layer at the bcc-(111) orientation also differs markedly from those at other orientations, as shown in Fig. 4. A region of perimeter typical of orientations other than the bcc-(111) exhibits a layer of reasonably constant thickness around the niobium single crystal with no evidence of Kirkendall voids [Easton *et al.*, 1979] or second phase [Dickey *et al.*, 1971]. When examined at higher magnification, the cylindrical surface of the exposed niobium shows a uniform distribution of "pockets" indicating short fingers of Nb3Sn lead the radially advancing Nb3Sn interface. At and in the vicinity of the (111) niobium substrate orientation, the interface, the Nb3Sn layer, and the exposed niobium surface all exhibit features that may arise from voids, second phases, or stress fields. (The existence of orientationally-dependent tin diffusion and Nb_xSn_{1-x} growth

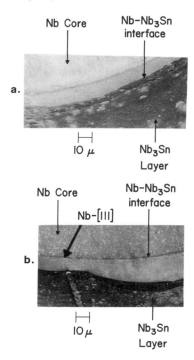

Nb Core   Nb–Nb$_3$Sn
               interface

a.

├─┤
10 μ   Nb$_3$Sn
               Layer

Nb Core   Nb–Nb$_3$Sn
               interface
        Nb–[111]

b.

├─┤
10 μ   Nb$_3$Sn
               Layer

FIGURE 4. Rod-shaped samples (magnification 800x) etched to expose the Nb$_3$Sn-Nb interface: a) a uniformity of thickness of Nb$_3$Sn is apparent around the niobium crystal core for this typical region of non-(111) orientations (blemishes are due to excessive etching); b) obvious interface and layer feature(s), and variable Nb$_3$Sn thickness, in the vicinity of the niobium-(111) substrate orientation.

kinetics should not be overlooked, either.) Moreover, the rapidly varying layer thickness, probably stemming at least in part from preferred etching by the preparation solution, may be understood as an angular profile indicative of an exceedingly limited range of anomalous layer character extending approximately 35μ to either side of this bcc-(111) direction, or across *only* ∿1° of orientation. A detailed examination of the Nb3Sn-Nb interface as a function of substrate orientation is currently in progress.

ACKNOWLEDGMENTS

The authors gratefully acknowledge the use of RED facilities at Northeastern University through the hospitality of W. Giessen. We would also like to thank S. Cogan and L. Vieland for providing the Nb3Sn samples used as reference standards in this work.

REFERENCES

DIADIUK, V., BOSTOCK, J. and MACVICAR, M. L. A., (1979). *IEEE Trans. MAG-15, 610.*

DICKEY, J. M., STRONGIN, M., and KAMMERER, O. F. (1971). *J. Appl. Phys. 42,* 5808; and STROZIER, J. A., MILLER, D. L., KAMMERER, O. F., and STRONGIN, M., (1976). *J. Appl. Phys. 47,* 1611.

EASTON, D. S., KROEGER, D. M., (1979). *IEEE Trans. MAG-15,* 178.

JACOBSON, B. E., HAMMOND, R. H., GEBALLE, T. H., and SALEM, J. R., (1978). *J. Less Comm. Metals 62,* 59.

EFFECTS OF ATOMIC ORDER ON THE UPPER CRITICAL
FIELD AND RESISTIVITY OF A15 SUPERCONDUCTORS

R. Flükiger[1]
S. Foner
E. J. McNiff, Jr.

Francis Bitter National Magnet Laboratory[2]
Massachusetts Institute of Technology
Cambridge, Massachusetts

Substantial variations in $H_{c2}(T)$ and $\rho(T)$ are observed in the A15 type compounds V_3Ga, Mo_3Os, Nb_3Pt and $Nb_3Au_{0.7}Pt_{0.3}$ for different degrees of atomic ordering produced by different heat treatment. The effect of atomic order on $H_{c2}(0)$ and $\rho(T)$ varies strongly from system to system.

I. INTRODUCTION

Experimental and theoretical investigations in the last decade confirm the strong influence of long-range atomic ordering on the superconducting properties of A15 type compounds. Here the effects of atomic order on the upper critical field, $H_{c2}(T)$, and on the electrical resistivity, $\rho(T)$ for well characterized V_3Ga, Mo_3Os, Nb_3Pt and $Nb_3Au_{0.7}Pt_{0.3}$ are presented. The changes in order were produced thermally. Because small changes in composition affect T_c and γ (Junod et al., 1971), ρ (Milewits et al., 1976) and H_{c2} (Flükiger et al., 1979a), the present measurements were made on pieces of the same materials on which measurements of the atomic order parameter, S, and the electronic specific heat, γ, had been measured previously at the University of Geneva. For V_3Ga, we refer to Flükiger et al.

[1]*Dept. Phys. Mat. Cond., University of Geneva, Switzerland.*
[2]*Supported by the National Science Foundation.*

265

(1976) and Junod et al. (1976), for Mo_3Os to Flükiger et al. (1974) and Paoli et al. (1975), and for Nb_3Pt and $Nb_3Au_{0.7}$-$Pt_{0.3}$ to Spitzli (1971) and Junod (1974). The materials discussed in this paper are characterized in Table I.

The strong dependence of H_{c2} on *composition*, illustrated in Fig. 1, suggests that the effects of order must be examined on well-characterized materials. If possible such studies should be made on the same piece of material for which comparisons are made.

II.   RESULTS AND DISCUSSION

The results for $H_{c2}(T)$ and $\rho(T)$ are shown in Figs. 2 and 3. All investigated systems exhibit large variations of $H_{c2}(T)$ in spite of the small changes in order parameter produced by heat-treatment. For V_3Ga and Nb_3Pt, increases of the order parameter $\Delta S = 0.03$ and 0.06 cause an increase in $H_{c2}(0)$ of 10

TABLE I.   *Some Physical Properties of the Compounds Studied in the Present Investigation*

Compound	Ordering treatment (time @ °C)	a (Å)	T_c (K)	S	γ (mJ/K^{2}cm^3)
V_3Ga	3 w. /610[a]	4.817	15.3		2.87
V_3Ga	1/2 h. /1250[a]	4.817	15.1	0.98±0.01	(2.72)[c]
V_3Ga	1/2 h. /1250[b]	4.817	13.8	0.95±0.01	2.37
Mo_3Os	2 w. /1100[a]	4.970	12.1	0.85±0.02	0.57
Mo_3Os	1/2 h. /1950[b]	4.970	11.4	0.79±0.02	0.56
Nb_3Pt	4 w. /900[a]	5.155	10.9	1.00±0.01	(0.58)[c]
Nb_3Pt	1/2 h. /1800[b]	5.155	8.6	0.94±0.02	0.52
$Nb_3Au_{0.7}Pt_{0.3}$	2 w. /750[a]	5.196	12.9		1.02
$Nb_3Au_{0.7}Pt_{0.3}$	1/2 h. /1550[b]	5.196	10.0		(0.62)[c]

[a] Slow cooled
[b] Cooled by argon jet quenching
[c] Interpolated values

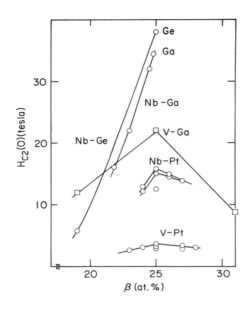

FIGURE 1. Upper critical field $H_{c2}(0)$ as a function of composition, β, in $A_{1-\beta}B_\beta$ compounds for the various A15 compounds. The variation of the points at $\beta = 25$ for Nb-Pt and V-Pt show the effects of atomic order (after Flükiger et al. 1979a)

and 16%, respectively. An even greater increase of $H_{c2}(0)$ ($\sim 40\%$) is observed for $Nb_3Au_{0.7}Pt_{0.3}$. (The different states of ordering are characterized by the corresponding heat treatments. The order parameter of a ternary system is not defined.) For Mo_3Os, however, an increase of $\Delta S = 0.06$ produced a decrease in $H_{c2}(0)$ from 8.3 to 7.1 tesla (Fig. 2a), in good agreement with earlier data (Fischer, 1975). The initial slope, $(dH_{c2}/dT)_{T=T_c}$, is also affected by ordering (see Table II). For Mo_3Os and Nb_3Pt, increasing disorder enhances the value of the initial slope from 0.85 to 1.05 and from 1.6 to 1.95 tesla/ K, respectively, in contrast to $Nb_3Au_{0.7}Pt_{0.3}$, for which the initial slope decreased. It should be noted (see Fig. 1) that the initial slope is difficult to determine for strongly Pauli paramagnetically limited materials such as V_3Ga. In this case, there is a noticeable variation of this slope very close to T_c and the slope cannot be determined accurately. The errors in this determination make detailed comparisons with S difficult. The upper critical fields of Mo_3Os and $Nb_3Au_{0.7}Pt_{0.3}$ agree with the predictions for a dirty type II superconductor; the value $h(0) = H_{c2}(0)/T_c(dH_{c2}/dT)_{T=T_c}$ for both systems is close to the

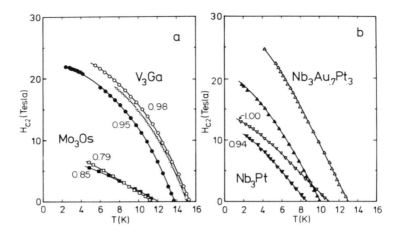

FIGURE 2. $H_{c2}(T)$ for different ordering parameters, S.
a) V_3Ga (S = 0.98 and 0.95); Mo_3Os (S = 0.85 and 0.79).
b) Nb_3Pt (S = 1.00 and 0.94). Heat treatment of $Nb_3Au_{0.7}$-
$Pt_{0.3}$: 2 weeks at 750°C (△), and 24 hours 1250°C (△), followed by quenching.

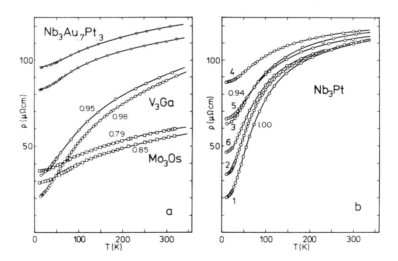

FIGURE 3. $\rho(T)$ for different treatments. a) V_3Ga (S =
0.98 and 0.95) and Mo_3Os (S = 0.85 and 0.79). $Nb_3Au_{0.7}Pt_{0.3}$ for
heat treatments given in Table I. b) Nb_3Pt after different
treatments. (1) 7 weeks at 750°C, (2) 3 weeks at 900°C,
(3) 10 hours at 1250°C and quenched, (4) 30 min at 1800°C and
quenched, (5) (4) + 48 hours at 1250°C, (6) (4) + 24 hours at
1000°C.

TABLE II. *High Field Parameters and Electrical Resistivity of V_3Ga, Mo_3Os, Nb_3Pt and $Nb_3Au_{0.7}Pt_{0.3}$ at Different Degrees of Atomic Ordering given in Table I*

Compound	Measured				Calculated	
	T_c (K)	$(dH_{c2}/dT)_{T=T_c}$ (tesla/K)	$H_{c2}(0)^a$ (tesla)	ρ_c ($\mu\Omega$cm)	$h(0)^b$	$H'_{c2}(0)^c$ (tesla)
V_3Ga	15.3	4.0	25	20.3	0.44	27
V_3Ga	15.1	4.0	23.1	24.2	0.38	30
V_3Ga	13.8	3.6	22	32.7	0.40	33
Mo_3Os	12.1	0.85	7	28.5	0.70	6
Mo_3Os	11.4	1.05	8	35.8	0.69	7
Nb_3Pt	10.9	1.6	15	20.1	0.80	4
Nb_3Pt	8.7	1.95	13	66.6	0.74	9
$Nb_3Au_{0.7}Pt_{0.3}$	12.9	3.3	30	60.1	0.70	24
$Nb_3Au_{0.7}Pt_{0.3}$	10.0	2.9	21	82.5	0.70	16
$Nb_3Au_{0.7}Pt_{0.3}$	9.0			96.5		

[a] Extrapolated
[b] Calculated from $h(0) = H_{c2}(0)/T_c(dH_{c2}/dT)_{T=T_c}$
[c] Calculated from $H'_{c2}(0) = 3.06 T_c \gamma \rho_c$

theoretical value, independent of the degree of order (Table II). It is interesting that $H_{c2}(T)$ for Nb_3Pt falls well above the predictions in agreement with Bongi et al. (1976). In addition, an increase of the order parameter from $S = 0.94$ to $S = 1$ enhances the value of $h(0)$ from 0.74 to 0.80. Recently, Flükiger et al. (1976) suggested that the high value of $h(0)$ for Nb_3Pt may be due to d-band overlap at the Fermi level.

An alternate determination of $H_{c2}(0)$ requires γ, T_c and the value of the electrical resistivity at 0 K, ρ_0. In this paper, this value was replaced by ρ_c, measured just above T_c. Figure 3 shows the variation of ρ for different degrees of ordering. For all compounds, ρ_c was found to increase with decreasing order parameter as expected. In order to exclude

an increase of ρ_c after argon jet quenching due to extraneous
effects such as strains, cracks or impurities, some samples
were remeasured after repeated heating cycles. Reversible
changes of both T_c and ρ_c were observed. The value of ρ_c for
bulk samples depends strongly on their metallurgical state.
Fully annealed, homogeneous, single-phased samples were found
to exhibit lowest ρ_c values.

Numerous metallurgical effects can influence the evaluation of the resistivity at low temperatures. Generally, these
effects increase the value of ρ_0 so that estimates of $H'_{c2}(0)$ in
Table II would be expected to give a value larger than measured. Thus it is difficult except under particularly favorable circumstances to determine $H'_{c2}(0)$ using resistivity data.
The estimates of $H'_{c2}(0)$ calculated with ρ_c gives values larger
than measured for V$_3$Ga which is consistent with the observed
strong paramagnetic limiting. The results for Mo$_3$Os are close
to the measured values, but the results for Nb$_3$Pt give a calculated value of $H'_{c2}(0)$ which is quite low. Even the relatively dirty Nb$_3$Au$_{0.7}$Pt$_{0.3}$ materials give a low calculated
value of $H'_{c2}(0)$.

The low resistivity and large residual resistance for
well-ordered Nb$_3$Pt suggests a long mean free path, ℓ_{tr}. Estimates of $\ell_{tr} \sim 100$ Å for sample 1 of Fig. 3b and $\xi_0/\ell_{tr} \simeq 0.7$
indicate that this material is a relatively clean supercon-
ductor.

It is interesting that by introducing 2% Pt atoms on the
Nb sites by quenching from 1800°C, corresponding to an order
parameter S = 0.92, ℓ_{tr} is reduced by almost 10, so that
Nb$_3$Pt then becomes a dirty type II superconductor. We emphasize that the alteration of the mean free path by thermal
methods is *reversible*. Similar estimates for the Mo$_3$Os materials in Table II yield $\ell_{tr} \sim 20$ Å and $\xi_0/\ell_{tr} \simeq 10$ so that
these are dirty. Calculations for the well ordered V$_3$Ga (T_c =
15.3 K) yield $\ell_{tr} \simeq 20$ Å and $\xi_0/\ell_{tr} \simeq 0.3$.

In summary, large variations of $H_{c2}(0)$ and the initial
slope with atomic ordering have been found on a series of A15
compounds with different high field characteristics: V$_3$Ga is
strongly paramagnetically limited, Mo$_3$Os and Nb$_3$Au$_{0.7}$Pt$_{0.3}$
show little paramagnetic limiting, and Nb$_3$Pt exhibits $H_{c2}(T)$
values which exceed the predictions for a dirty or clean type
II superconductor. The behavior of well-ordered Nb$_3$Pt shows a
large variation of resistivity with temperature and should behave as a clean type II superconductor.

REFERENCES

Bongi, G. H., Flukiger, R., Treyvaud, A., and Fischer, Ø.
    (1976). *J. Low Temp. Phys. 23*, 543.
Fischer, Ø. (1975). Proc. LT14, Helsinki, ed. M. Krusius and
    N. Vuorio, Vol. V, p. 172.
Flükiger, R., Padi, A., and Muller, O. (1974). *Solid State
    Commun. 74*, 443.
Flükiger, R., Staudemann, J. L., and Fischer, Ø. (1976).
    J. Less-Common Met. 50, 253.
Flükiger, R., Foner, S., McNiff, Jr., E. J., and Schwartz,
    B. B. (1979a). *J. Magn. Magnetic Mat. 11*, 186.
Flükiger, R., Foner, S., and McNiff, Jr., E. J. (1979b).
    Solid State Commun. 30, 723.
Junod, A., Staudemann, J. L., Muller, O., and Spitzli, P.
    (1971). *J. Low Temp. Phys. 5*, 25.
Junod, A. (1974). Thesis No. 1661, University of Geneva.
Junod, A., Flukiger, R., Treyvaud, A., and Muller, J. (1976).
    Solid State Commun. 19, 265.
Milewits, M., Williamson, S. J., and Taub, H. (1976).
    Phys. Rev. 13B, 5199.
Paoli, A., and Flükiger, R. (1975). Unpublished results.
Spitzli, P. (1971). *Phys. kondens. Materie 13*, 22.

POSITRON ANNIHILATION EXPERIMENT ON V_3Si

Manuel, A.A., Samoilov, S., Sachot, R., Descouts, P., Peter, M.

Department of Condensed Matter, Geneva University

I. INTRODUCTION

The A-15 intermetallic compound V_3Si belongs to the class of materials characterized by a high Tc (17.1°K) and anomalies in other electronic properties. Its band structure is complicated because of the large number of atoms per unit cell, yielding 38 conduction electrons. The common feature to all band structure calculations (Weger 1973; Mattheiss 1975; Van Kessel 1978; Klein 1978; Jarlborg 1979) is the presence of very flat 3d bands in the close vicinity of the Fermi level. Therefore, experiments in which $\vec{k}$ dependent properties are measured, are needed for the accurate determination of the band structure. A positron annihilation measurement is such an experiment.

In this technique (West 1973; Berko 1977) the momentum distribution is measured. In the independent particle approximation, at T=0 K, it is given by:

$$(1) \quad \rho(\vec{p}) = \sum_{\vec{k},\ell} n(\vec{k},\ell) \left| \int \exp(-i\vec{p}\cdot\vec{r}) \psi_+(\vec{r}) \psi_{\vec{k},\ell}(\vec{r}) d^3\vec{r} \right|^2$$

$$= \sum_{\vec{k},\ell} n(\vec{k},\ell) \sum_{\vec{G}} \delta(\vec{p}-\vec{k}-\vec{G}) \cdot \left| A_{\vec{G}}(\vec{k},\ell) \right|^2$$

where $\psi_+(\vec{r})$ is the positron ground state wave function, $\psi_{\vec{k},\ell}(\vec{r})$ is the electron wave function for wave vector $\vec{k}$ and band index ℓ. $n(\vec{k},\ell)$ is the occupation number of the state $(\vec{k},\ell)$. The $\vec{G}$ are reciprocal lattice vectors and the $A_{\vec{G}}(\vec{k},\ell)$ are the Fourier components:

$$(2) \quad A_{\vec{G}}(\vec{k},\ell) = \int_{cell} \exp(-i(\vec{G}+\vec{k})\cdot\vec{r}) \psi_+(\vec{r}) \psi_{\vec{k},\ell}(\vec{r}) d^3\vec{r}; \quad \sum_{\vec{G}} \left| A_{\vec{G}} \right|^2 = 1$$

Thus, each electron with wave vector k contributes to the anni-
hilation rate at every $\vec{p}=\vec{k}+\vec{G}$ with relative intensity $|A_{\vec{G}}(\vec{k},\ell)|^2$.
In V_3Si, these Fourier components are appreciable in quite a
number of Brillouin zones (BZ), since, the 3d wave function is
far from being a plane wave.

The major contributions to $\rho(\vec{p})$ are due to the annihilation
of the core electrons and of the valence electrons. The latter
is continuous up to every $\vec{k_F}+\vec{G}$, where a discontinuity is expected. From those breaks, which manifest themselves most
clearly in the derivative, information concerning the Fermi
surface (FS) topology, may be obtained.

Preliminary positron measurement (Berko 1970) on V_3Si demonstrated an anisotropy in the momentum distribution. Subsequent work suggested that planar sections in the FS were observed experimentally (Samoilov 1978) within the frame of the
coupled linear chains model (Weger 1973). In the last years,
bidimensional (2D) measurements (Berko 1975; Manuel 1979) gave
further experimental information which however was not yet
given a theoretical interpretation.

In a 2D measurement, two components of the pair momentum
are resolved, namely p_x and p_y. Therefore the measured angular
correlation function is given by:

(3) $F(p_x,p_y) = \int \rho(\vec{p})\,dp_z$ $\qquad\qquad p_x = mc\theta_x$; $p_y = mc\theta_y$

where θ_x and θ_y and the resolved angular components.

II. MEASUREMENTS

The experimental set-up used in this work (Manuel 1978a,
1978b; Manuel, to be published) is equipped with 2D high density proportional chambers (Jeavons 1976, 1978). The angular
resolution is equivalent to $0.31\pi/a$ x $0.26\pi/a$ (where is the
lattice constant of V_3Si),at 77K, including the e^+ thermal
motion and the sample width in one direction.

The V_3Si single crystals were grown by the zone melting
technique (Manuel 1978a). Specific heat measurement showed a
martensitic transition at 21.6K and a very sharp superconducting transition at 16.8K. The samples were cut by spark erosion, etched, annealed at 1400°C for 24 hours, slowly cooled
and etched again.

In this work we are mainly interested in the FS topology. Therefore, the small structures (1-3%), caused by the $n(\vec{k},\ell)$ term in (1), are the relevant information. The broad distribution underneath, representing the Fourier transform in (1), is considered as a "background". The latter is determined by computing the average value of the experimental distribution on concentric circles of increasing radius centered at the origin (Farmer, 1979). In fig.1 the angular correlation spectrum for the (110) plane is shown. The statistical error at the top is 0.2%. In fig.2 the residual topology is shown. The symmetry of order 2, typical for this plane, appears clearly. The statistical error on the residual structures is about 20%. In order to improve the signal to noise ratio, the spectrum is folded into the first BZ. This procedure has also the advantage of cancelling to a great extent the contribution of the wave function $(\Sigma |A_{\vec{G}}|^2 = 1$ eq. (2)), if a large number of BZ are considered (Lock 1973). Furthermore, for determining more accurately the FS crossings, the partial derivatives in various directions are computed and also folded into the first BZ.

The reduced zone scheme representation for the (100) plane, "background" substracted and folded over 36 BZ, is shown in fig. 3. The statistical error for the function is 2% and for the partial derivatives 5%. The symmetry of order 4 for the (100) plane, is obtained unambiguously. The observed structure consists of a small hole box and a bigger electron box centered at Γ(or X) and an ellipsoïdal hole depression along the MR line of the BZ. From a 2D measurement in the (100) plane it is not possible to distinguish between the ΓXMX and XMRM planes.

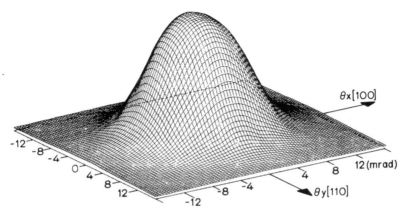

FIGURE 1.   Angular correlation spectrum for the (110) measurement in V₃Si.

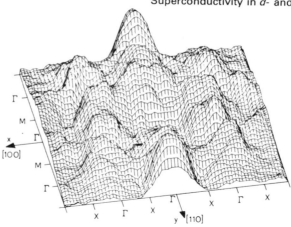

FIGURE 2. *Residual structures (see text) for the (110) measurement in V_3Si.*

The (110) measurement is more complex. The principal axes [100] and [110] were aligned along θ_x and θ_y respectively, thus in the third dimension (θ_z), the integration is over the [110] direction. Therefore, information concerning the planes ΓXMR and its mirror inverted MRΓX is obtained successively along the axis of integration. Fig.4 shows the reduced zone represen-tation for the (110) measurement, "background" substracted and folded over 96 BZ. The striking feature is the almost complete symmetry of the function with respect to the L line in fig.4. The meaning being obviously that the contribution from the ΓXMR and MRΓX planes is almost the same. This near symmetry in-duces antisymmetrical partial derivative $\partial/\partial y$, with respect to the L line. The structures appearing in the function and the partial derivatives indicate that there are holes at Γ <u>and</u> M, electrons at X <u>or</u> R and a big electron box centered at Γ.

From both measurements, a model for the FS topology may be extracted within the experimental resolution. As shown in fig. 5, it consist of:
- a hole box centered at Γ with 0.25 π/a edge
- an electron box centered at Γ with 0.75 π/a edge
- a hole ellipsoïd like structure at M ($a/2=b/2\simeq.35\pi/a;c/2\simeq.5\pi/a$)
- an electron structure at R, the shape of which is not yet determined.

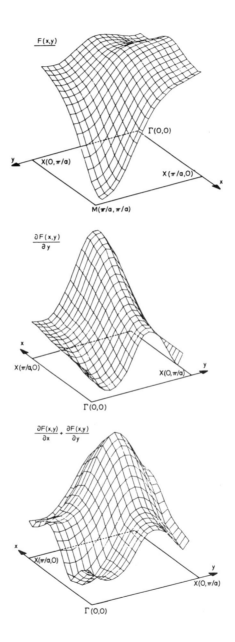

FIGURE 3. *Reduced zone representation for the (100) measurement in V₃Si.*

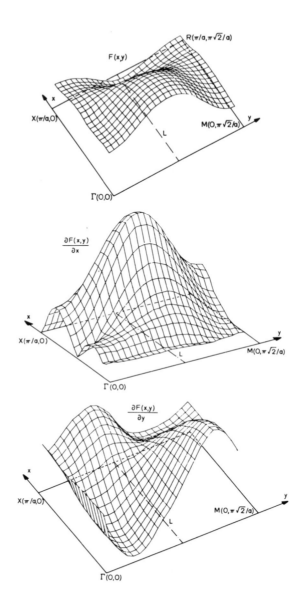

FIGURE 4.   *Reduced zone representation for the (110) measure-ment in V₃Si.*

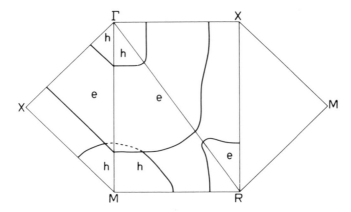

FIGURE 5. *Sections of the Fermi Surface of V₃Si consistent
with the present experiment.*

ACKNOWLEDGMENTS

This work benefited from a close collaboration with Dr. A.
P. Jeavons in the construction of the detectors. It also pro-
fited from valuable suggestions and stimulating discussions
with Prof. Ø. Fischer.

REFERENCES

Berko S. (1977). In "Compton Scattering" *(B. Williams, ed.)*
    p. 273. Mc Graw-Hill
Berko S. and Mader J.J. (1975). *Appl. Phys. 5,287.*
Berko S. and Weger M. (1970). *Phys. Rev. Lett. 24,55.*
Farmer W.S., Sinclair F., Berko S., Braedsley G.M. (1979).
    Proc. of Fifth Int. Conf. on Positron, April 7-11, Japan.
Jarlborg T. (1979). *J. Phys. 9,283.*
Jeavons A.P. and Cate C. (1976). *IEEE Trans. Nucl. Sci. 23,640.*
Jeavons A.P., Townsend D.W., Ford N.L., Kull K., Manuel A.A.,
    Fischer Ø. and Peter M. (1978). *IEEE Trans.Nucl.Sci.25,164.*
Klein B.M., Boyer L.L., Papaconstantopoulos D.A.,Mattheiss L.F.
    (1978). *Phys. Rev. B 18,6411.*
Lock D.G., Crisp V.M.C. and West R.N. (1973). *J. Phys. F3, 561.*
Manuel A.A. (1978a). *In Ph.D. Thesis No 1889, Geneva University*
Manuel A.A., Fischer Ø., Peter M. and Jeavons A.P. (1978b).
    Nucl. Inst. Meth. 156,67.
Manuel A.A., Samoilov S., Fischer Ø., Peter M. and Jeavons A.P.
    (1979). *Helv. Phys. Acta 52,37.*
Manuel A.A., Samoilov S., Fischer Ø., Peter M. and Jeavons A.P.
    to be published in Helv. Phys. Acta.

Mattheiss L.F. (1975). *Phys. Rev. B 12,2161.*

Samoilov S. and Weger M. (1977). *Solid State Comm. 24,821.*

Samoilov S., Ashkenazi J., Weger M. and Goldberg I.B. (1978). *J. de Phys. CG, 39,421.*

Van Kessel A.T., Myron M.W. and Mueller F.M. (1978). *Phys. Rev. Lett. 41,181.*

Weger M. and Goldberg I.B. (1973). *Solid State Phys. 28,1.*

West R.N. (1973). *Adv. in Phys. 23,266.*

STUDIES OF THE FERMI SURFACE OF V_3Si
BY 2D ACAR MEASUREMENTS*

S. Berko, W.S. Farmer, and F. Sinclair

Department of Physics
Brandeis University
Waltham, Massachusetts

Using a two-dimensional angular correlation of annihilation radiation apparatus, projections of the electronic momentum distribution of V_3Si (sampled by a thermalized positron) have been obtained. The three dimensional momentum distribution is reconstructed, and a zone-folding technique yields the density in the reduced zone $\rho(\underline{k})$, which reflects the gross features of the Fermi surface. Nested cylindrical hole surfaces along the edges of the zone that join at the corner points R are observed, and also additional electron density on the zone faces centered about X. Comparisons are made to recent APW calculations.

V_3Si belongs to the family of A-15 intermetallic compounds that have been intensively studied during the last decade. (See reviews by Weger and Goldberg (1973) the Testardi (1973).) Theoretical calculations for these compounds usually predict a Fermi energy in the vicinity of several flat bands of "d" character, leading to a complex Fermi surface (FS) consisting of several sheets. Experimental indications of the shape of the FS in some A-15 compounds have been provided by magneto thermal data and de Haas-van Alphen (dHvA) experiments (Graebner and Kunzler, 1969; Arko, et al 1977, 1978) suggesting a FS consisting of several nested cylindrical or ellipsoidal hole surfaces aligned along the <100> directions.

The applicability of the positron technique to the study of electronic band structure and particularly of the FS is well

Work supported by the National Science Foundation.

281

documented – see for example recent reviews by Mijnarends (1978)
and Berko (1978). In the independent particle model (IPM) the
momentum distribution $\rho(\underline{p})$ of the annihilation photons from a
periodic lattice at zero temperature is given by

$$\rho(\underline{p}) = \text{const} \sum_{\underline{k},n} \left| \int d^3r \, \exp(-i\underline{p}\cdot\underline{r})\psi^+(\underline{r})\psi^-_{\underline{k},n}(\underline{r}) \right|^2 =$$

$$(1)$$

$$\text{const} \sum_{\underline{k},n} \sum_{\underline{G}} |A_{\underline{G}}(\underline{k},n)|^2 \delta(\underline{p}-\underline{k}-\underline{G})$$

where the summation is over all occupied states in the first
zone, $\psi^+(\underline{r})$ is the ground-state ($\underline{k}^+ = 0$) positron Bloch wave
function, $\psi^-_{\underline{k},n}(\underline{r}) = u_{\underline{k},n}(\underline{r})\exp(i\underline{k}\cdot\underline{r})$ is the electron wave function with crystal momentum $\underline{k}$ and band index n, $\underline{G}$ is a recipro-
cal lattice vector, $h \equiv 1$, and

$$\psi^+(\underline{r})u^-_{\underline{k},n}(\underline{r}) = \sum_{\underline{G}} A_{\underline{G}}(\underline{k},n)\exp(i\underline{G}\cdot\underline{r})$$

One obtains from (1) a density in momentum space $\rho(\underline{p})$ exhibiting breaks not only at the FS in the first zone ($\underline{p} = \underline{k}$) but
also in higher zones ($\underline{p} = \underline{k} + \underline{G}$) due to the high momentum components (HMC) of the electron and positron wavefunctions ("um-
klapp annihilations"). One therefore predicts a density bound-
ed by a set of FS in $\underline{p}$ space modulated at each $\underline{p} = \underline{k} + \underline{G}$ by
$|A_{\underline{G}}(\underline{k},n)|^2$. The measurement of $\rho(\underline{p})$ by the angular correlation
of annihilation radiation (ACAR) relfects the size and shape of
the FS as well as the nature of the wave functions via the
Fourier coefficients $A_{\underline{G}}(\underline{k},n)$. By the use of symmetry arguments
it is possible to show that not all parts of the FS have to
appear in every zone in $\underline{p}$-space.

ACAR measurements are usually performed with the "long-
slit" geometry, measuring planar integrals in $\rho(\underline{p})$, $N_{\hat{n}}(p_z) =$
$\iint \rho_{\hat{n}}(\underline{p})dp_x dp_y$, where $\hat{n}$ indicates the crystal orientation of
the sample and p_z is related to the measured angle between the
two photons by $p_z = \theta mc$; thus momenta can be measured in milli-
radian units, 1 mrad = mc $\times 10^{-3}$, for V_3Si $\pi/a = 2.57$ mrad. More
recently, full 2D ACAR surfaces have been measured by Berko et
al (1977) using "point-slit" geometry, $N_{\hat{n}}(p_y,p_z) = \int \rho_{\hat{n}}(\underline{p})dp_x$.
Clearly these yield more unambiguous information about $\rho(\underline{p})$.

Long slit measurements on oriented crystals of V_3Si were
reported by Berko and Weger (1970, 1972) for several orientations $\hat{n}$. Fine structure in the derivative dN/dp_z was attributed to planar sheets in the FS normal to <100> as predicted by
the linear chain model (Barak et al, 1975). Recently, Samoilov
and Weger (1977) have performed a similar experiment yielding a
derivative structure again attributed to planar FS sheets. 2D

ACAR measurements on V_3Si were performed by Berko and Mader (1975) for a few fixed p_y values. Recent reports by Manuel et al (1978, 1979) also include 2D V_3Si data.

Full 2D ACAR surfaces have now been obtained and analyzed at Brandeis University (Farmer, et al, 1979a, 1979b). Our angular cross-correlation apparatus using 64 NaI detectors has been described elsewhere (Berko, et al, 1977). The distributions $N(p_y,p_z)$ are sampled in steps of 0.6 and 0.2 mrad in p_y and p_z respectively. The "point-slit" geometrical resolution is Δp_y = 1.5 mrad and Δp_z = 0.5 mrad FWHM. Full ACAR surfaces have been measured for three crystal orientations: $N_{001,100}$, $N_{011,100}$ and $N_{011,0\bar{1}1}$. The first index gives the good resolution direction p_z and the second the integration direction p_x. All these distributions exhibit a broad gaussian shape with very small structures. This is to be expected because of the many filled bands of V_3Si (Berko and Weger, 1970, 1972). Subtracting a rotationally averaged distribution $R(|\underline{p}|) = (1/2\pi)\int N(|\underline{p}|,\theta)d\theta$ from the data, a distribution is produced which displays the small deviations from isotropic behavior. Fig. 1 shows a contour map of this anisotropy distribution for the [001][100] orientation. The minima at the $(\pm \pi/a, \pm 3\pi/a)$ points are clearly commensurate with the reciprocal lattice, indicating a concentration of hole density projecting onto the corners of the [100] projection of the repeated zone in p-space. Interpretation of the anisotropy along other projections is more difficult because of the projection down a lower symmetry direction. The direct analysis of these anisotropies requires care because the rotational averaging may cause aliasing of a sharp feature around a

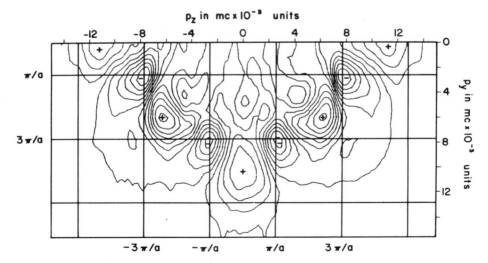

FIGURE 1. Contour map of the anisotropic part of $N_{001,100}$ (p_y,p_z).

circle of the same radius. In Fig. 1, the sharp minima can generate a ring of compensatory maxima. For these reasons we decided to carry out a transformation from p-space to k-space developed originally for long-slit data by Lock et al (1973).

From equation (1) it is clear that an electron with crystal momentum $\underline{k}$ contributes to $\rho(\underline{p})$ at all $\underline{p} = \underline{k} + \underline{G}$ where $\underline{G}$ is any reciprocal lattice vector. To reassemble these effects we calculate

$$\rho'(\underline{p}) = \sum_{\underline{G}} \rho(\underline{p} + \underline{G}) \tag{2}$$

In the approximation that $\sum_{\underline{G}} |A_{\underline{G}}(\underline{k},n)|^2 = 1$, $\rho'(\underline{p})$ becomes a repeated zone scheme representation of $\rho(\underline{k})$, thus leading to a constant density for filled bands, and fluctuations that are only due to the FS. For a constant positron wave function in the IPM this condition is identically satisfied. The extent to which this approximation is accurate in real situations has been studied by Lock and West (1975) and Beardsley et al (1975).

This folding technique can be applied directly to the measured 2D distribution $N(p_y,p_z) = \int \rho(\underline{p}) dp_x$ to obtain the equivalent projection of $\rho'(\underline{p})$. In two dimensions we calculate

$$N'_{\hat{n}}(p_y,p_z) = \sum_{\underline{H}} N_{\hat{n}}(p_y + H_y, \; p_z + H_z) \tag{3}$$

where H_y and H_z are the components of the 2D vectors $\underline{H}$, produced when the full reciprocal lattice is projected onto the p_y, p_z plane. For the case of a simple cubic reciprocal lattice projected down [100] this projection gives a simple square lattice, and the folding technique yields a projection of a single zone with no overlap. In all but a few special cases, however, the projections of the different zones in the repeated zone scheme overlap, making it more difficult to interpret directly the resulting distribution $N'_{\hat{n}}(p_y,p_z)$.

Fig. 2 shows a contour map of $N'_{001,100}$. There are strong minima ($\sim$5% amplitude) at the corners of the projected zone, with shallow troughs running along the edges. The lack of four-fold symmetry is due to the rectangular p_y,p_z resolution of our apparatus. We find that a simple FS model consisting of a set of empty cylinders running along the cube edges of joining at the corners in a six-pronged cross ("cubic jungle gym") reproduces the observed distribution remarkably well. To produce $N'_{\hat{n}}(k_y,k_z)$ from this model, we assign a constant value to $\rho(\underline{k})$ outside the jungle gym and a zero value inside. We then integrate along k_x and fold in the actual two-dimensional resolution function. Finally the results of the calculation are

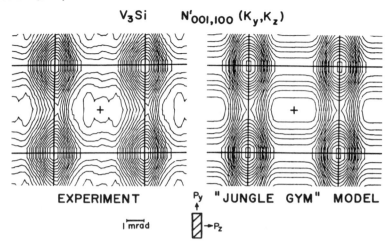

$$V_3Si \qquad N'_{001,100}\,(K_y,K_z)$$

EXPERIMENT P_y "JUNGLE GYM" MODEL

$\overline{1\ mrad}$ ▨→P_z

FIGURE 2. Comparison between the 2D ACAR data folded according to eqn. (3) and the simple jungle gym model.

scaled so that the range of amplitude equals that in the experimental $N'_{\hat{n}}(k_y,k_z)$. This computation was performed with several
cylinder radii r and we find that $r = 1$ mrad $= 0.39$ π/a gives the
best fit. The resulting model surface is also shown in Fig. 2.
This comparison was also carried out for the orientations [011]
[100] and [011][0$\bar{1}$1], which gave hints at the deficiencies of
this over-simple model. The finite momentum resolution and the
reducing technique used could conceal the detailed features of
the multiple sheets of the FS, and average them to yield an effective $\rho(\underline{k})$ resembling the jungle gym.
    Since it is impossible to completely determine $\rho(\underline{p})$ from
projections in only two directions, and the 2D folding algo-
rithm in (3) is only useful for a few high-symmetry directions,
we decided to reconstruct $\rho(\underline{p})$ in 3D and then apply the folding
in (2). Thus $\rho(p_x,p_y,p_z)$ was reconstructed in independent p_x,
p_z planes from data taken in six different orientations rotated
about the $p_y = [010]$ direction in 9° steps. Data was obtained
for each orientation in the form of a 2D ACAR distribution with
steps of 1.2 and 0.2 mrad in p_y and p_z respectively. For each
p_y value, single curves $N_{\hat{n}}(p_z)$ for the six orientations were
used to reconstruct $\rho(\underline{p})$ in this p_x, p_z plane. The 4 mm symmetry along [010] permits us to back-project the available data
from 40 orientations around 360°. We use the filtered back-
projection technique described by Chesler and Riederer (1975).
Using the 3D extensions of the folding technique of Lock et al.
(1973) given in (2) we then produce $\rho(\underline{k})$, containing information from all bands in a single zone. In Fig. 3 (experiment)
we show the results of this reconstruction for sections through

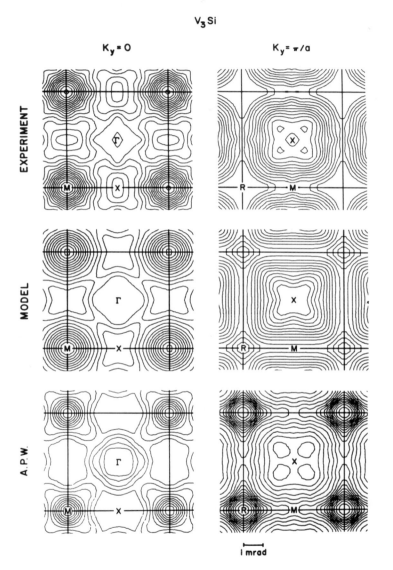

FIGURE 3. Comparison between sections through k-space for the reconstructed positron annihilation data, the model in Fig. 4, and the APW calculation of Klein et al (1978a, 1978b). For all three the lowest electron density in the $k_y = 0$ section is at the M point, and in the $k_y = \pi/a$ section at the R point.

the center ($k_y = 0$) and through the face ($k_y = 2.57$ mrad) of the
zone.   The effective resolution in the k_x, k_z plane is dependent on the p_z resolution in the ACAR data (0.5 mrad) and the
angle between adjacent projections (9°).   Once again the low
electron density along the zone edges is revealed, reconfirming
the overall topology of the jungle gym model.   We have refined
this model to fit the reconstructed $\rho(\underline{k})$ more closely, and as
a result we can draw the following conclusions: a) The hole
structure along the zone edges is made up of at least two nested cylindrical surfaces ($r_1 = 0.2$ π/a and $r_2 = 0.4$ π/a provides
a good fit).   b) This hole structure extends to the R-point,
where we see a low electron density.   c) The structure observed
in the center of the zone in the 2D results (Fig. 2) is due to
extra electron density centered about the X point on the zone
face.   Our best FS model to date that incorporates these features is shown in Fig. 4.   The electron density centered at X
is represented by a square box, extending .34 π/a along X-M and
.44 π/a along X-Γ.   Joining the corners of these boxes by cylinders that meet along the Γ-R line with a radius of 0.2 π/a as
indicated by the dashed lines in Fig. 4 again improved the fit
to the data.   This is the model FS which has been used to calculate the density in Fig. 3 (model), with the weights of each
sheet as indicated in Fig. 4.   There still remain some differences between the model and the data, particularly around the
R-point, where the model has too much hole density.   It is possible that the inner jungle gym does not continue all the way

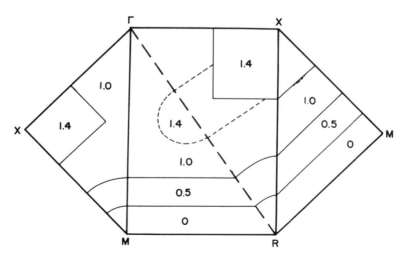

FIGURE 4.   FS model, made up out of simple geometrical surfaces, which was used to calculate the distributions in Fig. 3
(model).   The numbers indicate the densities assigned to the
various parts of k-space.

to the corner, but terminates before reaching R, yielding closed ellipsoids along the cube edges.

We have compared these results with several FS calculations.
From these we produce theoretical contour maps in the k_x, k_z
planes by assigning to each partially filled band an equal
weight and convoluting the theoretical FS with the experimental
k_y resolution. An early APW calculation by Mattheiss(1975) has
hole surfaces along the edges, but they do not extend to the R
point. The agreement with FS of Nb_3Sn calculated by van Kessel
et al (1978) is better, although again the calculation does not
show enough hole density around R. We find that the V_3Si FS
obtained from a self-consistent APW calculation by Klein et al
(1978a, 1978b) provides the best overall agreement. Fig. 3
(APW) shows the contour maps based on this FS. The main difference is once again around the R point, where now the theory
shows more hole surfaces than observed.

ACAR data with better resolution and statistics will be required to study more details of the complex V_3Si FS, in particular possible surfaces centered on Γ as well as the structure
of hole orbits around R. Theoretical calculations of $\rho(\underline{p})$ incorporating the positron wave functions will be necessary for a
complete $\underline{p}$ space comparison.

ACKNOWLEDGMENTS

We are grateful to B.M. Klein and F.M. Mueller for sending
us their FS results prior to publication, to J.H. Wernick of
Bell Laboratories for providing us with single crystal V_3Si
samples, and to G.M. Beardsley for help in the early stages of
this work.

REFERENCES

Arko, A.J., Fisk, Z., and Mueller, F.M. (1977). *Phys. Rev. B16,*
1387.
Arko, A.J., Lowndes, D.H., Muller, F.A., Roeland, L.W., Wolf-
rat, J., van Kessel, A.T., Myron, H.W., Mueller, F.M., and
Webb, G.W. (1978). *Phys. Rev. Lett. 40,* 1590.
Barak, G., Goldberg, I.B., and Weger, M. (1975). *J. Phys.*
Chem. Solids 36, 847.
Beardsley, G.M., Berko, S., Mader, J.J., and Shulman, M.A.
(1975). *Appl. Phys. 5,* 375.
Berko, S. (1978). *J. de Physique 39,* C6-1568.
Berko, S., and Mader, J.J. (1975). *Appl. Phys. 5,* 287.
Berko, S., and Weger, M. (1970). *Phys. Rev. Lett. 24,* 55.

Berko, S., and Weger, M. (1972). *In* "Computational Solid State Physics" (F. Herman and T. Koehler, eds.), p. 59. Plenum Press, New York.

Berko, S., Haghgooie, M., and Mader, J.J. (1977). *Phys. Lett.* *63A*, 335.

Chesler, D.A., and Riederer, S.J. (1975). *Phys. Med. Biol.* *20*, 632.

Farmer, W.S., Sinclair, F., Berko, S., and Beardsley, G.M. (1979a). Proc. 5th Int. Conf. Positron Annihilation (Japan).

Farmer, W.S., Sinclair, F., Berko, S., and Beardsley, G.M. (1979b). *Solid State Comm.* *31*, 481.

Graebner, J.E., and Kunzler, J.E. (1969). *J. Low Temp. Phys.* *1*, 443.

Klein, B.M., Boyer, L.L., Papaconstantopoulos, D.A., and Mattheiss, L.F. (1978a). *Phys. Rev.* *B18*, 6411.

Klein, B.M. (1978b). Private communication.

Lock, D.G., Crisp, V.H.C., and West, R.N. (1973). *J. Phys.* *F3*, 561.

Lock, D.G., and West, R.N. (1975). *Appl. Phys.* *6*, 249.

Manuel, A.A., Samoilov, S., Fischer, O., Peter, M., and Jeavons, A.P. (1978). *Nucl. Instr. and Methods* *156*, 67.

Manuel, A.A., Samoilov, S., Fischer, O., Peter, M., and Jeavons, A.P. (1979). *Helvetica Physica Acta* *52*, 37.

Mattheiss, L.F. (1975). *Phys. Rev.* *B12*, 2161.

Mijnarends, P.E. (1978). *In* "Positrons in Solids" (P. Hautojarvi, ed.). Springer-Verlag, Berlin.

Samoilov, S. and Weger, M. (1977). *Solid State Comm.* *24*, 821.

Testardi, L.R. (1973). *In* "Physical Acoustics" vol. 10 (W.P. Mason and R.N. Thurston, eds.), p. 193. Academic Press, New York.

van Kessel, A.T., Myron, H.W. and Mueller, F.M. (1978). *Phys. Rev. Lett.* *41*, 181.

Weger, M. and Goldberg, I.B. (1973). *In* "Solid State Physics" vol. 28 (H. Ehrenreich, F. Seitz and D. Turnbull, eds.), p. 2. Academic Press, New York.

THEORY OF RESISTIVITY "SATURATION"

*Philip B. Allen**

Department of Physics
State University of New York
Stony Brook, New York, USA

*High T_c superconductors at room temperature and above
have resistivities rising less rapidly with T than Bloch-
Grüneisen theory predicts. Critical discussion of seven
possible mechanisms is given. Three mechanisms (anharmonicity, Fermi smearing, T-dependent energy bands) are
consistent with Boltzmann theory, and four (localization,
Debye-Waller effects, Phonon drag and ineffectiveness, and
non-classical conduction channels) are not. It is argued
that the last mechanism is the most plausible.*

I. THE SATURATION PHENOMENON VERSUS NORMAL METALLIC
    RESISTIVITY

Figure 1 shows the electrical resistivity of Nb_3Sn as a
function of temperature (Woodard and Cody, 1965.). The
shape of $\rho(T)$ departs strongly from Bloch-Grüneisen theory
(Ziman, 1960). It was pointed out by Fisk and Lawson (1973)
that the same shape of $\rho(T)$ is seen in nearly all high T_c
d-band metals (and other highly resistive metals as well),
whereas low T_c materials tend to have a more normal behavior.
Sometimes the effect is called a "bulge" - implying that
the anomalous feature is excess resistivity for temperatures
near θ_D. This seems to me incorrect; if there were only a
bulge, then $\rho(T)$ above θ_D should be normal, i.e., (a) nearly

*Supported in part by National Science Foundation Grant No.
DMR79-00837.*

linear in T and (b) extrapolating back through ρ_o, the residual resistance. This second constraint – that the intercept is fixed – is often forgotten. It follows from the basic variational solution of the Bloch-Boltzmann equation (Allen, 1978),

$$\rho_{ideal}(T) = \rho_o + [(n/m)_{eff} \, e^2 \tau_{ph}(T)]^{-1} \tag{1}$$

$$1/\tau_{ph}(T) = (2\pi/\hbar) \, k_B T \, \lambda_{tr}(T) \tag{2}$$

$$(n/m)_{eff} = N(o) \, \langle v_x^2 \rangle \tag{3}$$

$$\lambda_{tr}(T) = 2 \int_o^\infty \frac{d\Omega}{\Omega} \, \alpha_{tr}^2 F(\Omega) \left[\frac{\hbar\Omega/2k_B T}{\sinh(\hbar\Omega/2k_B T)} \right]^2 \tag{4}$$

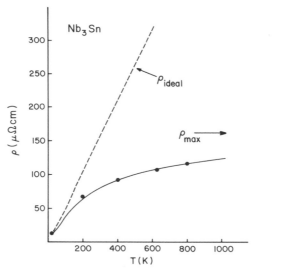

FIGURE 1.   Resistivity of Nb_3Sn versus temperature.   Dots are data of D.W. Woodard and G.D. Cody (Phys. Rev. 136, A166 (1964).)   The solid line is a fit using eq. (6) with $\rho_o = 13.5\mu\Omega cm$ and $\rho_M = 163\mu\Omega cm$.   Assuming $\lambda_{tr} = \lambda = 1.7$, the fitted value of $(n/m)_{eff}$ gives a Drude plasma frequency 3.7 eV which compares well with 3.4 eV calculated by L.F. Mattheiss et al. (Phys. Rev. B17, (1978).)   The dashed curve is ρ_{ideal}, the Boltzmann resistivity calculated from eqs. (1-4) using the phonon density of states of P. Schweiss et al. (in "Superconductivity in d- and f-Band Metals," D.H. Douglass, ed., Plenum, New York, 1976; p. 189) for the shape of $\alpha_{tr}^2 F(\Omega)$.   The corresponding mean free path at 300K is 7Å, using the Fermi velocity of Mattheiss et al.

where $N(o)$ is the density of states at the Fermi surface,
$<v_x^2>$ is the mean square x-component of electron velocity at
the Fermi surface, and $\alpha^2 F(\Omega)$ is a function very similar to
$\alpha^2 F(\Omega)$ in superconductivity, with a slightly different
weighting used in the Fermi surface average. At high temperatures ($T \gtrsim \theta_D$) the factor in brackets in eq. (4) becomes unity and $\lambda_{tr}(T)$ achieves its maximum value, a
constant, λ_{tr}, very similar to λ in superconductivity. A
simplified version of equation (1) valid at T=0 and at
$T \gtrsim \theta_D$ is

$$\rho_{ideal}(T) \underset{\sim}{} \rho_o + \rho_1 T \tag{5}$$

where ρ_1 involves two coupling constants, λ_{tr} and $(n/m)_{eff}^{-1}$.
From eq. (5) it follows that $\rho(T)$ of Nb_3Sn (fig. 1)
is badly amiss at high T. One cannot tell (without a
detailed theory) whether there is excess resistivity in the
"bulge" region, but no detailed theory at all is required to
see that at high T, $\rho(T)$ falls below the required linear
behavior. A better name for the phenomenon is "deviation
from linearity" (d.f.ℓ.), which carries no prejudice as to
the cause. Fisk and Webb (1976) have introduced a more
descriptive term, "saturation." This word conveys the
prejudice (which I think is correct) that the downward turn
in $\rho(T)$ represents the approach to Mott's (1971) maximum
metallic resistivity (or minimum metallic conductivity),
$\rho_{max} = 1/\sigma_{min}$, and is associated with very short electron
mean free paths (ℓ). Unfortunately we have no clear microscopic picture for the conduction mechanism in Mott's
regime. Mooij (1973) has assembled much evidence that in
d-band alloys, ρ saturates at a value $\rho_{max} \underset{\sim}{} 150\mu\Omega$ cm.

II. PARALLEL RESISTOR FORMULA

An empirical formula used by Wiesmann et al. (1977) fits
a great deal of data.

$$1/\rho(T) = 1/\rho_{ideal}(T) + 1/\rho_{max} \tag{6}$$

If eq. (5) is used for ρ_{ideal}, then there are three adjust-
able parameters. For A15 metals, ρ_{max} is found to be
$\sim 150\mu\Omega$ cm. Only ρ_o varies significantly when eq. (6) is
fit to data for the same material with varying degrees of
radiation damage (Gurvitch, 1978). The electron-phonon
parameter ρ_1 derived from fitting (6) to experiment, agrees
with independent theoretical estimates of ρ_1 by Allen et al.

(1978a) for $Nb_3A\ell$ and Nb_3Ge. These estimates consist of calculating $(n/m)_{eff}$ from band theory and estimating $\lambda_{tr} \approx \lambda$, taking λ from T_c. This procedure was used success-fully by Chakraborty *et al.* (1976) for elements like Nb where "saturation" is not seen at 300K and below. Infrared experiments (Mattheiss *et al.*, 1978) confirm the value of $(n/m)_{eff}$. Detailed theoretical calculations (Pinski *et al.* 1978) of λ and λ_{tr} for Nb and Pd show that these two numbers differ by only 10%. From the theory used to estimate ρ_1 it is found that if Boltzmann theory worked, ℓ at 300K would be 3-4Å in Nb_3Ge and $Nb_3A\ell$.

The success of eq. (6) suggests that it may be more than an accident. Perhaps an extra conduction channel parallel to the Boltzmann channel is physically real. If so, it should be a general occurrence, available in all metals. This does not contradict the data. In elements not showing saturation, ρ_{ideal} is sufficiently small compared to $150\mu\Omega$ cm that it is difficult or impossible to detect the "parallel resistor."

III.   WHAT IS BLOCH-BOLTZMANN THEORY?

The first step is to understand thoroughly the ordinary theory of metallic conduction formulated by Felix Bloch (1928). The Bloch-Boltzmann equation for electrons in steady-state in a dc external electric field E is

$$- e \underset{\sim}{E} \cdot \underset{\sim}{v}_k \; (-\partial f/\partial\varepsilon_k) = \underset{k'}{\Sigma} \; Q_{kk'}\phi_{k'} \tag{7}$$

where the distribution function F_k for electrons in state $|k>$ with velocity $\hbar v_k = \partial\varepsilon_k/\partial\underset{\sim}{k}$ is $f(\varepsilon_k) + \phi_k \, (-\partial f/\partial\varepsilon_k)$ and f is the Fermi function. The scattering operator $Q_{kk'}$ is the sum of terms for impurity scattering, phonon scattering, and electron-electron (Coulomb) scattering. This equation has been linearized, which means that it only gives the Ohmic part of the current. This equation rests on four assumptions:

(A)   The E field and the currents are described by the semiclassical theory. That is, the E field "accelerates" electrons according to $\underset{\sim}{\dot{k}} = -e\underset{\sim}{E}/\hbar$, and the current is caused by the resulting excess of electrons with a band velocity pointing along $\underset{\sim}{E}$,

$$\underset{\sim}{j} = -e \underset{k}{\Sigma} \; \underset{\sim}{v}_k \; \phi_k \; (-\partial f/\partial\varepsilon_k) \tag{8}$$

(B) Scattering events are statistically independent of each other; they are separated by enough wavelengths that the electron "recovers" from previous collisions before experiencing the next one. For free electrons the criterion is $k_F \ell \gg 1$.

(C) The phonon part of the collision operator is calculated using only the first term in the Taylor series

$$\mathcal{K}_{ep} = \sum_{\ell} [\underset{\sim}{u}_{\ell} \cdot \underset{\sim}{\nabla}_{\ell} V + \frac{1}{2} \underset{\sim}{u}_{\ell} \underset{\sim}{u}_{\ell'} : \underset{\sim}{\nabla}_{\ell} \underset{\sim}{\nabla}_{\ell'} V + \ldots] \tag{9}$$

where u_{ℓ} is the displacement of the ℓth atom and V is the crystal potential. Specifically, the phonon part of Q is second order in $(u \cdot \nabla V)$. Higher powers of $(u \cdot \nabla V)$ would give effects higher order in $(k_F \ell)^{-1}$ and so are omitted according to (B). Higher order terms in the series (8), treated to second order in perturbation theory, will give corrections of order $(u/a)^2$ and higher; these are omitted.

(D) The phonons are assumed to be in thermal equilibrium.

In spite of the four assumptions, this theory is remarkably sophisticated and includes many processes to infinite order of perturbation. This sometimes surprises theorists unfamiliar with the details. Feynman diagrams provide a classification scheme which is helpful (to some) in understanding the content and limitations of eq. (7). Figure 2 shows representative graphs which are completely included in eq. (7). It is surprising that no renormalization effects occur to alter eq. (7) at the level of the graphs depicted in fig. 2. The most careful proof is by Holstein (1964). He found that for ac fields, when $T < \theta_D$ and $\omega \lesssim \omega_D$, renormalization effects do occur, but they cancel as $\omega \to o$.

Representative graphs which are not summed by eq. (7) are shown in fig. 3. Each graph has been classified according to which of the four assumptions A-D is violated. Graphs A will be discussed in Sec. XI, B in VII, C in VIII, and D in IX.

IV. ANHARMONICITY

There are two effects within Boltzmann theory which can cause the "ideal" resistivity to depart from linearity at high temperature. Unfortunately, they don't seem sufficient to explain saturation. They are described in this section and the next. The simplest is the fact that phonon frequencies change with temperature partly because of thermal expansion. For convenience we call this "anharmonicity,"

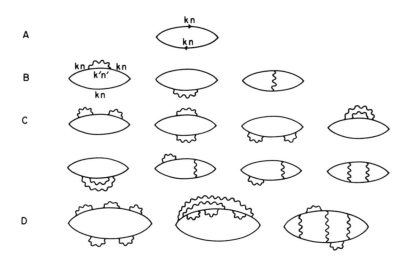

FIGURE 2.   Representative Feynman graphs summed in
Bloch-Boltzmann theory.   Subsets A,B,C are zeroth, first,
and second-order graphs respectively; D shows a few of the
fifth-order graphs which are summed.

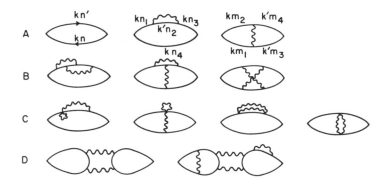

FIGURE 3.   Feynman graphs omitted in Bloch-Boltzmann
theory.   Subsets A,B,C,D are the lowest-order graphs
omitted because of assumptions A,B,C,D.   In the special
cases (n'=n) or (n_1=n_3=n_4) or (m_1=m_2 and m_3=m_4), graphs A
reduce to graphs of fig. 2.

even though the mechanism is possibly not anharmonic
coupling. To estimate the importance of anharmonicity,
Allen *et al.* (1976) followed McMillan (1968) and wrote

$$\lambda_{tr} = \eta/M<\omega^2> \tag{10}$$

Thus the effect of anharmonicity on $\rho(T)$ at high T is
through a factor $<\omega^2>^{-1}$. Normal behavior is for $<\omega^2>$ to
diminish with increasing T, causing $\rho(T)$ to rise above
linearity. However, many A15 metals have anomalous
anharmonic behavior, and $<\omega^2>$ is known to increase. It has
been estimated that at most 1/3 of the d.f.ℓ. can be
explained this way (Bader and Fradin, 1976).

V.   FERMI SMEARING

Eq. (8) can be written

$$\sigma(T) = \int_\infty^\infty d\varepsilon \ \sigma(\varepsilon,T) \ (-\partial f/\partial \varepsilon) \tag{11}$$

$$\sigma(\varepsilon,T) = \sum_k e^2 v_{kx}^2 \ \tau_k \ \delta(\varepsilon_k - \varepsilon) \tag{12}$$

where τ_k is *defined* by the relation $\phi_k = -e \ \underset{\sim}{v}_k \cdot \underset{\sim}{E} \ \tau_k$ and $\underset{\sim}{E}$ is
in the x direction. Mott (1936) pointed out that in d-band
systems, $\sigma(\varepsilon,T)$ may vary with ε on a scale $k_B T$, so that
$(-\partial f/\partial \varepsilon)$ in (11) may *not* be replaced by $\delta(\varepsilon)$. The resulting
corrections are called "Mott Fermi smearing." The effect
on the T-dependence of ρ can be significant when energy bands
are narrow and the Fermi velocity changes rapidly with ε.
Many authors have made model calculations purporting to show
how saturation in A15 metals can be explained this way
(Cohen, Cody, and Halloran, 1967; Bader and Fradin, 1976;
Nakayama and Tsuneto, 1978; Ting, Synder, and Williamson,
1979). Unfortunately it is extremely difficult to calculate
this effect, and most of the existing literature on the
subject is wrong. A reasonable procedure (Allen, 1976)
starts by defining a density of state $N(\varepsilon)$, mean square
velocity $v_x^2(\varepsilon)$, and scattering rate $\tau(\varepsilon)$ through the
relations

$$N(\varepsilon) \equiv \sum_k \delta(\varepsilon_k - \varepsilon) \tag{13}$$

$$N(\varepsilon) v_x^2(\varepsilon) \equiv \sum_k v_{kx}^2 \ \delta(\varepsilon_k - \varepsilon) \tag{14}$$

$$N(\varepsilon)v_x^2(\varepsilon)\tau(\varepsilon) \equiv \sigma(\varepsilon,T)/e^2 \qquad\qquad (15)$$

Calculating $\tau(\varepsilon)$ requires solving Boltzmann's equation. It is not permissible to use the standard variational solution which assumes $\tau(\varepsilon)$ = const. Roughly speaking, $1/\tau(\varepsilon)$ behaves as $N(\varepsilon)$ with some additional factors which (hopefully!) vary less rapidly with ε. Note that $N(\varepsilon)$ *cancels* out of $\sigma(\varepsilon,T)$ leaving $v_x^2(\varepsilon)$ as the parameter to be smeared. Attempted calculations along these lines (Laubitz *et al.*, 1977; Allen and Chakraborty, 1978; Pinski *et al.*, 1978) have been moderately successful for the elements. There is no doubt that the effect on ρ_{ideal} can be significant. However, it is hard to believe that Fermi smearing – a complex and band-structure-dependent effect, can account for the simple and regular behavior of so many materials. Why should $\rho(T)$ not sometimes bend upwards, or else bend too far in the downwards direction?

VI.   ALTERATIONS OF BAND STRUCTURE WITH T

Two suggestions are sometimes made based on intuitive plausibility: (i) putting a T-dependent band structure into eq. (7), or (ii) introducing a phenomenological lifetime broadening (Testardi and Mattheiss, 1978; Wiesmann *et al.*, 1978; Greig and Morgan, 1973) in addition to the thermal broadening of eqs. (11-15). The latter receives some support from a CPA theory by Chen *et al.* (1972); also Brouers (1978), but this theory omits electron-hole interactions and deals only with a single band. A rigorous theory (making neither of these simplifications, but valid only for weak disorder) has been given by Chakraborty and Allen (1978, 1979). Their theory can account for T dependent shifts in $\sigma(\omega)$ at optical frequencies in semiconductors, and gives a generalization of Boltzmann theory in the dc limit (discussed further in sec. XI.). Unfortunately their theory is in no way equivalent to a modified version of eq. (7). No justification is found for either intuitive procedure, except that for the trivial effects of thermal expansion, a T-dependent band structure can be used in (7). This latter is a small effect, less significant than the thermal shifts of phonon frequencies already discussed in sec. IV.

VII. CONNECTION WITH ANDERSON LOCALIZATION

The remaining sections deal with attempts to explain saturation outside the framework of Bloch-Boltzmann theory. Of the omitted graphs in fig. 3, the hardest to deal with are those of 3B. They contribute to ρ terms which are higher order in $(k_F \ell)^{-1}$. It is known (Langer and Neal, 1966) that a straightforward expansion of σ in powers of $(k_F \ell)^{-1}$ has logarithmically divergent coefficients in third order and higher. There is as yet no accepted method for solving this problem. It seems likely that the difficulties with perturbation theory relate to an actual physical singularity, the onset of Anderson localization (Anderson, 1958; Thouless, 1974). This subject remains difficult and controversial. Only simple models involving a single band have been studied, whereas A15 metals have many overlapping bands. At T=0 it is generally agreed that σ will vanish above a critical value of the disorder, d_c. There is some controversy about whether σ vanishes continuously (as $(d_c-d)^P$ for some positive P) or discontinuously (as $\theta(d_c-d)$). Older evidence suggested the latter, with the value of σ just below d_c being σ_{min}. This strongly suggests a connection between saturation and localization. More recent evidence suggests the former, which seems incompatible with saturation, i.e. there is no hint of an *excess* conductivity before localization occurs.

The situation is even less clear at finite temperature. For $d>d_c$, either thermally assisted hopping or activation above a mobility edge will allow σ to be finite. For $d<d_c$, little work has been done. The data of fig. 1 suggest the question whether vibrational disorder can *cause* localization, and if so how will $\sigma(T)$ behave above and below the threshold. In summary, there is yet not much evidence for a close connection between saturation and localization.

VIII. DEBYE-WALLER FACTORS

It has been proposed at various times that when an electron scatters from the atomic displacement, a Debye-Waller factor should enter, weakening the strength of the scattering potential, just as for x-ray scattering. This is undoubtedly true, and is described diagramatically by dressing the electron-phonon vertex by additional closed phonon loops which represent virtual emission and reabsorption of the same phonon. (fig. 3C, first two). Clearly this will give a diminished high-T resistivity, as observed

experimentally (Rossiter, 1977; Visscher, 1978). There are
two difficulties. First, the Debye-Waller expansion
parameter is (u/a), a number which behaves roughly the same
in all metals, approaching the Lindemann limit at melting.
There is nothing to discriminate strong-scattering metals
where saturation is seen from weak-scattering metals like
Cu or Aℓ where it is not. Second, there is a theoretical
objection (Sham and Ziman, 1963) that multi-phonon
scattering enters in the same order of perturbation theory
as Debye-Waller effects, (fig. 3C, second two) and clearly
enhance the resistivity. The degree to which the two
effects cancel is unknown. Speculations of exact cancella-
tion seem unfounded. It is difficult to estimate the
effects well and the matter awaits further theoretical
development.

IX. PHONON DRAG

    It is known that in a metal carrying a current, the
phonons are not exactly in thermal equilibrium. In an
idealized case one can imagine that the phonons are dragged
with the current, so that in the rest-frame of the drifting
electrons, the phonons and electrons are both close to
thermal equilibrium. Fewer current-degrading collisions
occur and the conductivity is enhanced (Ziman, 1960). This
picture applies only if phonons decouple from everything
except electrons. In actuality, phonons couple to each
other (by anharmonic coupling) and to defects, allowing
relaxation toward equilibrium. Thus phonon drag is
expected to be significant only in pure metals at low
temperatures where anharmonic and defect effects are small.
A theory is achieved by generalizing eq. (7) to two
coupled Boltzmann equations, one for electrons and one
for phonons. Holstein (1964) has shown that these coupled
Boltzmann equations can be derived by summing up the
additional graphs of fig. 3D.

X. PHONON INEFFECTIVENESS

    Morton *et al.* (1978) and Cote and Meisel (1978) have
independently proposed the hypothesis that phonons with
$Q\ell < 1$ become ineffective scatterers. Little justification
has been offered except that it gives an economical explanation for the resistivity data and for the degradation of T_c
by radiation damage. Two things are bothersome. First, in

the spirit of Mott's ρ_{max}, saturation should be independent of which mechanism causes the principal scattering (defects, phonons, or Coulomb scattering, whereas "phonon ineffective-ness" treats only the phonon mechanism. Second, amorphous s-p metals have T_c values comparable to their crystalline counterparts and tunneling shows excess strength in $\alpha^2 F(\Omega)$ at low Ω. If phonon ineffectiveness applies, one would expect the small Ω phonons with small Q to have a deficiency of weight in $\alpha^2 F$.

Both groups proposing phonon ineffectiveness have invoked Pippard's (1955) work on ultrasonic attenuation. Pippard showed that if the impressed phonon has $Q\ell < 1$, then electrons are ineffective in degrading the ultrasonic energy. This is because electrons are dragged by the impressed phonon wave. The connection between Pippard's ideas and saturation seems obscure. The thermal phonons present during electrical conduction are incoherent and cannot all drag electrons. It seems to me that if there is truth in the idea of "phonon ineffectiveness," it is contained in the phonon drag processes of sec. IX.

XI. BEYOND SEMICLASSICAL THEORY

Finally we turn to the graphs of fig. 3A. These graphs were summed in the generalized Boltzmann equation of Chakraborty and Allen (1978, 1979), as a solution to the problem of how to include temperature-dependent electron bands into conductivity theory. Both the semiclassical acceleration equation and the semiclassical current (eq. 8) must be supplemented by non-classical terms which give the interband dipole transitions and the interband currents. These graphs by themselves would give significant contribu-tion to $\sigma(\omega)$ only for values of ω in the interband energy range. However, when summed to high order along with the graphs of fig. 2, a new channel for d.c. conduction is found. The resulting theory, when solved to first order in the new effects, has exactly the form of the parallel resistor model, eq. (6). The "shunt resistor" ρ_{max} represents the new dc current channel available when inter-band currents, interband excitation by the $\underset{\sim}{E}$ field, and interband scattering are allowed to mix with the usual semiclassical processes. Disorder both allows and inhibits these processes. For example, a virtual interband excitation by the E field could not affect the dc current unless a virtual interband scattering event were available to restore energy conservation. On the other hand, the amount of current carried in this fashion is limited by collisions just as in the semiclassical case. The expression obtained for $\sigma_{min} = 1/\rho_{max}$ is quite complicated. Collisions appear in both numerator and denominator, leaving a result which is formally independent of the strength of the disorder. The order of magnitude of the term is $\sigma_{min} \sim nhe^2/mE_B$ where E_B is a band separation. This has the right size to account for saturation. Comparing with Boltzmann conduction the ratio is $\hbar/\tau E_B$, similar to $1/k_F \ell$. However, this is only a first order solution. It is not clear why the phenomenological eq. (6) should be successful in the range where the second term dominates. Why do higher order terms not appear?

XII. FINAL COMMENTS

On studying "saturation" we have the double misfortune of a complicated physical parameter (conductivity) and a complicated system (A15 metals). However, we have the good fortune of a clean experiment where disorder can be tuned in a simple reversible way (temperature) and where the results, although unexplained, are simple (eq. 7). It is my prejudice that none of the proposed explanations, except the non-classical channel of sec. XI, have the requisite combination of universality and microscopic validity. If the theory of sec. XI is the correct explanation , we are still lacking a complete or adequately simple picture of the phenomenon.

There is a simple verbal description (V. Heine, private communication) of conduction in the Mott regime, even though there is still no simple mathematical description. Electrons in d-band compounds spend quite a lot of time circulating around the transition metal atom they happen to be associated with, before moving on to the next one. When disorder is weak, the process of motion from one unit cell to the next has phase coherence, and Bloch states are formed which carry currents (until interrupted by scattering.) However, if the disorder is high, phase memory may be lost within a unit cell (Ohkawa 1978) and the resulting eigenstate carry no current. Collisions now play a dual role of helping create the current through transitions between states, and preventing the current from continuing too far in the new state. Thus it is possible for ρ_{max} to be independent of the strength of the disorder.

ACKNOWLEDGMENTS

It is a pleasure to thank the many collaborators who
have worked with me on this subject, especially W.H. Butler,
B. Chakraborty, V. Heine, F.S. Khan, W.E. Pickett,
F.J. Pinski and M. Strongin.

REFERENCES

Allen, P.B. (1976). *Phys. Rev. Lett.* *37*, 1638.
Allen, P.B. (1978). *Phys. Rev.* *B17*, 3725.
Allen, P.B. and Chakraborty, B., (1978). *In* "Thermo-
    electricity in Metallic Conductors" (F.J. Blatt and
    P.A. Schroeder, eds.), p. 125. Plenum Press, New York.
Allen, P.B., Hui, J.C.K., Pickett, W.E., Varma, C.M., and
    Fisk, Z. (1976). *Sol. State Commun.* *18*, 1157.
Allen, P.B., Pickett, W.E., Ho, K.M., and Cohen, M.L. (1978).
    Phys. Rev. Lett. *40*, 1532.
Anderson, P.W. (1958). *Phys. Rev.* *109*, 1492.
Bader, S.D., and Fradin, F.Y. (1976). *In* "Superconductivity
    in d- and f-Band Metals" (D.H. Douglass, ed.), p. 567.
    Plenum Press, New York.
Bloch, F. (1928). *Z. Phys.* *52*, 555.
Brouers, F. (1978). *J. Phys. (Paris) Lett.* *39*, L323.
Chakraborty, B., Pickett, W.E., and Allen, P.B. (1976).
    Phys. Rev. *B14*, 3227.
Chakraborty, B., and Allen, P.B. (1978). *Phys. Rev.* *B18*,
    5225.
Chakraborty, B., and Allen, P.B. (1979). *Phys. Rev. Lett.* *42*,
    736.
Chen, A-B., Weiss, G., and Sher, A. (1972). *Phys. Rev.* *B5*,
    2897.
Cohen, R.W., Cody, G.D., and Halloran, J.J. (1967). *Phys.
    Rev. Lett.* *19*, 840.
Cote, P.J., and Meisel, L.V. (1978). *Phys. Rev. Lett.* *40*,
    1586.
Fisk, Z., and Lawson, A.C. (1973). *Sol. State Commun.* *13*,
    277.
Fisk, Z., and Webb, G.W. (1976). *Phys. Rev. Lett.* *36*, 1084.
Greig, D., and Morgan, G.J. (1973). *Phil. Mag.* *27*, 929.
Gurvitch, M., (1978). Ph.D. Thesis, S.U.N.Y. Stony Brook.
Holstein, T.D. (1964). *Ann. Phys. (N.Y.)* *29*, 410.
Langer, J.S., and Neal, T. (1966). *Phys. Rev. Lett.* *16*, 984.
Laubitz, M.J., Leavens, C.R., and Taylor, R. (1977).
    Phys. Rev. Lett. *39*, 225.

Mattheiss, L.F., Testardi, L.R., and Yao, W.W. (1978).
    Phys. Rev. B17, 4640.
McMillan, W.L. (1968). *Phys. Rev. 167,* 331.
Mooij, J.H. (1973). *Phys. Stat. Sol. (a)17,* 521.
Morton, N., James, B.W., and Wostenholm, G.H. (1978).
    Cryogenics 18, 131.
Mott, N.F. (1936). *Proc. Roy. Soc. (London) A156,* 368.
Mott, N.F., and Davis, E.A. (1971). "Electronic Processes
    in Non-Crystalline Materials," Oxford University Press,
    Oxford.
Nakayama, I., and Tsuneto, T. (1978). *Prog. Theor. Phys. 59,*
    1418.
Ohkawa, F.J. (1978). *J. Phys. Soc. Jpn. 44,* 1105 and 1112.
Pinski, F.J., Allen, P.B., and Butler, W.H. (1978), *Phys.
    Rev. Lett. 41,* 431; and to be published.
Pippard, A.B., (1955). *Phil. Mag. 41,* 1104.
Rossiter, P.L. (1977). *J. Phys. F: Metal Phys. 7,* 407.
Sham, L.J., and Ziman, J.M. (1963). *In* "Solid State Physics"
    (F. Seitz and D. Turnbull, eds.) vol. 15. Academic
    Press, New York.
Testardi, L.R., and Mattheiss, L.F. (1978). *Phys. Rev.
    Lett. 41,* 1612.
Thouless, D.J. (1974). *Phys. Reports (Phys. Lett. C) 13,* 93.
Ting, C.S., Snyder, T.M., and Williamson, S.J. (1979).
    J. Low Temp. Phys., in press.
Visscher, W.M. (1978). *Phys. Rev. B17,* 598.
Wiesmann, H., Gurvitch, M., Lutz, H., Ghosh, A., Schwarz, B.,
    Strongin, M., Allen, P.B., and Halley, J.W. (1977).
    Phys. Rev. Lett. 38, 782.
Wiesmann, H., *et al.,* (1978). *J. Low Temp. Phys. 30,* 513.
Woodard, D.W., and Cody, G.D. (1964). *Phys. Rev. 136,* A166.
Ziman, J.M. (1960). "Electrons and Phonons," Oxford
    University Press, Oxford.

DENSITY OF STATES
AND T_C OF DISORDERED A15 COMPOUNDS[1]

A. K. Ghosh and Myron Strongin
Brookhaven National Laboratory
Upton, New York

In this paper various data for the depression of T_C and
N(O) are presented for a wide class of A15 materials. The
question of disorder and the limits on T_C in these materials
are discussed.

I. INTRODUCTION

It is a well-established fact that the transition temperature, T_C, of the A15 superconductors is drastically altered by
particle irradiation. In this paper we primarily focus on the
question of the behavior of the density of states (DOS) at the
Fermi level, N(O) and its relationship to T_C and disorder,
where the disorder is characterized by the residual resistivity ρ_O. In detail we realize that ρ_O alone cannot exactly define T_C. Low temperature irradiations and anneals have shown
that T_C is not necessarily a unique function of ρ_O (Wiesmann,
1978; Brown *et al.*, 1978). However we feel that the approximation is adequate for the purposes of most discussions.

It was noted (Hulm and Blaugher, 1972) that for bcc alloys
of neighboring d-band transition metals, the electron phonon
coupling parameter λ as determined from McMillan's equation
was proportional to N(O). Since $\lambda/N(O) = \langle I^2 \rangle / M \langle \omega^2 \rangle$, it implied
that $\langle I^2 \rangle / M \langle \omega^2 \rangle$ (all the terms have their usual meaning) was
constant for a particular alloy system. More recently Varma
and Dynes, (1976) speculated that for the A15's, disorder would
not change the ratio $\lambda/N(O)$.

[1]*Work supported by DOE.*

305

The question of how disorder affects the DOS and hence T_c was considered by Crow *et al.*, (1969) in explaining the T_c depression of thin film niobium with disorder. In that paper they argued that the DOS of niobium $N(\epsilon)$ would be smeared out because of the decreasing mean free path ℓ or the decreasing electron lifetime τ owing to the energy uncertainty relation $(\Delta\epsilon)\tau \sim \hbar$, where $\Delta\epsilon$ is the width of the $N(\epsilon)$ peak in energy. This smearing effectively decreases $N(0)$ as τ decreases and thereby T_c drops. Experimental support for this viewpoint is found for the case of vanadium thin films where $N(0)$ was determined (Teplov *et al.*, 1976). For the case of the A15 where sharp structure in the DOS is expected, a similar smearing argument would indicate a decreasing $N(0)$ with disorder (either defect or thermally induced).

In this paper we shall present new data on disordered Nb_3Al, V_3Si, and Mo_3Ge. We briefly outline the way the DOS is estimated from critical field measurements and show how $N(0)$ varies with the residual resistivity ρ_o. Based on these data we shall examine the assumption of $\lambda \propto N(0)$, and also the question of sharp structure in high T_c A15's. Finally we discuss the question of disorder and the maximum attainable T_c among the transition metal based A15 compounds. We argue that recent estimates (Ho *et al.*, 1978) of the effect of phonon-disorder are too pessimistic.

II. EXPERIMENTAL DETAILS

Samples of A15 superconductors were prepared by e-beam co-evaporation onto heated single crystal sapphire substrates, which were masked to give an appropriate geometry for resistivity measurements. These samples were analyzed by x-ray diffraction and microprobed to determine the composition and the phases present. Table I lists the properties of the "as deposited" samples whose data are presented in the subsequent section. Defect disorder was introduced in the samples by means of 2.5 MeV α^{++}-particle irradiation at the BNL small Van de Graaff. After each irradiation the T_c, ρ_o, and the critical field slope $(dH_{c2}/dT)_{T_c} \equiv H'_{c2}(T_c)$ was measured.

In earlier publication (Ghosh *et al.*, 1978b) we have shown how one can estimate the electronic specific heat coefficient γ from the critical field slope near T_c. The expression for γ is:

$$\gamma = \frac{1}{\rho_o \eta}\left(2.2 \times 10^4 \; H'_{c2} - \frac{7.17 \times 10^{17} \; T_c \eta}{v_F^{*2}}\right) \tag{1}$$

TABLE I. Some Parameters of Evaporated A15

Material	$T_c(K)^d$	$\Delta T_c(K)$	$\rho_o(\mu\Omega cm)$	$H'_{C2}(KOe/K)$	η	$<\hbar\Omega_p>(ev)$
Nb_3Sn	17.88	0.09	14.6	21.6	1.21	--
Nb_3Ge	20.9	0.9	50.0	25.2	1.23	3.7^a
Nb_3Al	16.3	0.4	53.8	26.8	1.19	3.7
V_3Si	16.0	0.3	8.4	21.5	1.14	3.1^b
Mo_3Ge	1.45	0.01	0.14	--	1.02	--
Mo_3Ge^c	1.76	0.2	9.9	0.8	1.02	--

[a]P. B. Allen et al., 1978.
[b]L. F. Matheiss et al., 1978.
[c]Sputtered sample.
[d]$T_C(K)$ indicates midpoint of resistive transition.

where H'_{C2} is in KOe/K, ρ_o is in $\mu\Omega$-cm, γ is in ergs/cm³-K², v_F^* is the renormalized averaged Fermi velocity in cm/sec, and η is a strong-coupling correction (Rainer and Bergmann, 1974) that depends on $T_C/<\omega>$, and is of the order of unity. It was shown (Ghosh et al., 1978) that for Nb_3Sn, making the assumption that either $<N(0)v_F>$ = constant or $<N(0)v_F^2>$ = constant leads essentially to similar results. In this paper we have assumed that $<N(0)v_F^2>$ is held constant as the sample is damaged. This implies that the plasma energy Ω_p^2 remains constant, which is a fairly good assumption since in the A15's its variation with energy near the Fermi surface is very much weaker than either $N(\epsilon)$ or v_F (Mattheiss et al., 1978; Pickett et al., 1979). Then eq. (1) reduces to

$$\gamma = \frac{2.2 \times 10^4 \ H'_{c2}}{\eta\rho_o} \left[1 + \frac{4.844 \ T_c(1+\lambda)}{\rho_o\Omega_p^2} \right]^{-1} \qquad (2)$$

where Ω_p is in ev. The term in the square brackets represents the correction term which becomes important when $\ell \gtrsim \xi_o$, where ξ_o is the B.C.S. coherence length. The values of Ω_p used in this paper were taken from the literature and is listed in Table I. Once γ is calculated from the data, the value of $N(0)=(3/2)(\gamma/\pi^2k_B^2)/(1+\lambda)$ is obtained, λ being calculated from the measured T_C using either the Allen-Dynes equation (Allen and Dynes, 1975) with f_1, f_2 equal to unity or the McMillan equation (McMillan, 1968).

III. RESULTS

A. T_C Variation with Disorder

The systematics of T_C depression for the case of Nb_3Sn and Nb_3Ge has been studied earlier (Ghosh et al., 1978a). In Fig. 1(a) the T_C behavior of α^{++}- and electron irradiated Nb_3Sn is shown (O) as a function of the residual resistivity ρ_o. That of Nb_3Al and V_3Si are also shown in Fig. 1(a). For the case of Mo_3Ge, irradiation is seen to increase T_C (Gurvitch et al., 1978) to saturated values >6 K, as shown in Fig. 1(b). We will return to this point later in the discussion.

B. Estimate of N(O)

To examine the behavior of N(O) as a function of disorder, we have used eq. (2) to determine γ from H'_{c2} and ρ_o measurements.

1. *Nb_3Sn.* In the unirradiated state the specific heat coefficient $\gamma \sim 50 \pm 2$ mJ/g-mol K^2. Using $\langle \omega_{log} \rangle \sim 125$ K, (Allen and Dynes, 1975) $\mu^* \sim 0.1$, λ is calculated from the Allen-Dynes equation and N(O) is estmated to be ~ 0.91 states/ev-atom. The data of Ghosh et al., (1978b) is replotted in Fig. 2(a) which shows the behavior of N(O) as a function of ρ_o.

It was shown (Ghosh et al., 1978b) that the behavior of N(O) with T_C is quantitatively explained by taking $\lambda \propto N(O)$ assuming $\langle \omega_{log} \rangle$ does not change with disorder. This indicates that $\langle I^2 \rangle / M \langle \omega^2 \rangle \sim 2.1-2.3$ remains relatively unchanged as T_C is changed by a factor of ~ 4.5. In the heavily irradiated state ($\rho_o > 130$ $\mu\Omega$-cm), the data do seem to indicate a slight drop in the ratio to $\sim 1.7-1.8$. This change could be attributed to a 5-10% increase in $\langle \omega \rangle$ (or θ_D) or a 5-10% change in $\langle I^2 \rangle$, some of which is expected owing to an increase in lattice parameter of the irradiated sample (Butler, 1977).

2. *Nb_3Al.* Using $\Omega_p \sim 3.7$ ev, the value of γ was calculated to be $\sim 30 \pm 2$ mJ/g-mol K^2. Using $\langle \omega_{log} \rangle \sim 153$ K (Allen and Dynes, 1975) and $\mu^* \sim 0.1$, the calculated $\lambda \sim 1.4$ indicates a value of N(O) ~ 0.7 states/ev-atom in the unirradiated state. With increasing disorder N(O) decreases and its behavior with ρ_o is shown in Fig. 2(b)...(+). In the case of Nb_3Al, the data show that the ratio $\lambda/N(O)$ changes from an initial value of 2.0-2.4 to $\sim 1.4-1.6$ as the T_C is changed from 16 K to <4 K. The uncertainty in the values arises from the uncertainty in the experimental value of γ and the value of μ^* (0.1-0.13) that one uses in calculating λ from the expression for T_C.

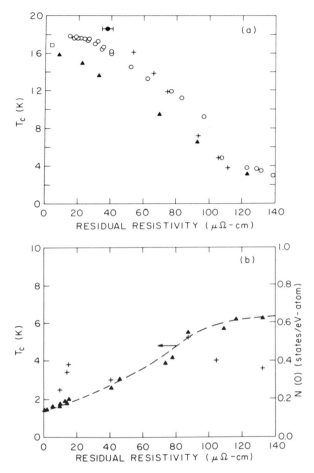

FIGURE 1. *Behavior of T_c as a function of residual defect resistivity ρ_o. (a) shows T_c vs ρ_o for Nb_3Sn (O), V_3Si (▲), V_3Si single crystal data (□), Nb_3Al (+), and Nb_3Al bulk (●). (b) shows data for Mo_3Ge: (△) T_c and (+) $N(O)$.*

3. Nb_3Ge. The γ value was calculated using $\Omega_p \sim 3.7$ ev. It is not crucial what value one use for Ω_p, since the contribution of the correction term at $\rho_o \sim 50$ $\mu\Omega$-cm is quite small $\sim 5\%$. The unirradiated value of $\gamma \sim 26\pm1$ mJ/g-mol K^2. λ was calculated from the Allen-Dynes equation with $\langle \omega_{log} \rangle \sim 153$ K, from which $N(O)$ was estimated to be ~ 0.5 states/ev-atom. The data of Wiesmann et al., (1978) is plotted (+) here in Fig. 2(a) to compare with the data of Nb_3Sn.

The high T_c Nb_3Ge behaves quite differently from $Nb3Sn$. It is found that the value of $\langle I^2 \rangle / M \langle \omega^2 \rangle$ changes drastically from

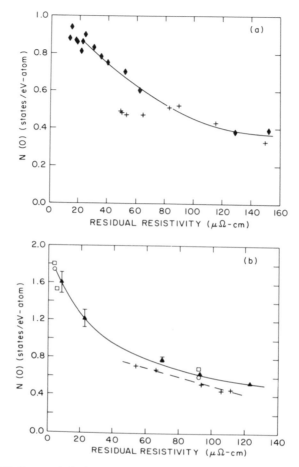

FIGURE 2. $N(0)$ behavior as a function of ρ_0.   (a) Nb$_3$Sn
(◆), Nb$_3$Ge (+).   (b) shows $N(0)$ for Nb$_3$Al (+), and V$_3$Si (▲,□,0).
See text for details.   Lines are drawn only to indicate trend
of data.

~3.7–3.9 to 1.8–1.9 as T$_C$ is changed from 21 K to ~11 K.   Below
that the ratio remains unchanged to a T$_C$ ~4 K.   Tsuei (1978)
has interpreted this to mean that for Nb$_3$Ge <I^2> changes a
factor ~4 as the T$_C$ is depressed to ~4 K.   Although this may be
the actual trend in Nb3Ge, we would like to point out that the
starting material may not be a homogeneous high T$_C$ Nb$_3$Ge but
may contain some fraction of disordered A15 as has been
measured by EXAFS (Brown *et al.*, 1977).

4. V_3Si. For this sample Ω_p was chosen to be $\sim$3.1 ev which
is intermediate to the calculated and the measured values for
V_3Si. Under this assumption eq. (2) gave a value of $\gamma\sim56\pm4$
mJ/g-mol K^2. Using McMillan's equation with $\theta_D \sim$400 K and μ^*
$\sim$0.13 (Viswanathan and Caton, 1978) λ is calculated to be $\sim$0.9
and hence N(0) is estimated to be $\sim$1.6 states/ev-atom. With
irradiation, the γ value drops rapidly and the behavior of N(0)
with ρ_o is shown ($\blacktriangle$) in Fig. 2(b). Also shown are the heat
capacity data (Viswanathan and Caton, 1978) ($\square$) on single
crystal V_3Si before and after neutron irradiation and the data
(o) on the same crystal using the reported values of $H'_{c2}(T_c)$
and ρ_o (Guha et al., 1978). The trend in the data is remarkably similar for the different samples and for the different
methods of measurement.

V_3Si shows a rather high value for the N(0) in the unirradiated state and decreases very rapidly with increasing ρ_o.
In this case we find that contrary to niobium compounds the
initial value of $<I^2>/M<\omega^2>$ $\sim$0.5 is low and rises to $\sim$1 for ρ_o
$>$65 $\mu\Omega$-cm and then remains fairly constant at the value for
increasing values of ρ_o.

5. Mo_3Ge. For this low T_c A15, γ has been calculated from
eq. (2) where the square bracket has been taken $\approx$1.0. This is
always the case for Mo_3Ge since ℓ<<ξ_0 for ρ_o $>$5 $\mu\Omega$-cm. The λ
has been calculated from McMillan's equation assuming $\theta_D \sim$400 K
and $\mu^* \sim$0.1. The values of N(0) thus estimated (+) are shown
in Fig. 1(a). While there is considerable scatter in the data,
the overall trend is one of increasing N(0) with increasing T_c,
λ/N(0) staying relatively constant at $\sim$1.3-1.6.

IV. DISCUSSION AND CONCLUSIONS

In this section we summarize the behavior of N(0) and
λ/N(0) with disorder and we discuss the question of the maximum
limit on T_c.

A. Behavior of N(0) with Disorder

For the A15 class of materials, it is expected that the DOS
will have structure on a scale much smaller than that for transition metals like Nb. Recent band structure calculations indicate that indeed the DOS has fine structure on the scale
$\sim$0.05-0.1 ev (Pickett et al., 1979; Van Kessel et al., 1978).
If the Fermi level falls near a peak in the DOS, then as T is

decreased due to increased scattering (either phonons or defects), the DOS is "smeared out" and $N(0)$ will drop as a consequence of the increasing width of the peak as per the uncertainity relation $(\Delta\epsilon)\tau\sim\hbar$. Since $\tau=4\pi/(\Omega_p^2\rho_0)$, then as ρ_0 increases, $\Delta\epsilon$ will increase. Typically, for $\Omega_p \sim 3.7$ ev, $(\Delta\epsilon) \sim 200$ K for a $\rho_0 \sim 10 \ \mu\Omega$-cm. In the case of Nb_3Sn, the data in Fig. 2(a) do not clearly indicate whether there is a sharp structure in DOS. If a fine structure does exist its $\Delta E > \theta_D$ which is ~ 250-280 K. However, for V_3Si, the data in Fig. 2(b) indicate a rather dramatic decrease in $N(0)$ as ρ_0 increases from about 4 to $\sim 30 \ \mu\Omega$-cm. The data show that for V_3Si at least, structure in the DOS ~ 100 K is possible. We also note that most of the structure is smeared out by a $\rho_0 \sim 65 \ \mu\Omega$-cm. One consequence of a structure which is finer than θ_D is that the $N(0)$ appearing in λ or T_c will have to be averaged over $\theta_D \sim 400$-500 K (Nettle and Thomas, 1977) which translates to an effective $\rho_0 \sim 25 \ \mu\Omega$-cm. This would indicate that the actual $<I^2>/M_\omega^2>=$ $=\lambda/<N(0)>\theta_D$ would be larger in the high T_c state. Recently it was shown by Testardi and Mattheiss, (1978) that by using the calculated band structure parameters the trend with disorder of $N(0)$, $<v_F>$ and $<\Omega_p^2>$ could be calculated by averaging these parameters over energy ranges $E_b=\hbar/\tau$. They calculated these parameters as a function of ρ, (this could be either defect or thermally induced), for a number of bcc alloys and V_3Si. The calculated behavior of $N(0)$ for V_3Si is very similar to Fig. 2(b). Moreover, by taking $N(0)$ averaged over θ_D, they showed that the experimental T_c behavior of V_3Si with ρ_0 could be accounted for by assuming $\lambda\propto<N>$.

In the case of Mo_3Ge, the Fermi level is most probably located in the region of low DOS between the bonding and antibonding peaks. As the material is disordered and the peaks smear out, it is expected that the $N(0)$ of the low DOS material will increase while that of a high DOS like $Nb3Sn$ will drop. The situation is analogous to the case of transition metals where the T_c of Nb drops on disordering while that of Mo increases (Collver and Hammond, 1973).

B.   Variation of $\lambda/N(0)$ with Disorder

The data show that for A15 materials with high DOS the behavior of T_c is largely governed by $N(0)$ and its variation with disorder. This is the case for both the well-ordered A15's Nb_3Sn and V_3Si and also the low T_c Mo_3Ge. For Nb_3Al the data indicate a decrease in $<I^2>/M_<\omega^2>$. However what causes this cannot be answered without some knowledge about the dependence of $<\omega^2>$ with disorder. The behavior is in the right direction since a decrease in $N(0)$ and $<I^2>$ would affect the phonon frequencies, making them higher.

The Nb_3Ge data show that $<I^2>/M<\omega^2>$ changes a factor of 2 from an initial high value of $\sim 3.7-3.9$. This could mean either (1) that the electron-phonon interaction in the metastable Nb_3Ge is substantially higher than that in Nb_3Sn or V_3Si, as has been mentioned by Tsuei (1978) or (2) that even for Nb_3Ge, $\lambda/N(0)$ behavior is similar to Nb_3Sn but that the actual $N(0)$ changes are incorrectly estimated owing to poor sample quality. The implication of a large $<I^2>$ is, that for A15 material the structural instability (or metastability) may be tied to a strong electron-phonon interaction, which in effect also softens the phonons by renormalization.

C. *Maximum Attainable T_C in A15's*

There is no *a priori* reason why λ and hence T_C cannot be higher than currently found. Recently, it was suggested (Ho et al., 1978) that temperatures ~ 25 K effectively acts to decrease the λ (T=0) value by the electron lifetime broadening of DOS peaks due to electron-phonon scattering. They estimate a reduction in $T_C \sim 2°K$ for Nb_3Ge. Our data on T_C and $N(0)$ vs ρ_0 show that this ΔT_C is an overestimate if one assumes that temperature disorder and defect disorder are equivalent in the sense that the smearing of $N(0)$ depends on τ which is affected by both kinds of scattering. Our data (Ghosh et al., 1978a) show that for a ρ_0 change of about 2.5 $\mu\Omega$-cm T_C changes by about 0.5°K. This change in ρ is equivalent to $\rho_{thermal}$ at 25°K. Hence we conclude that the lowering of T_C would be about 0.5°K at 25°K due to thermal smearing of $N(0)$. Furthermore, for real materials like Nb_3Ge or possibly Nb_3Si the major contribution to the resistivity at low temperature is not from ρ_{el-ph}, but from the defect disorder ρ_0. Hence the DOS smearing is more strongly dependent on the residual defect scattering than on phonon disorder.

In conclusion, it should be mentioned that $N(0)$ appears to approximately correlate with the residual defect structure. For well-ordered compounds like V_3Si and Nb_3Sn, $N(0)$ is high and the $\rho_0 \sim 2-10$ $\mu\Omega$-cm (Taub and Williamson, 1974) and ~ 10 $\mu\Omega$-cm respectively. $N(0)$ for Nb_3Al is $\sim 40\%$ that of Nb_3Sn and its $\rho_0 \sim 35-40$ $\mu\Omega$-cm (Webb et al., 1977), while Nb_3Ge has a low $N(0)$ and $\rho_0 \sim 40-50$ $\mu\Omega$-cm (Harper et al., 1975). These materials are disordered as evidenced from x-ray and neutron diffraction studies (Sweedler et al., 1978), however, their T_C is high and the λ values are all ~ 1.7. We speculate that for niobium based A15 materials whose Fermi energy lies near a peak in the DOS, then as one increases the electron phonon coupling strength, the structural instability inherent in the high T_C materials (Matthias, 1970) causes the defect structure to increase thereby reducing $N(0)$. We feel this is a possible mechanism by which the λ values are limited. In Nb_3Ge, the trade off in stability and $N(O)$ may be just optimum to give a $T_C \sim 23°K$. Based on this speculation, the chances of Nb_3Si (which will have a higher ρ_0 then Nb_3Ge) being a high T_c superconductor may be a mere theoretical possibility.

ACKNOWLEDGMENTS

The work described here was done in collaboration with many other colleagues at Brookhaven. Dr. H. Lutz prepared most of the samples, Dr. H. Wiesmann did the earlier work on Nb_3Ge, Dr. M. Gurvitch was closely connected with most of the work on Nb_3Sn and as usual O. F. Kammerer provided expert metallurgical support.

REFERENCES

Allen, P. B. and Dynes, R. C. (1975). Phys. Rev. B 12, 905
Allen, P. B., Pickett, W. E., Ho, K. M., and Cohen, M. L. (1978). Phys. Rev. Letts. 40, 1532.
Brown, G. S., Testardi, L. R., Wernick, J. H., Hallack, A. B., and Geballe, T. H. (1977). Solid State Commun. 23, 875.
Brown, B. S., Birtcher, R. C., Kampwirth, R. T., and Blewitt, T. H. (1978). J. of Nucl. Mater. 72, 76.
Butler, W. H. (1977). Phys. B15, 5627.
Collver, M. M. and Hammond. R. H. (1973). Phys. Rev. Lett. 30, 92.
Crow, J. E., Strongin, M., Thompson, R. S., and Kammerer, O. F. (1969). Phys. Lett. 30A, 161.
Dynes, R. C. and Varma, C. M. (1976). J. Phys. F6, L215.
Ghosh, A. K., Wiesmann, H., Gurvitch, M., Lutz, H., Kammerer, O. F., Snead, C. L. Jr., Goland, A., and Strongin, M. (1978a). J. of Nucl. Mater. 72, 70.
Ghosh, A. K., Gurvitch, M., Wiesmann, H., and Strongin, M. (1978b). Phys. Rev. B18, 6116.
Guha, A., Sarachik, M. P., Smith, F. W., and Testardi, L. R. (1978). Phys. Rev. B18, 9.
Gurvitch, M., Ghosh, A. K., Gyorrfy, B. L., Lutz, H., Kammerer, O. F., Rosner, J. S., and Strongin, M. (1978). Phys. Rev. Letts. 41, 1616.
Harper, J. M. E., Geballe, T. H., Newkirk, L. R., and Valencia, F. A. (1975). J. of Less Common Metals 43, 5.
Ho, K. M., Cohen, M. L., and Pickett, W. E. (1978). Phys. Rev. Letts. 41, 815.

Hulm, J. K. and Blaugher, R. D. (1972). In "Superconductivity
    in d- and f-Band Metals" (D. H. Douglass, ed.), p. 1, AIP,
    New York.
Mattheiss, L. F., Testardi, L. R., and Yao, W. W. (1978).
    Phys. Rev. B17, 4640.
Matthias, B. T. (1970). Comments Solid State Phys. 3, 93.
McMillan, W. L. (1968). Phys Rev. 167, 331.
Nettle, S. J. and Thomas, H. (1977). Solid State Commun. 21,
    683.
Pickett, W. E., Ho, K. M., Cohen, M. L. (1979). Phys. Rev. B19,
    1734.
Rainer, D. and Bergmann, G. (1974). J. of Low Temp. Phys. 14,
    501.
Sweedler, A. R., Cox, D. E., and Moehlecke, S. (1978). J. of
    Nucl. Mater. 72, 50.
Taub, H. and Williamson, S. J. (1974). Solid State Commun. 15,
    181.
Tepov, A. A., Mikheeva, M. N., Golyanov, V. M., and
    Gusev, A. N. (1976). Sov. Phys. JETP 44, 587.
Testardi, L. R. and Mattheiss, L. F. (1978). Phys. Rev. Letts.
    41, 1612.
Tsuei, C. C. (1978). Phys. Rev. B18, 6385.
Varma, C. M. and Dynes, R. C. (1976). In "Superconductivity in
    d- and f-Band Metals" (D. H. Douglass, ed.), p.507, Plenum,
    New York.
Viswanathan, R. and Caton, R. (1978). Phys. Rev. B18, 15.
Van Kessel, A. T., Myron, H. W., and Mueller, F. M. (1978).
    Phys. Rev. Letts. 41, 181.
Webb, G. W., Fisk, Z., Engelhardt, J. J., and Bader, S. D.
    (1977). Phys. Rev. B15, 2624.
Wiesmann, H. (1978). Ph.D. thesis.
Wiesmann, H., Gurvitch, M., Ghosh, A. K., Lutz, H.,
    Kammerer, O. F., and Strongin, M. (1978). Phys. Rev. B17,
    122.

RESISTIVITY OF A15 COMPOUNDS[*]

Michael Gurvitch[+]

Brookhaven National Laboratory
Upton, New York

There is more to A15s than high critical fields and
transition temperatures. In particular, resistivity studies
of A15s, although related to superconducting properties, have
a life of their own. As it is well known, most properties of
A15s are rather anomalous, and so is the resistivity (Sarachik
et al. 1963; Woodard and Cody, 1964; Webb *et al.*, 1977;
Gurvitch *et al.*, 1979). Yet anomalous as it appears when
compared to normal, well-behaved metals, the form of the
resistivity of A15s is not unique to these compounds. In fact
both high temperature "saturation" of the resistivity curve
$\rho(T)$ and low temperature proportionality to T^2 are found in
other systems, as will be discussed below. Hence there is a
need for sufficiently general explanations. I believe that at
present the behavior of $\rho(T)$ at high temperatures is understood better than at low temperatures.

I. HIGH TEMPERATURES (T $\gtrsim$ 100 K): SATURATION

The most obvious feature of the $\rho(T)$ curve of all high T_c
A15s is the strong bending towards the temperature axis.
Resistivity tends to "saturate" to a value of the order of
$\sim$ 150 $\mu\Omega$cm. The reader will find a comprehensive discussion
of the saturation phenomena in Allen's paper in this volume. So
I will make only a few points on saturation before moving on
to discuss the low temperature behavior of $\rho(T)$.

---

[*]*Preparation of the paper and part of the analysis were
done in Bell Laboratories.*
[+]*Present address: Bell Laboratories, Murray Hill, New Jersey*

One way to study resistivity behavior of A15s is to
introduce controllable amounts of disorder by irradiating
films of these materials with fast particles (Poate *et al.*,
1975; Lutz *et al.*, 1976; Testardi *et al.*, 1977; Wiesmann *et
al.*, 1977; Gurvitch *et al.*, 1978). In Fig. 1a our data on
α-irradiated films of Nb_3Sn are presented. One can see that the
curves have a tendency to merge or saturate at high temperatures and because of that Matthiessen's rule is disobeyed.
This behavior is by no means unique to A15 compounds: one
finds in the literature much evidence that many metallic
systems and elements with comparably high values of ρ resemble
A15s in their resistivity behavior. For example, on Fig. 1b
data on Ti – Aℓ alloys (Mooij, 1973) are presented along with
the data for some elements: Np (Lee, 1961) on Fig. 1c and
self-irradiated Pu (King *et al.*, 1965) on Fig. 1d. Similar
"saturation" behavior was recently found in the ternary compound $ErRh_4B_4$ (Rowell *et al.*, this volume). Numerous examples
can be also found in the work of Fisk and Lawson (1973).
Unfortunately, in the latter paper resistivity is given in
arbitrary units, normalized to the room temperature values, so
it seems at first sight that the occurrence of the nonlinear
resistivity correlates strongly with superconductivity, which
was the conclusion of the authors. I suspect strongly that
the real correlation is with the value of resistivity, which
is simply higher in superconductors (i.e., materials with high
electron-phonon coupling λ) than in nonsuperconductors,
creating the nonlinear "saturation" effect. Should superconductivity be suppressed by the presence of magnetic moments or
for any other reason, resistivity, provided it is high enough,
would still deviate from linearity, like in nonsuperconducting
Np and Pu (Fig.1c,d). One finds that in metallic systems resistivities rarely exceed values $\sim$ 200–300 $\mu\Omega$cm. As was pointed
out on several occasions (Fisk and Webb, 1976; Ioffe and Regel,
1960; Mott, 1974; Mooij, 1972; Allen, 1976) on a simple intuitive level this general behavior can be understood by noting
that the mean free path (m.f.p.) ℓ of electrons calculated from
the conventional theory, in these systems comes out to be of
the order of interatomic spacing a: presumably the natural lower
limit for ℓ. In a classical theory which is not concerned with
this problem, at high temperatures the m.f.p. $\ell_{e-ph} \propto \frac{const}{T}$,
so that $\ell_{e-ph} \to 0$ as $T \to \infty$. The easiest way to limit ℓ_{e-ph}
to a lowest value of a is to write $\ell_{e-ph} \propto \frac{const}{T} + a$, so
that $\ell_{e-ph} \to a$ as $T \to \infty$. Hopefully this simple form may be a
reasonable first approximation. (Similar forms can be written
for $\ell_{defects}$ and for ℓ_{total}). From that immediately follows
the "parallel resistor" formula

$$\rho^{-1}(T) = \rho_{ideal}^{-1}(T) + \rho_{max}^{-1} \tag{1}$$

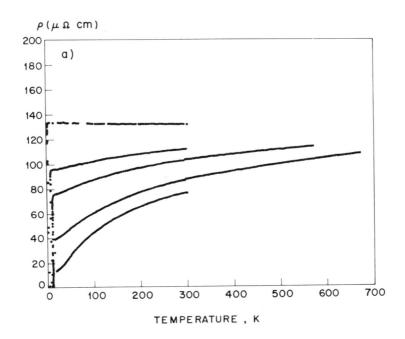

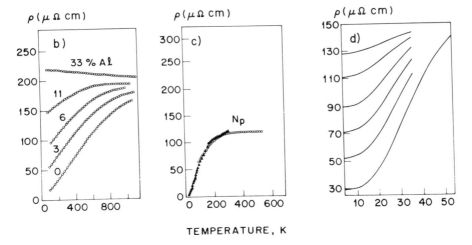

FIGURE 1a. Series of resistivity curves taken on α-irradiated Nb₃Sn. b. Resistivity of Ti-Aℓ alloys (Mooij, 1972), numbers on the curves indicate % of Aℓ. c. Resistivity of Np (Lee, 1961). d. Series of resistivity curves taken on self-irradiating Pu (King et al., 1965).

which turned out to describe the data fairly well (Wiesmann
et al., 1977). The simplistic argument given above is by no
means a proof. The reader is again referred to Allen's paper
for a discussion of the "parallel resistor" mechanism.

Other explanations were proposed in the past for the
resistivity saturation. One of them is Mott's Fermi smearing
(Mott, 1936). This approach was applied by Cohen, Cody and
Halloran (1967) and Bader and Fradin (1976) to A15s. For us it
is important now that in the framework of these ideas, if the
density of states $N(E_F)$ increases with disorder or with temperature smearing, then the resistivity should rise faster than
linear, and vice versa. Now, it turns out that in the low T_c
A15 material Mo_3Ge the density of states increases with disorder, as is evident from the increase of the T_c with disorder
(Gurvitch *et al.* 1978). In fact the density of states itself
was estimated from magnetic measurements and was indeed found
to be increasing. Figure 2 shows our data on two α-irradiated
samples of Mo_3Ge fitted to the parallel resistor formula with
$\rho_{max} = 120$ $\mu\Omega cm$. The slope of $\rho_{id}(T)$ at high temperatures is
assumed to be proportional to λ and λ is calculated from measured values of T_c, assuming that the Debye temperature is a constant. So here we have a situation when $N(E_F)$ is rising, the
slope of $\rho_{id}(T)$ is rising, and yet saturation overcomes this
effect and forces damaged Mo_3Ge to behave like all the other
metallic systems with high values of ρ - in the direct contra-
diction to the predictions of the Fermi smearing approach.

II. LOW TEMPERATURES $(T_c < T \stackrel{\sim}{<} 50K)$: T^2 DEPENDENCE

At low temperatures resistivity of A15s continues to be
anomalous in a sense that it does not follow predictions of
the classical theory of conductivity (see Ziman, 1960).
Instead of being proportional to T^5 or T^3 as is the case for
normal and transition metals, resistivity of high T_c A15s
Nb_3Sn, Nb_3Al, Nb_3Ge and V_3Si was found to be proportional to
T^2:

$$\rho(T) - \rho(o) = AT^2 \quad \text{(Webb } et\ al.\text{, 1977; Marchenko, 1973)} \quad (2)$$

We studied the resistivity of these compounds and of Mo_3Ge
as a function of temperature and, in addition, as a function
of disorder which was introduced by electron and α-particle
irradiations. Our data were taken with the use of a computer-
ized system which permitted us to obtain up to several
hundreds of points along each resistivity curve and to determine the power index n in the temperature dependence

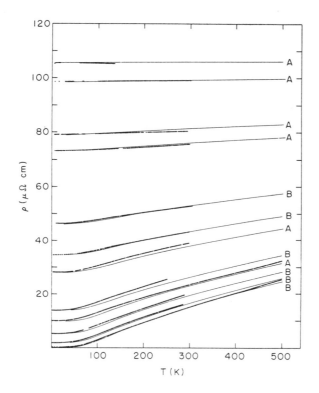

FIGURE 2.   Resistivity of two Mo₃Ge samples: A, sputtered film; B, electron-beam-evaporated film.   Solid lines are fitted by Eq. (1) using $\rho_{max} = 120$ μΩcm and $\rho_{id} = \rho_{oid} + L \cdot T \cdot G(\frac{\theta_D}{T})$, where $G(\frac{\theta_D}{T})$ is the Gruneisen function and $\theta_D = 435K$. Coefficient L is taken to be proportional to the electron-phonon coupling λ (Gurvitch et al., 1978).

$\rho(T) - \rho(o) = AT^n$ with high precision.   The method of data analysis that we used is a modification of the well-known way of determining n by plotting resistivity vs. T^z and finding the index z = n which produces the straight line plot. It is rather difficult to judge graphically just how well a straight line fits the data.   Instead we are connecting two points on the ρ vs. T^z plot with the straight line and taking deviations of the ρ-curve from that straight line (Fig. 3b). These deviations then can be plotted separately (Fig. 3c). The value of z which produces smallest deviations is the closest to the true index n.   The deviation plot (DP) method, as we call it, is discussed in more detail elsewhere (Gurvitch, 1978; Gurvitch *et al.*, 1979).   Typical DP for

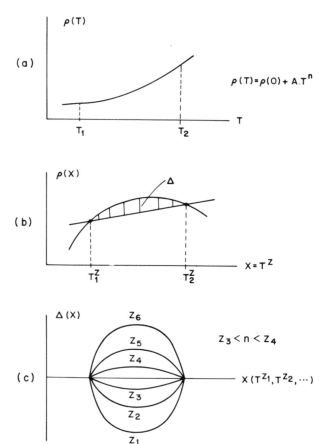

FIGURE 3. Explanation of the deviation plot. a. Resistivity curve between temperatures T_1 and T_2. b. The same curve plotted vs. T^z between T_1^z and T_2^z. Straight line in these coordinates is drawn between points (T_1^z, ρ_1) and (T_2^z, ρ_2). c. The difference (deviation) of the curve $\rho(T^z)$ from the straight line plotted separately for different values of z. Ends of all curves are kept at the same points for convenience.

unirradiated samples of Nb_3Sn, Nb_3Al and Mo_3Ge are shown on Figs. 4a,d,e. From these and other DP we can estimate the indices n (Table 1).

In the case of Nb_3Ge only one sample was analyzed and we do not insist on the obtained value of n = 2.3. (If it is indeed n = 2.3, perhaps it has something to do with the

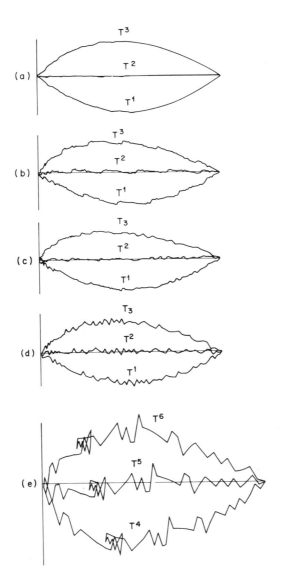

FIGURE 4. Low temperature data presented on deviation
plots. Data points are connected by straight lines. a. As-
made Nb₃Sn, ρ(o) = 11.4 μΩcm, 18.4K < T < 28.1K; b. α-irradiated
Nb₃Sn, ρ(o) = 75.7 μΩcm, 16.3K < T < 33.4K; c. α-irradiated
Nb₃Sn, ρ(o) = 95.7 μΩcm, 14.9K < T < 43.6K; d. As-made Nb₃Aℓ,
ρ(o) ≈ 55 μΩcm, 18.6K < T < 35.8K; e. As-made Mo₃Ge,
ρ(o) = 1.9 μΩcm, 14K < T < 34.5K.

TABLE 1

Compound	n	Temperature Range $T_1 < T < T_2$
Nb_3Sn	2.0 ± 0.05	$18.4 < T < 28 - 30$
Nb_3Al	2.0 ± 0.05	$18.6 < T < 35 - 40$
V_3Si	2.0 ± 0.2	$17\ \ \ < T < 26 - 28$
Nb_3Ge	2.3 ± 0.2	$21\ \ \ < T <\ \ \sim 35$
Mo_3Ge	5.0 ± 0.1	$10\ \ \ < T <\ \ \sim 35$

inevitable presence of a second phase with low T_c in Nb3Ge (Brown *et al.*, 1977). Apart from that, all the high T_c A15s seem to have an integer power index n = 2.0.

However, the low T_c A15 material Mo3Ge has n = 5)Fig. 4e), behaving as if it were a good Bloch–Gruneisen metal (see Ziman, 1960). Indices n remain constant from T_1 T_c to T_2 (see Table 1), at which point they start to deviate from their values. The way the DP method was used, values of T_2 could be determined only approximately by trying DP for different upper boundaries in temperature. Furthermore, a DP analysis of the data show that n stays equal to 2 with increasing $\rho(o)$ for both electron and α–particle irradiations. Figure 4b and 4c illustrate that point for α–irradiated Nb3Sn with $\rho(o)$ up to 95.7 $\mu\Omega$cm. At the same time it was found that the region in temperature in which n = 2 becomes wider as $\rho(o)$ goes up. Figure 5 summarizes the T_2 dependence on the value of $\rho(o)$ for different A15s. Let us leave any speculation about that behavior until the discussion. Finally, we studied the behavior of the coefficient A in (2) as a function of disorder (Fig. 6). The decrease in the value of A with increasing $\rho(o)$ was very reproducible for different samples, as is evident from Fig. 6.

In summary, we found that for high T_c A15s n fairly precisely equals 2 in the temperature range $T_c \gtrsim T_1 < T < T_2$, where T_2 is material and $\rho(o)$ dependent, that n = 2 even for very highly damaged samples, and that T_2 increases and coefficient A decreases with disorder in a consistent manner.

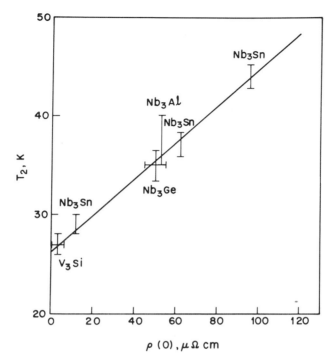

FIGURE 5.   Dependence of the upper limit in temperature T_2 to which resistivity is proportional to T^2 on the residual resistance $\rho(o)$ for different A15s.   Nb3Sn sample was α-irradiated.

III. DISCUSSION

a.   The Problem of n = 2

Webb *et al.* (1977) observed that both resistivity and lattice heat capacity (data on heat capacity of Knapp *et al* (1976) of high T_c A15s follows a T^2 dependence at low temperatures. They argued that it makes it plausible that both dependences result from the particular phonon properties of A15s, namely that the phonon density of states $F(\omega)$ of the A15s differs from the Debye approximation $F(\omega) \propto \omega^2$ in such a way as to reduce the power index n = 3, which occurs in Debye's model for the lattice heat capacity $C_L(T)$ and Wilson's model for the resistivity $\rho(T)$, to the observed value of n = 2.

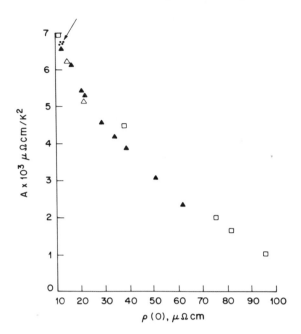

FIGURE 6.   A vs. ρ(o) for three Nb₃Sn samples:
□ - *α-irradiations, 1st sample.*
□ - *α-irradiations, 2nd sample.*
▲ - *electron irradiations.*
Dots indicated by the arrow: scatter in the data for the
unirradiated reference part of the electron-irradiated sample.

In the light of the data presented in the last section,
we think this explanation is not plausible. First, if the
observed value of n is due to the pecularities of a phonon
spectrum, then it is hard to understand why n is an integer,
and in more than one material. Secondly, if the explanation
of Webb *et al.,* is correct, then n should change as the phonon
spectrum is altered. However even in highly disordered
samples with $\rho(o) \sim 96$ μΩcm n remains the same.

Let us look more closely at the analysis. The formulae
for $\rho(T)$ and $C_L(T)$ are written in the forms

$$\rho(T) \propto \int \frac{x}{[\sinh x]^2} F(\omega)\, d\omega \tag{3}$$

$$C_L \propto \int \left(\frac{x}{\sinh x}\right)^2 F(\omega)\, d\omega \tag{4}$$

where $x = \frac{\hbar\omega}{2K_BT}$. In the absence of the actual $F(\omega)$ Webb *et al.* deduce it from the $\alpha^2F(\omega)$ obtained from superconducting-tunneling data. The model $F(\omega)$ is inserted in (3) and (4) and the results for $C_L(T)$ and $\rho(T)$ are plotted as a function of T^2 between 20 and 50K. For both $C_L(T)$ and $\rho(T)$ straight line plots are obtained, (at least up to 40K), which are reproduced here on Fig. 7a. Notice the negative intercepts, which were interpreted by Webb *et al.* as a manifestation of the fact that there is no "T^2 law" which would go to the lowest temperatures. Now, (3) and (4) are generalizations of the Debye's and Wilson's formulae for C_L $_{Debye}(T)$ and $\rho_{Wilson}(T)$ with $F(\omega)$ instead of the Debye ω^2 (Allen, 1971). It is instructive to compare the behavior of C_L and ρ as calculated by Webb *et al.* with the behavior of C_L $_{Debye}$ and ρ_{Wilson}. Following Webb *et al.*, we used the Debye temperature $\theta_D = 265K$. Results of our calculation are plotted between 20 and 50K on Fig. 7b. It can be seen that the two graphs are rather similar. One can hardly base an argument on the difference in shape of C_L and ρ curves on these two graphs.

The reason why C_L $_{Debye}$ and especially ρ_{Wilson} fit reasonably well to the T^2 between 20 and 50K is that the temperature range taken is fairly wide and T^2 rather large: with the value of $\theta_D = 265K$ we have $0.075 \leq \frac{T}{\theta_D} \leq 0.19$. Figure 8 shows a DP made for $\rho_{Wilson}(T)$ in this range. As we see, an average $n \cong 2.4$. Plotted as ρ vs. T^2 it looks pretty much like a straight line and, in any case, pretty much like the model calculation with the realistic $F(\omega)$ of Webb *et al.* A negative intercept results from the fact that the true index $n > 2$. Finally, there are reports that the heat capacity of Nb_3Al, V_3Si (Junod *et al.*, 1971) and Nb_3Ge (Harper *et al.*, 1975) actually follow Debye theory reasonably well up to $T \approx 30-32K$.

To summarize, given the facts on n being very precisely 2 and on the T^2 behavior in disordered samples, plus having realized that the analysis of Webb *et al.* is not as conclusive as it may seem at first, I think we have to look for another explanation for the T^2 resistivity in A15s.

The possibility of electron-electron (e-e) scattering (Marchenko, 1973) has to be ruled out in the A15s because the coefficient A is two orders of magnitude larger than it would be from this mechanism (Webb *et al.*, 1977; Gurvitch *et al.*, 1979).

There are also magnetic or nearly magnetic metals in which the large T^2 term in $\rho(T)$ is present. However there is no reason to expect free magnetic moments in, say, Nb_3Sn.

Apparently the T^2 term is commonly seen in the resistivity of disordered alloys (see references in the work of Cote and Meisel, 1977), where, at least in some cases, e-e scattering and the possibility of spin fluctuations can be ruled out

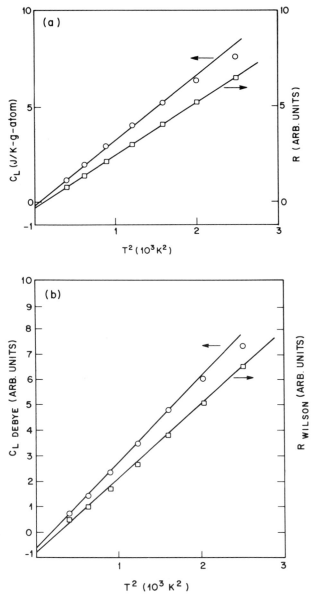

FIGURE 7.   a. Calculated C_L and R for the model $F(\omega)$ plotted vs. T^2 between 20 and 50K, taken from the work of Webb et al. (1977). b. Calculated $C_{L\ Debye}$ and R_{Wilson}, $\theta_D = 265K$, plotted vs. T^2 between 20 and 50K.

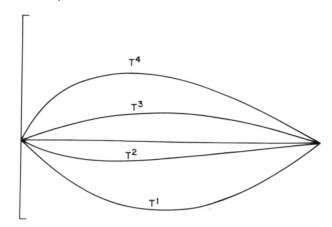

FIGURE 8.   Deviation plot for Wilson's resistivity

$$\rho_W \propto \int\limits_{O}^{\omega_D} \frac{x}{(sinhx)^2}\, \omega^2 d\omega, \quad \omega_D = \frac{K_B \theta_D}{\hbar} \; ; \; temperature \; range$$

$$0.075 \le \frac{T}{\theta_D} \le 0.19.$$

(Hasegawa,1971). Cote and Meisel make use of the modified
Ziman's liquid-metal theory to account for the T^2 term in
these systems. On the other hand, Hasegawa (1971) speculated
that the T^2 term in disordered transition metal alloys can
result from the violation of the momentum conservation in the
electron-phonon scattering, as can be the case in "dirty"
systems (Campbell *et al.*, 1971; Mills, 1971; Gubanov, 1957).
The question is whether these ideas are applicable to A15s,
where $\rho(o)$ is initially much smaller than that of disordered
and amorphous transition metal alloys, and the coefficient A
of the T^2 resistivity term is much larger. But we will come
back to the question of comparing values of A.
    It is interesting that the T^2 behavior in A15s was so far
seen only in compounds with fairly high values of λ: even in
the most damaged sample of Nb_3Sn studied ($\rho(o) = 95.7$ $\mu\Omega cm$,
$T_c = 9.15K$), $\lambda = 0.97$. At the same time in Mo_3Ge we found $n = 5$
with $\rho(o)$ values up to 45 $\mu\Omega cm$ ($T_c = 3K$). In this case λ
calculated from the McMillan's equation with $\mu^* = 0.1$ and
$\theta_D = 435K$ comes out to be $\lambda = 0.46$. Perhaps this observation
suggests that the T^2 term is associated with the strong
electron-phonon scattering. On the other hand in Nb $\lambda \overset{\sim}{\sim} 1$ and
yet $\rho(T) - \rho(o) \propto T^3$ (Webb, 1969). Is the simultaneous presence
of high λ and high $\rho(o)$ essential?

b. The Problem of T_2 Increase With Disorder

The fact that the upper temperature limit T_2 of the T^2 region increases with disorder (Fig. 5) can be interpreted in two ways. One way is to say that this result is another indication that the T^2 law is related to having disorder (strong scattering) in the system. However, on the other hand, let us not forget that according to the "parallel resistor" formulation, ρ_{max} should influence the resistivity behavior at low as well as at high temperatures provided resistivity is high enough to "feel" the presence of ρ_{max}. It is really the $\rho_{ideal}(T)$ branch that we should speak about when considering the low temperature behavior. We can calculate $\rho_{id}(T) = \dfrac{\rho(T)}{1 - \dfrac{\rho(T)}{\rho_{max}}}$ from the measured resistivity $\rho(T)$ provided we know the ρ_{max} value from the high temperature fits. Then using DP for the ideal branch we can find its low temperature behavior. It turns out that $\rho_{id}(T)$ also follows the T^2 law $\rho_{id}(T) = \rho_{o\ id} + A_{id}\ T^2$. On Fig. 9 both $\rho(T)$ and $\rho_{id}(T)$ are shown with corresponding T^2 fits. The reader may notice that the ideal branch follows the T^2 law to higher T_2's than the measured $\rho(T)$. This can be further established with DP; in fact it turns out that within the error bars T_2^{ideal} is independent of $\rho(o)$ ($T_2^{ideal} = 35 \pm 3K$ for Nb₃Sn), while, as we have seen, T_2 is increasing with $\rho(o)$. Hence from the point of view of the "parallel resistor" model the increase in T_2 with disorder is simply a result of approaching the ρ_{max}.

c. Behavior of the Coefficient A With Disorder and the Link Between High- and Low-Temperature Resistivity of A15s

Not only the increase of T_2, but also the behavior of A with disorder can be understood in terms of the "parallel resistor" model. Namely, we assume that A_{ideal} does not change drastically with disorder, i.e., that Matthiessen's rule holds for the ideal branch. However the measured resistivity curve $\rho(T)$ will not obey Matthiessen's rule since it is being "squeezed" between $\rho(o)$ and ρ_{max}. So, for higher values of $\rho(o)$, the $\rho(T)$ curve has to become flatter than it was initially, i.e., the coefficient A will decrease with disorder. Now, the following question can be asked: what value of ρ_{max} is needed to explain the observed variation of A with $\rho(o)$ assuming that A_{ideal} stays constant? One way to answer this question is to assume a value of ρ_{max} and to calculate a standard deviation for the number of values of A_{ideal} corresponding to a set of curves with different $\rho(o)$

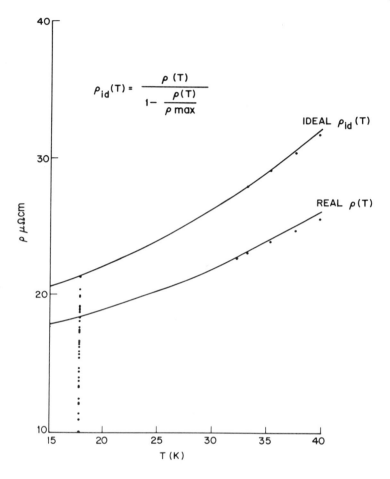

$$\rho_{id}(T) = \frac{\rho(T)}{1 - \frac{\rho(T)}{\rho \, max}}$$

IDEAL $\rho_{id}(T)$

REAL $\rho(T)$

FIGURE 9. *Nb₃Sn resistivity data fitted to*
$\rho(T) = \rho(o) + AT^2$ *and calculated points* $\rho_{id}(T) = \dfrac{\rho(T)}{1 - \dfrac{\rho(T)}{\rho_{max}}}$ *with*

$\rho_{max} = 140$ μΩcm *fitted to* $\rho_{id}(T) = \rho_{oid} + A_{id}T^2$.

values. This was done for a series of α-irradiations of
Nb₃Sn and Mo₃Ge films. (In case of Mo₃Ge n = 5 and so A, A_{id}
are in this case coefficients in front of T^5.)

The results are shown on Fig. 10. The minima on these
curves indicate the values of ρ_{max} which provide the smallest
variation in A_{ideal} or, in other words, these values of ρ_{max}
are needed to explain the variation of A with disorder
assuming that this variation comes from approaching ρ_{max}.

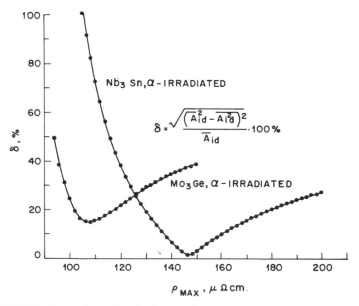

FIGURE 10. *Standard deviations of* A_{id} *values vs.* ρ_{max} *for* Nb_3Sn *and* Mo_3Ge.

As the reader sees, this procedure does not assume the knowledge of ρ_{max} from the high temperature fits. Here we are operating with the data taken at $T < 40K$, and yet the values of ρ_{max} come out to be very close to those found at $T > 100K$. This is another success of the "parallel resistor" formula.

We have mentioned before that in transition metal alloys values of A are considerably smaller (more than 10 times) than in clean Nb_3Sn. However we know now that besides all other things, the value of A depends on how close is $\rho(o)$ to ρ_{max}. In those alloys $\rho(o)$ is fairly high, of the order of 100 $\mu\Omega$cm (Hasegawa, 1971), so A can be small for that reason. Hence the big difference in A values does not necessarily rule out the possibility that the T^2 behavior has a common origin in both A15s and disordered alloys.

ACKNOWLEDGMENTS

    I wish to thank collaborators on this work, especially Myron Strongin, A. K. Ghosh, H. Lutz, H. Wiesmann and P. B. Allen. I also thank D. F. Moore for the critical reading of the manuscript and good suggestions.

REFERENCES

Allen, P. B. (1971). *Phys. Rev. B3*, 305.

Allen, P. B. (1976). *Phys. Rev. Lett. 37*, 1638.

Bader, S. D., and Fradin, F. Y. (1976). *In* Superconductivity in d- and f-Band Metals, edited by D. H. Douglass, Plenum Press.

Brown, G. S., Testardi, L. R., Wernick, J. H., Hallack, A. B., and Geballe, T. H. (1977). *Solid State Comm. 23*, 875.

Cohen, R. W., Cody, G. D., and Halloran, J. J. (1967). *Phys. Rev. Lett. 19*, 840.

Cote, P. J., and Meisel, L. V. (1977). *Phys. Rev. Lett. 39*,2.

Fisk, Z., and Lawson, A. C. (1973). *Solid State Comm. 13*, p. 277-279.

Fisk, Z., and Webb, G. W. (1976). *Phys. Rev. Lett. 36*, 1084.

Gubanov, A. I. (1957). *Soviet Physics - JETP 4*, No. 4, p. 465-473.

Gurvitch, M. (1978). Ph.D. Thesis, S.U.N.Y. at Stony Brook.

Gurvitch, M., Ghosh, A. K., Gyorffy, B. L., Lutz, H., Kammerer, O. F., Rosner, J. S., Strongin, Myron (1978). *Phys. Rev. Lett. 41*, 23, p. 1616.

Gurvitch, M., Ghosh, A. K., Lutz, H., Strongin, Myron, to be published in Phys. Rev. B.

Harper, J. M. E., Geballe, T. H., Newkirk, L. R., Valencia, F. A. (1975). *Journ. of the Less-Comm. Metals 43*, 5-11.

Hasegawa, R. (1971). *Phys. Letters 36A*, 5, P. 425.

Joffe, A. F., Regel, A. R. (1960). *Prog. Semicond. 4*, 237.

Junod, A., Staudenmann, J. L., Muller, J., and Spitzli, P. (1971). *J. Low Temp. Phys. 5*, 1.

King, E., *et al.*, (1965). *Proc. Roy. Soc. A284*, 325.

Knapp, G. S., Bader, S. D., and Fisk, Z. (1976). *Phys. Rev. B13*, 3783.

Lee, J. A. (1961). Progress in Nuclear Energy, Vol. 3 (International Conference on the Peaceful Uses of Atomic Energy, Series 5, Geneva), p. 453.

Lutz, H., Wiesmann, H., Kammerer, O. F., Strongin, Myron (1976). *Phys. Rev. Lett. 36*, 1576.

Marchenko, V. A. (1973). *Sov. Phys-Solid State 15*, 1261.

Mooij, J. H. (1972). *Phys. Status Solidi (a)17*, 521.

Mott, N. F. (1974). Metal-Insulator Transitions (Taylor and Francis, London).

Mott, N. F. (1976). *Proc. Roy. Soc. (London) Ser. A153*, 699.

Poate, J. M., Testardi, L. R., Storm, A. R., and Augustyniak, W. M. (1975). *Phys. Rev. Lett. 35*, 19.

Rowell, J. M., Dynes, R. C., Schmidt, P. H., this volume.

Sarachik, M. P., Smith, G. E., Wernick, J. H. (1963). *Can. J. Phys. 41*, 1542.

Testardi, L. R., Poate, J. M., Levinstein, H. J. (1977). *Phys. Rev. B15*, 5.

Webb, G. W. (1969). *Phys. Rev. 181*, 3, 1127–1135.

Webb, G. W., Fisk, Z., Engelhardt, J. J., Bader, S. D. (1977). *Phys. Rev. B15*, 5, p. 2624.

Wiesmann, H., Gurvitch, M., Lutz, H., Ghosh, A. K., Schwarz, B., Strongin, M., Allen, P. B., Halley, J. W. (1977). *Phys. Rev. Lett. 38*, 782.

Woodard, D. W., Cody, G. D. (1964). *RCA Rev. 25*, 392

Ziman, J. M. (1960). "Electrons and Phonons", Oxford University Press, Oxford.

SUPERCONDUCTIVITY AND ATOMIC ORDERING IN
NEUTRON-IRRADIATED Nb_3Ge.[1]

D. E. Cox

Brookhaven National Laboratory
Upton, New York

A. R. Sweedler

Brookhaven National Laboratory
and
Department of Physics
California State University
Fullerton, California

S. Moehlecke

Brookhaven National Laboratory
and
Unicamp
Campinas, Brazil

L. R. Newkirk
F. A. Valencia

Los Alamos Scientific Laboratory
Los Alamos, New Mexico

*The effects of high energy (E > 1 MeV) neutron irradiation
on the properties of Nb_3Ge prepared by chemical vapor deposition have been studied. Irradiation to a fluence of 3.4×10^{19}
n/cm^2 produces a decrease in the superconducting transition*

[1]*Work at Brookhaven supported by the U.S. Department of
Energy under contract No. EY-76-C-02-0016*

temperature T_c from 20.9 to 4.4 K and in the long range order parameter S from 0.86 to 0.46, and an increase in the lattice parameter a_o from 5.142 Å to 5.174 Å. For a fluence of 6.5 × 10^{19} n/cm², most of the Nb_3Ge is transformed into a noncrystalline phase. The A15 phase is recovered after annealing

INTRODUCTION

It is clear from a variety of recent experiments on high-T_c A15 compounds[2] that defects play an important role in the stability of the lattice and the attainment of high T_c's (Testardi et al., 1977a; Sweedler et al., 1978). In particular, irradiation with energetic particles has been shown to result in substantial decreases in T_c, which have been associated with various kinds of defects such as site-exchange disorder (Sweedler et al., 1974; Sweedler and Cox, 1975), static displacements (Meyer and Seeber, 1977; Testardi et al., 1977b; Meyer, 1978), and highly disordered regions 20 - 40 Å in size coherent with a matrix of ordered material (Karkin et al., 1976; Pande, 1977 and 1978). For some time, it has been widely believed that exact stoichiometry and a high degree of order are crucial in the attainment of maximum T_c in these materials, but this viewpoint has been questioned in recent work on Nb-Ge films (Testardi et al., 1975; Poate et al., 1976; Ilonca, 1977). This system is of particular interest in view of the high T_c's of up to 23 K and the fact that low temperature methods of synthesis are required to produce stoichiometric or near-stoichiometric material. Previous diffraction studies (Cox et al., 1976) have indicated the presence of a small amount of site-exchange disorder in Nb_3Ge prepared by chemical vapor deposition, and the present paper discusses this effect in more detail and describes some further studies of neutron-irradiated samples in which it is shown that the previously observed depressions in T_c (Sweedler et al., 1976) are accompanied by a substantial decrease in the degree of long range order.

---

[2]*The formula A_3B is used to denote the A15 phase in the A-B system regardless of the actual composition. The term "stoichiometric A_3B" denotes the ideal 3:1 composition.*

EXPERIMENTAL

The Nb_3Ge samples were prepared by chemical vapor deposition as previously described (Newkirk et al., 1975). Two different batches of material were used, the first (377/8 M6-12) for order measurements as a function of fluence, and the second (127M5-10) for annealing experiments after irradiation. The properties of both materials are quite similar to those of the sample (Nb_3Ge-II) used in the earlier study (Sweedler et al., 1976), in which the A15 phase was found to contain 76 ± 1 at % Nb from wet chemical and x-ray diffraction analysis. The A15 phase in the present samples was assumed to have the same Nb content. Details of the characterization of all three materials are summarized in Table 1.

Experimental details describing the irradiation procedure, annealing studies and T_C measurements have been described previously (Sweedler and Cox, 1975; Sweedler et al., 1976). The fast neutron flux has recently been measured more accurately and is now known to be 1.3×10^{14} n/cm²/sec (E > 1 MeV) and 5.0×10^{14} n/cm²/sec (E > 0.1 MeV). The figures quoted for E > 1 MeV neutrons in previous papers should be scaled by a factor of 1.3.

Data were collected for an unirradiated sample of batch

Table 1. *Characterization of Nb_3Ge samples described in test. S_A and S_B, the order parameters for A and B sites respectively, defined as in earlier study (Cox et al., 1976). Figures in parentheses are least-squares esd's and refer to the least significant digit. Errors in a_0 are ± 0.001 Å.*

	Sample		
	377/8 M6-12	*127 M5-10*	*Nb_3Ge-II*
T_C onset (K)	20.9	20.2	19.7
a_0 (Å)	5.141₅	5.143	5.142
T-Nb_5Ge_3 (wt %)	10(1)	34(3)	35(2)
H-Nb_5Ge_3 (wt %)	8(1)	3(1)	2(0.3)
NbO (wt %)	<1	1(0.3)	3.5(3)
NbO_2 (wt %)	---	4(1)	5.5(2)
x	0.12(1)	0.06(2)	0.10(2)
B(Å²)	0.7(2)	0.2(3)	0.8(3)
S_A	0.83(3)	0.92(3)	0.86(3)
S_B	0.79(3)	0.87(3)	0.82(3)

377/8 M6-12, and three irradiated samples with T_C onsets of
14.2, 9.1 and 4.4 K, respectively, out to the (421) reflection
from the A15 phase ($2\theta \simeq 85°$). The sample holders were
covered with a piece of Be foil 0.025 mm thick to prevent contamination by the radioactive samples. The observed 2θ values
of the A15 peaks were corrected with the use of Si as an
internal standard, and the integrated intensities were corrected for attenuation by the Be cover.

Least squares refinement of the corrected intensity data
was carried out with six variable parameters; an Nb-site
occupation factor x as represented by the formula ($Nb_{3-x}Ge_x$)
[$Ge_{0.96}Nb_{0.04+x}$], an overall temperature factor B for the A15
phase, and scale constants for this phase and the three impurity phases present, tetragonal and hexagonal Nb_5Ge_3, and NbO.
The atomic positional parameters and temperature factors of
the impurity phases were fixed at previously determined values
(Cox et al., 1976).

It is important to recall that it is the intensities of
the A15 "difference" peaks which are most sensitive to the
order parameters. Since in Nb_3Ge the scattering factors f_{Nb}
and f_{Ge} are quite similar, the difference term is small. In
the present experiments, S is determined largely by the intensity of the (110) reflection, which occurs at a 2θ value of
about 24° and is by far the strongest of the difference peaks.
The estimated errors obtained for x in the refinement are
strongly dependent upon the precision of the measurement of
this intensity. Considerable care was taken to determine this
to a statistical precision of 5% or better, involving repeated
scans over periods of up to 24 hours. It should also be emphasized that there is no overlap of the (110) peak with any of
the possible reflections from the impurity phases. The (220)
reflection of tetragonal Nb_5Ge_3 is situated about 0.3° higher,
but has completely negligible intensity.

RESULTS

A number of the experimental results are summarized in
Table 2. The depression of T_C and the increase in a_O follow
curves similar to those reported previously (Sweedler et al.,
1978). Figure 1 shows some of the diffraction data obtained
for samples before irradiation and after irradiation to a
fluence of 3.4×10^{19} n/cm^2. The ordinate scales have been
chosen so that the relative intensities are directly comparable.
A very significant decrease in the intensity of the (110) peak
is seen to have occurred at this fluence.

A careful check for line-broadening as a function of

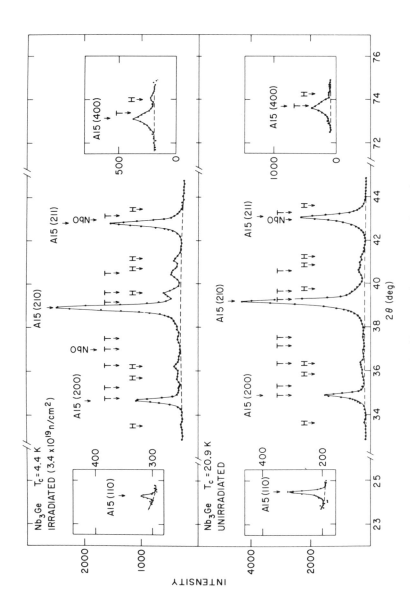

FIGURE 1. Part of x-ray (Cu Kα) diffraction scans for Nb₃Ge. Bottom: unirradiated. Top: irradiated to a fluence of 3.4 x 10¹⁹ n/cm². T and H denote tetragonal and hexagonal Nb₅Ge₃, respectively.

irradiation was made by scanning the (321) reflection with Cu
Kβ radiation, with the results shown in Fig. 2. No increase in
the full-width at half-maximum (Γ) is seen up to a fluence of
2.1×10^{19} n/cm^2 (T_C = 9.1 K), but a small increase is apparent
at the highest fluence. The increase in a_0 is reflected in
the systematic shift in Bragg angles.

A much more drastic change occurs in more heavily irradiated samples. Figure 3 shows a part of the diffraction pattern for a sample of the second batch of material 127 M5-10
before irradiation (bottom), after irradiation to a fluence of
6.5×10^{19} n/cm^2 (middle), and with a subsequent anneal at 750°C
for two hours (top). In the middle pattern, degradation has
progressed to the point where most of the A15 Bragg peaks are
no longer visible. At the same time, the appearance of a broad
diffuse peak indicates that most of the Nb$_3$Ge phase has been
transformed to a non-crystalline state. T_C and a_0 in this case
are 3.4 K and 5.195 Å, respectively. However in the top pattern, the A15 phase is seen to be recovered after an anneal at
750°C for two hours.

The recovery of T_C as a function of annealing time and
temperature has been described elsewhere (Sweedler et al., 1976
and 1978) and attributed to recovery of long range order. The
intensity of the (110) reflection in the top part of Fig. 3
confirms that most of the long range order has indeed been
recovered after the 750°C anneal.

The results of the least-squares refinement for the unirradiated materials are summarized in Table 1. As previously
noted (Cox et al., 1976) a significant amount of disorder is
found in all three as-deposited materials, as shown by the x
values of around 0.1 (S_A = 0.86, S_B = 0.82). If the Nb content

Table 2.   *Data for irradiated samples of Nb$_3$Ge, sample*
            377/8 M6-12. Errors in a_0 are ±0.002 Å.

	Neutron fluence (10^{19} n/cm^2, E > 1 MeV)			
	0.0	0.75	2.1	3.4
T_C onset (K)	20.9	14.2	9.1	4.4
T_C midpoint (K)	20.3	13.0	8.3	4.2
a_0 (Å)	5.142	5.152	5.158	5.174
x	0.10(1)	0.18(2)	0.25(3)	0.39(1)
B (Å^2)	0.5(1)	0.7(3)	0.2(4)	1.1(2)
S_A	0.86	0.75	0.65	0.46
S_B	0.82	0.71	0.62	0.43

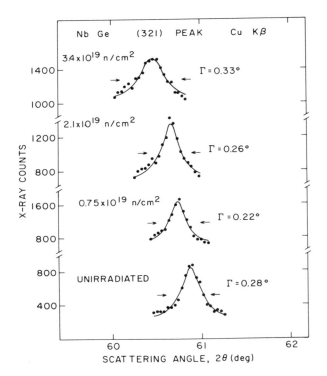

FIGURE 2. X-ray (Cu Kβ) scans of the (321) reflection from Nb₃Ge as a function of fluence. Solid curves are least-squares fits to Lorentzian peaks. Γ is the full-width at half-maximum, and has an esd of 0.02°.

were assumed to be at the limiting values of 75 and 77 at %, this value would change to about 0.13 ($S_A = S_B = 0.83$) and 0.07 ($S_A = 0.90$, $S_B = 0.81$), respectively.

Results for the measurements on successively irradiated samples of batch 377/8 M6-12 are summarized in Table 2. There appear to be no serious errors introduced by the use of Be-covered sample holders, as shown by the good agreement between the results for the unirradiated material in Tables 1 and 2.

The depressions in T_C and increases in a_O as a function of irradiation are seen to be accompanied by a systematic increase in x, or a decrease in the order parameters S_A or S_B. Figure 4 shows that a_O is an approximately linear function of x and hence follows Vegard's law, while T_C has an approximately exponential dependence. From the linear dependence of a_O, it can be inferred that the Al5 phase remaining after irradiation to a fluence of 6.5×10^{19} n/cm² (Fig. 3) is heavily disordered

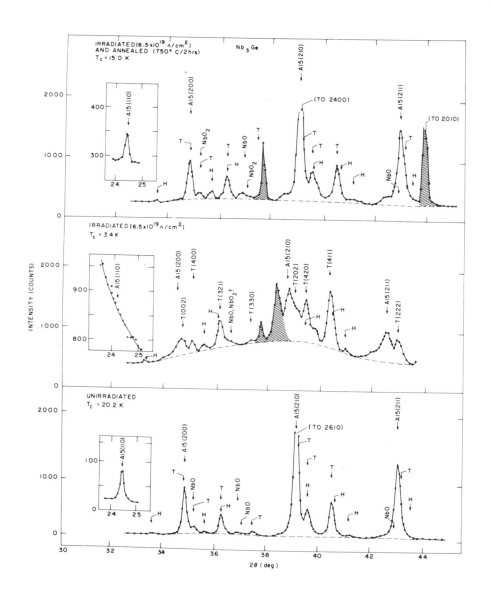

FIGURE 3. Part of x-ray (Cu Kα) diffractometer scans for Nb₃Ge. Bottom: unirradiated. Middle: irradiated to a fluence of 6.5 x 10¹⁹ n/cm². Top: irradiated to a fluence of 6.5 x 10¹⁹ n/cm² and annealed at 750°C for two hours. The Al peaks (shaded) are from the sample holder.

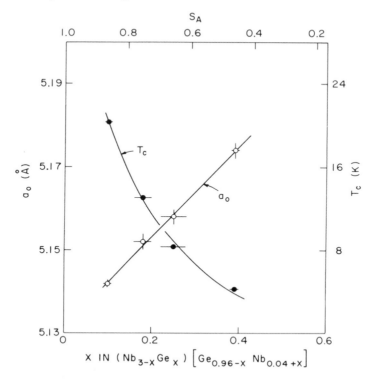

FIGURE 4. *Variation of a_O (open circles) and T_C midpoints (solid circles) for Nb₃Ge as a function of x (defined in text) and S_A. The straight line is a guide to the eye, the curve represents the relationship $T_C = T_{CO} \exp[-3.6(1 - S_A/S_{AO})]$, where T_{CO} is 20.3 K and S_{AO} is 0.86.*

($S_A \sim 0.2$). It is interesting to note that the lattice parameters of the tetragonal and hexagonal Nb₅Ge₃ phases also increase as a function of fluence, but at only about half the rate of that of the A15 phase.

DISCUSSION

The analysis of the diffraction data provides strong evidence that significant amounts of disorder exist on both Nb and Ge sites in three separate samples of CVD Nb₃Ge (Table 1). However, there are some assumptions made in this analysis which need to be examined. First of these is the starting composition. The effect of changing this within the experimentally

determined limits of 75 – 77 at % Nb has been shown above to
lead to an increase in the order on one site, but only at the
expense of a decrease on the other site. The assumption of an
overall isotropic temperature factor appears to be a good
approximation, based on published data for a number of A15
compounds (Flükiger and Staudenmann, 1976). Furthermore the
(110) peak lies at a sufficiently low angle ($\sin \theta/\lambda \simeq 0.14$)
that errors introduced in this way cannot plausibly account for
more than a small fraction of the difference between the calculated intensity for the fully-ordered state and that observed.
The third assumption of neutral atom scattering factors is more
questionable. Reference to scattering factor tables (International Tables for X-ray Crystallography, 1974) shows that
charge transfer could have a significant effect on the (110)
intensity, but very little on the others. Available information on this point is rather limited. From x-ray single crystal studies of V_3Si (Staudenmann et al., 1976) and Cr_3Si
(Staudenmann, 1976) it has been deduced that there is charge
transfer of roughly two electrons from each Si to the three
transition metal atoms, although this interpretation has been
questioned by Ho et al. (1978). Mössbauer studies on Nb_3Sn
indicate a reduction in the 5s electron density at the tin
nucleus of about 0.4 electrons (Herber and Kalish, 1977). A
similar trend in Nb_3Ge would result in a larger calculated
intensity for the (110) reflection and hence an even smaller
order parameter in the samples studied.

     One effect of irradiation is clearly to produce additional
site exchange disorder. It is interesting to compare the
associated decreases in T_C with those produced by increasing
the Nb content, and hence the amount of B-site disorder, in
Nb-Ge films grown epitaxially on Nb_3Ir substrates (Dayem et al.,
1978). For example, Table 2 shows that after irradiation to a
fluence of 2.1×10^{19} n/cm², the state of order may be expressed
as $(Nb_{2.75}Ge_{0.25})[Ge_{0.71}Nb_{0.29}]$ and the T_C onset is 9.1 K.
Reference to Fig. 16 of Dayem et al. shows a T_C of 10.5 K for
a sample containing 17.9 at % Ge. If the Nb sites are assumed
to be fully ordered, this corresponds to $Nb_3[Ge_{0.716}Nb_{0.284}]$.
It is not possible to make a similar comparison for A-site
disorder, since the A15 single phase region was found by Dayem
et al. to extend only to about 26 at % Ge. Furthermore, there
is no reliable way to estimate the combined effect of A-site
and B-site disorder on T_C, but it is clear that the latter
alone is capable of accounting for a large part of the observed
T_C decrease in the irradiated samples.

     At first sight, the Vegard's law dependence of a_o on order
in the irradiated samples is not particularly surprising, and
is consistent with the simple radius model previously described
(Moehlecke et al., 1977). This is an extension of the model
proposed by Geller (1956) who was the first to note that the

lattice parameters of Al5 compounds could be accounted for strikingly well with a self-consistent set of radii for atoms on A and B sites. However, lattice parameter data for a variety of Nb systems with an appreciable range of stoichiometry on the Nb-rich side make it clear that the Geller radius for Nb differs according to whether it occupies its normal A site or is substituted on to a B site. The commonly accepted Geller radii for Nb and Ge are 1.51 Å and 1.36 Å, respectively, (Johnson and Douglass, 1974), and there is now abundant evidence that the B-site Geller radius of Nb must be 1.42 - 1.44 Å (Vieland and Wicklund, 1974; Moehlecke et al., 1977; Tarutani and Kudo, 1977). With these values, the calculated a_o for the unirradiated sample of Nb$_3$Ge would be between 5.140 - 5.145 Å, in good agreement with the observed value of 5.142 Å (Table 2). For Al5 material containing 83 at % Nb, the calculation gives 5.168 - 5.180 Å, against an observed a_o of 5.179 Å, (Sweedler et al., 1976), indicating a value of about 1.44 Å for r_{Nb}^B. This leads to calculated a_o's of 5.149, 5.153 and 5.161 Å, respectively, for the three irradiated samples, and thus accounts for only about 60% of the observed increase. The remaining 40% must then be attributed to the effect of more complex defects. Defects of this type could also be the origin of the smaller increases in a_o observed for the Nb$_5$Ge$_3$ phases, which are much more stable than Nb$_3$Ge, and presumably much more resistant to site-exchange disorder.

The arguments above and previous studies on both compositionally and radiation-disordered high-T_C Al5 materials lead us to the conclusion that the predominant factor in the depression of T_C is the presence of atomic disorder on either or both of the A and B sites. Other kinds of defects undoubtedly exist, particularly in irradiated samples, and most likely act to depress T_C as well, but we believe their role is less important. The mechanism by which atomic disorder produces this degradation is still an unanswered question, however.

The presence of significant amounts of disorder in the as-deposited materials may be essential for their preparation, in that the Al5 phase can thereby form with a slightly larger a_o than that of fully-ordered, stoichiometric Nb$_3$Ge, which is generally predicted to be around 5.13 Å. The tendency towards instability and higher T_c's in the Nb$_3$X systems (X = Al, Ga, Si, Ge and Sn) as the X atom sizes and lattice parameters decrease has been noted by many authors and discussed recently in some detail by Ashkin and Gavaler (1978). The limiting value of about 5.14 Å in these systems may correspond to the minimum Nb-Nb separation in the chains which can be tolerated before repulsive forces dominate.

ACKNOWLEDGMENTS

    We would like to acknowledge many helpful discussions with
G. W. Webb and J. J. Engelhardt of the University of California,
San Diego.

REFERENCES

Ashkin, M. and Gavaler, J. R. (1978). J. Low Temp. Phys. 31,
    285.
Cox, D. E., Moehlecke, S. and Sweedler, A. R. (1976). "Super-
    conductivity in d- and f-Band Metals" (D. H. Douglass, ed.)
    p. 461. Plenum Publishing Corp., New York.
Dayem, A. H., Geballe, T. H., Zubeck, R. B., Hallak, A. B. and
    Hull, G. W. (1978). J. Phys. Chem. Solids 39, 529.
Flükiger, R. and Staudenmann, J.-L. (1976). J. Less Common
    Metals 50, 253.
Geller, S. (1956) Acta Cryst. 9, 885.
Herber, R. H. and Kalish, R. (1977). Phys. Rev. B16, 1789.
Ho, K. M., Pickett, W. E. and Cohen, M. L. (1978). Phys. Rev.
    Lett. 41, 580.
Ilonca, G. (1977). Phys. Stat. Solidi a43, 387.
International Tables for X-ray Crystallography (1974). Vol. 4,
    The Kynoch Press, Birmingham.
Johnson, G. R. and Douglass, D. H. (1974). J. Low Temp. Phys.
    14, 565.
Karkin, A. E., Arkhipov, V. W., Goshchitskii, B. N. Romanov,
    E. P. and Sidorov, S. K. (1976). Phys. Stat. Solidi a38,
    433.
Meyer, O. (1978). J. Nucl. Mater. 72, 182.
Meyer, O. and Seeber, B. (1977). Solid State Commun. 22, 603.
Moehlecke, S., Cox, D. E. and Sweedler, A. R. (1977). Solid
    State Commun. 23, 703.
Newkirk, L. R., Valencia, F. A., Giorgi, A. L., Sklarz, E. G.,
    and Wallace, T. C. (1975). IEEE Transactions on Magnetics,
    MAG-11, 221.
Pande, C. S. (1977). Solid State Commun. 24, 241.
Pande, C. S. (1978). J. Nucl. Mater. 72, 83.
Poate, J. M., Dynes, R. C., Testardi, L. R. and Hammond, R. H.
    (1976). "Superconductivity in d- and f-Band Metals"
    (D. H. Douglass, ed.) p. 489. Plenum Publishing Corp.,
    New York.
Sweedler, A. R., Schweitzer, D. G. and Webb, G. W. (1974).
    Phys. Rev. Lett. 33, 168.
Sweedler, A. R. and Cox, D. E. (1975). Phys. Rev. B12, 147.

Sweedler, A. R., Cox, D. E., Moehlecke, S., Jones, R. H.,
    Newkirk, L. R. and Valencia, F. A. (1976). J. Low Temp.
    Phys. $\underline{24}$, 645.
Sweedler, A. R., Cox, D. E. and Moehlecke, S. (1978). J. Nucl.
    Mater. $\underline{72}$, 50.
Staudenmann, J.-L. (1976). Thesis No. 1735, University of
    Geneva.
Staudenmann, J.-L., Coppens, P. and Muller, J. (1976). Solid
    State Commun. $\underline{19}$, 29.
Tarutani, Y. and Kudo, M. (1977). Japan. J. Appl. Phys. $\underline{16}$,
    509.
Testardi, L. R., Meek, R. L., Poate, J. M., Royer, W. A.,
    Storn, A. R., and Wernick, J. H. (1975). Phys. Rev. $\underline{B11}$,
    4304.
Testardi, L. R., Poate, J. M. and Levinstein, H. J. (1977a).
    Phys. Rev. $\underline{B15}$, 2570.
Testardi, L. R., Poate, J. M., Weber, W. and Augustyniak, W. M.
    (1977b). Phys. Rev. Lett. $\underline{39}$, 716.
Vieland, L. J. and Wicklund, A. W. (1974). Phys. Lett. $\underline{49A}$,
    407.

DIRECT OBSERVATION OF DEFECTS IN A15 COMPOUNDS
PRODUCED BY FAST NEUTRON IRRADIATION[1]

C. S. Pande
Brookhaven National Laboratory
Upton, New York

The nature of defect or defect complexes produced in superconducting compounds Nb_3Sn, Nb_3Pt, and V_3Si by high energy ($E \gtrsim 1$ MeV) neutron irradiation is investigated by transmission electron microscopy. The newly developed technique of superlattice reflection imaging is used whereby the regions of reduced long range order are directly imaged. Unlike metals these regions were found in general not to collapse into dislocation loops. The size and the volume fraction of these disordered regions are obtained for fluences ranging from 10^{17} neutrons/cm^2 to 3×10^{19} neutrons/cm^2. The size ranges from 20Å to 60Å. Typical volume fraction for 10^{18} neutrons/cm^2 is over 1%.

I. INTRODUCTION

Marwick (1975) has estimated that more than 95% of the displacements produced in a metal undergoing neutron irradiation in a fast reactor occur within displacement cascades. In intermetallic compounds like A15s the cascades will also contain replacements (i.e. antisite defects). (In addition there is a possibility of the creation of antisite defects by the operation of long range collision sequences.) These cascades or disordered regions were extensively observed in Nb_3Sn using transmission electron microscopy (Pande, 1979), employing the newly developed superlattice reflection technique (Jenkins et al., 1976). For sufficiently low fluences, the specimens contain volumes of cascades where the long range order S is

[1]This work was performed under the auspices of the U.S. Department of Energy.

much smaller than the surrounding matrix. In the superlattice
reflection technique regions of low S are mapped as dark dots.
The contrast arises due to the difference in structure factor
between the disordered zones and the more or less ordered
matrix. So far the evidence for the disordered regions were
obtained only in Nb_3Sn, by transmission electron microscopy
(Pande 1977, 1979) and low angle scattering (Karkin *et al.*,
1976). It is of interest to know whether other A15 compounds
such as Nb_3Pt, V_3Si show similar behavior. The chief aims of
the present work are:

1) to use the superlattice reflection technique to investigate irradiated Nb_3Sn, Nb_3Pt, and V_3Si.

2) to estimate the volume fraction of the disordered regions in Nb_3Sn and to compare with what would be expected
theoretically, if all antisite defects are confined to these
disordered regions.

II. SPECIMEN PREPARATION

Nb_3Sn specimens were prepared by solid state diffusion
("bronze") process. The grain size of such a sample is
usually of the order of 1μ and is suitable for transmission
electron microscopy in the diffraction contrast modes, although detailed Burgers vector determination is very difficult.
V_3Si and Nb_3Pt were prepared by arc melting. Both of them had
a grain size of over 10μ.

Discs of these materials ~3 mm in diameter were sealed in
quartz tubing. The quartz tube was then placed in an aluminum
container and surrounded with fine mesh aluminum powder and
held in place with aluminum foil. The irradiation was carried
out in the Brookhaven High Flux Beam Reactor (HFBR). The
specimen temperature during the irradiation is expected to be
150°-250°C. The fluences mentioned in this paper refer only
to neutrons with energy $E \gtrsim 1$ MeV, i.e. high energy neutrons
with a flux of 1.3×10^{14} n/cm^2-sec. However neutrons of lower
energy as well as thermal neutrons are also present.

For obtaining specimens suitable for transmission electron
microscopy, ion thinning procedure was used. The iron milling
was done at a grazing angle of 12-15° and Argon ions were
accelerated with voltage of 4.5-5 kV. Ion current was 25-50
microamperes and gun current ranging from 0.4 to 2 milli-
amperes. In some cases the specimens were washed in HF to remove damage layers due to ion milling. For a discussion on
the possibility of damage due to ion milling, see Pande (1979).

III. EXPERIMENTAL OBSERVATIONS AND RESULTS

   The irradiated specimens were examined using diffraction contrast technique employing both fundamental and superlattice reflections. In Nb_3Sn and Nb_3Pt 'small black dots' (disordered regions) were observed in specimens irradiated to various fluences. For example Fig. 1 shows a single Nb3Sn grain of the irradiated specimen imaged in superlattice reflection (g=011) showing a high density of 'black dots'. As stated previously we identify the 'black dots' in dark field superlattice reflection as arising from the disordered regions created by the high energy neutrons and that the image is a simple projection of the disordered regions on the imaging plane [(100) in this case]. The size of the contrast figures under these conditions was taken to be the size of the disordered regions as they are close to spherical shape, although some of the disordered regions appear to be slightly elliptical. Figure 2 shows similar disordered regions in Nb3Pt.
   In the case of V_3Si however this technique has so far not been successful. We have not observed disordered regions in V_3Si on neutron irradiation after repeated attempts. This result is in agreement with that of Küpfer and Manuel (1979) and probably indicates annealing of defects in V_3Si at the reactor temperature. Unlike metals, very few of the disordered regions were found to collapse into dislocation loops.

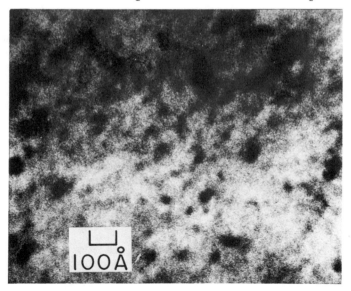

FIGURE 1. *Disordered regions (black dots) in irradiated* Nb_3Sn.
(Fluence $\sim 10^{19}$ *neutrons/cm^2).*

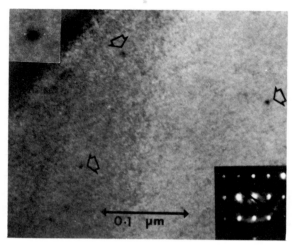

FIGURE 2. *Disordered regions in irradiated Nb_3Pt.* (*Fluence* $\leq 10^{17}$ *neutrons/cm^2.*) *Diffraction conditions are shown on bottom right hand corner. Top left hand corner shows a magnified view of one of the disordered regions.*

The volume F_v of the disordered regions is an important parameter in calculating T_c degradation of the material on irradiation. This was done by measuring the disordered regions and measuring the thickness of the foil. It should be emphasized that F_v so obtained are rough estimates only. The error in F_v is estimated to be ± 30% at low fluences. At high fluences only the lower limit could be estimated (see 4th point in Figure 3). Figure 3 is a plot of the volume fraction of the disordered regions as a function of fluence for irradiated Nb_3Sn. It is clear that F_v is not insignificant as sometimes claimed. The theoretical curve is obtained on the assumption that all the antisite defects are present in the disordered regions. Experimental value of k as obtained by Sweedler and Cox (1975) is used to estimate theoretical F_v (see Pande (1977) for details).

IV. CONCLUSIONS

1) Disordered regions in two A15 compounds, viz. Nb_3Sn, and Nb_3Sn, and Nb_3Pt produced by fission neutron irradiation at reactor temperature are directly imaged.

2) Volume fraction of the disordered regions is measured from the electron micrographs. It is shown that the majority of defects in Nb_3Sn are present in these disordered regions.

3) No disordered regions were observed in V_3Si irradiated at reactor temperature indicating the possibility of annealing of the defects at this temperature.

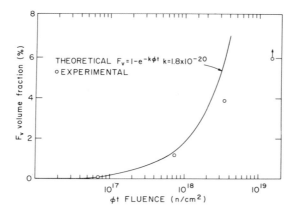

FIGURE 3. *Volume fraction F_V of the disordered regions as a function of fluence in Nb3Sn. (See text for details.)*

ACKNOWLEDGMENTS

The author is grateful to Dr. M. Suenaga for initiating this work and for many valuable suggestions and comments. Superb technical assistance by Messers. Bob Jones and Frank Thomsen is also gratefully acknowledged.

REFERENCES

Jenkins, M. L., Katerbau, K. H., and Wilkens, M. (1976). Phil. Mag. 34, 1141.
Karkin, A. E., Arkhipov, V. E., Ghoshchitskii, B. N., Romanov, E. P., and Sidorov, S. K. (1976). Phys. Stat. Sol. (a) 38, 433.
Küpfer, H. and Manuel, A. A. (1979). Phys. Stat. Sol. (a) (in press).
Marwick, A. D. (1975). J. Nucl. Mater. 55, 259.
Pande, C. S. (1979). Phys. Stat. Sol. (a) 52, 687.
Pande, C. S. (1977). Solid State Commun. 24, 241.
Sweedler, A. R. and Cox, D. E. (1975). Phys. Rev. B12, 147.

RECOVERY OF T_C BY ANNEALING IN DISORDERED Nb_3Pt[1]

David Dew-Hughes
Randall Caton[2]
Robert Jones
David O. Welch

Metallurgy and Materials Science Division
Department of Energy and Environment
Brookhaven National Laboratory
Upton, New York

I. INTRODUCTION

The superconducting critical temperature, T_C, of the A15 compound Nb_3Pt, can be depressed by rapid 'splat' quenching from the melt and by fast neutron irradiation (Moehlecke, Cox, and Sweedler 1977, 1978). In both cases the effect on T_C is believed to be due to anti-site disorder. Annealing allows the recovery of T_C to its original value. Studies of the kinetics of recovery can yield information about atom movements in disordered compounds, and in particular give estimates of activation energies for the reordering processes. Such studies in neutron irradiated material are complicated by the additional presence of an excess of radiation-induced vacancies, and other defects. These complications should be absent in thermally-disordered and rapid-quenched samples, and a comparison between their behavior and that of irradiated samples is especially valuable.

II. EXPERIMENTAL PROCEDURES AND RESULTS

A button of Nb_3Pt was prepared by arc-melting Nb and Pt in ultra-purity argon. The cast sample was homogenized at

[1]*Work supported by the Department of Energy.*
[2]*Present address: Department of Physics, Clarkson Technical College, Potsdam, New York.*

1800°C for 12 h, followed by an ordering anneal of 10 d at
900°C. This treatment produced material with superconducting
critical temperatures from 10.5-11.0 K. Some pieces were cut
from the button, the remainder was crushed to small lumps.
These lumps were individually remelted in the arc furnace on a
copper hearth, and 'splat' quenched by striking with a spring-
loaded copper hammer. The resulting flakes were ground to
powder and encapsulated in 2 mm diameter quartz tubing with 0.5
atm. of high purity helium gas. Their T_c was found to be ~6.6 K.
Two of the cut pieces were irradiated in the Brookhaven HFBR to
a dose of 6×10^{18} nvt of fast (E>1 MeV) neutrons. Superconducting T_cs (midpoints) before and after irradiation were 10.55 K
and 6.15 K.

The 'splat' quenched and irradiated specimens were annealed, in high-purity helium, at 750°C and 900°C for times up
to 1000 h. At intervals the specimens were removed from the
furnace and their superconducting T_C was measured inductively;
the temperature was monitored with a germanium resistance
thermometer accurate to ±0.1 K. Only one specimen of the
'splat' quenched material, 2A, was studied. It was first annealed at 750°C for up to 770 h, and subsequently annealed at
900°C for up to 440 h. Of the two irradiated samples, 3A was
first annealed at 750°C for 1000 h and then at 900°C for 250 h;
3B was annealed at 900°C for up to 400 h.

The results of the isothermal anneals, T_C versus annealing time t, are plotted in Figure 1. The T_C values quoted all
correspond to the mid-point of the inductive transition. Extrapolation of the 900°C curves suggest that the final T_C
values (T_{CO}) for the various samples will be: 2A, 9.85 K; 3A,
10.9 K; 3B, 11.1 K. The low value for 2A may be due to
either a loss of one component, or some compositional inhomogeneity, introduced by the remelting. Both 3A and 3B have
higher values then that for the starting material; diffusion
enhanced by the radiation induced vacancies may have removed
some slight inhomogeneities in the initial material.

III. RECOVERY KINETICS

From an analysis of isochronal reordering data in several
neutron irradiated A15 compounds, Dew-Hughes, Moehlecke, and
Welch (1978) concluded that reordering in these compounds follows second order kinetics at a rate given by:

$$\frac{dy}{dt} = - K f_v e^{-U_R/kT} y^2 \tag{1}$$

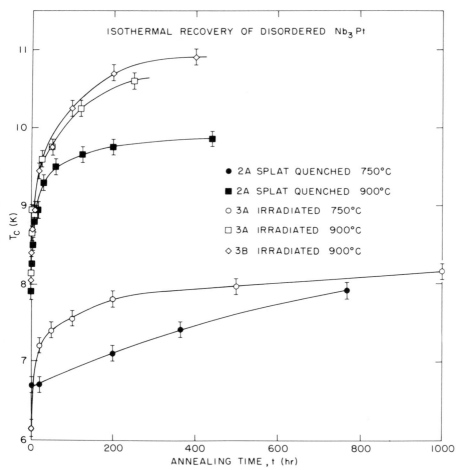

FIGURE 1. *Superconducting critical temperature* $T_c(K)$ *versus annealing time t for 'splat' quenched are irradiated samples of Nb₃Pt annealed at 700°C and 900°C.*

where $y=(1-T_C/T_{CO})$. Over the range of T_C and S for Nb₃Pt in this study $y \propto (1-S)$, where S is the Bragg-Williams order parameter (Moehlecke et al. 1978). U_R is the activation energy for reordering by a Nb atom on a Pt site jumping into an adjacent vacancy on a Nb site. f_V is the fraction of vacancies; in the absence of radiation induced vacancies it is the equilibrium vacancy concentration at a temperature T given by:

$$f_{V_e} = V_0 \, e^{-U_f/kT} \qquad (2)$$

where U_f is the activation energy for the formation of a vacant lattice site. If the excess vacancies introduced by irradiation diffuse to sinks where they are annihilated the vacancy fraction decreases with time as:

$$f_v = f_{v_O} \ e^{-t/\tau} \tag{3}$$

where f_{v_O} is the initial non-equilibrium concentration, and the relaxation time $\tau = (\alpha D)^{-1}$ where the diffusion constant at a temperature T is:

$$D = D_O \ e^{-U_m/kT} \tag{4}$$

U_m is the activation energy for diffusion. It must be stressed that the reordering process itself does not affect the vacancy concentration, the vacancies merely move from one sublattice to the other to effect the ordering. Substituting in equation 1:

$$\frac{dy}{dt} = - K \ e^{-U_R/kT} \ y^2 \ (f_{v_O} \ e^{-t/\tau} + V_O \ e^{-U_f/kT}) \tag{5}$$

and integrating with respect to time

$$\frac{1}{y} - \frac{1}{y_O} = K \ e^{-U_R/kT} \ [f_{v_O} \ \tau \ (1-e^{-t/\tau}) + V_O \ e^{-U_f/kT} t] \tag{6}$$

A plot of $1/y$ versus t should show an exponential portion at short times, the plot becoming linear at long times. The initial slope is given by:

$$\left[d(1/y)/dt \right]_{t\approx o} = K \ (f_{v_O} + f_{v_e}) \ e^{-U_R/kT} \tag{7}$$

and the final, linear slope is

$$\left[d(1/y)/dy \right]_{t\to\infty} = K \ V_O \ e^{-(U_R+U_f)/kT} \tag{8}$$

In the absence of radiation induced excess vacancies, the curve should be a simple straight line.

IV. ANALYSIS OF RESULTS

$1/y$ for the 'splat' quenched specimen 2A for both 750°C and 900°C anneals is plotted versus time t in Figure 2. A straight line plot is clearly revealed at both temperatures. The 900°C data are also plotted as $\ln(y)$; this would yield a straight line if first order kinetics were followed. It does not, and the use of second order kinetics in the above analysis

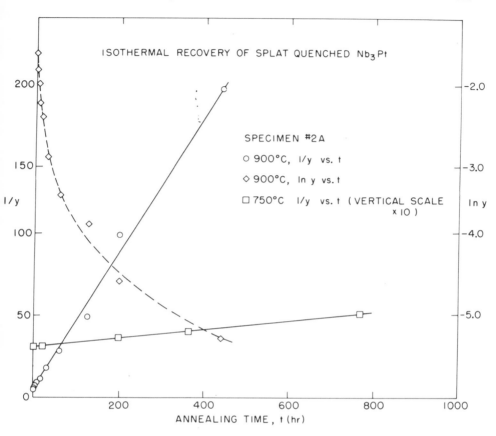

FIGURE 2. 1/y versus t for 'splat' quenched Nb$_3$Pt annealed at 750ºC and 900ºC. The 900ºC results are also plotted as ℓn(y) versus t.

is thus justified. From the ratio of the slopes at the two temperatures an activation energy (U_R+U_f) ~3.6 eV is deduced. This is close to that obtained from the change in slope of the T$_c$ versus t curves when the annealing temperature was raised from 750ºC to 900ºC, ~3.7 eV.

1/y versus t is plotted for the irradiated samples 3A and 3B in Figure 3. The curves for 3A at 750ºC and 3B at 900ºC both show an initial portion which can be approximated to an exponential, followed by a linear portion. The points for 3A at 900ºC fall on the linear portions of the curve for 3B at 900ºC, as is to be expected. The ratio of the linear slopes gives an activation energy (U_R+U_f) ~3.8 eV, in good agreement

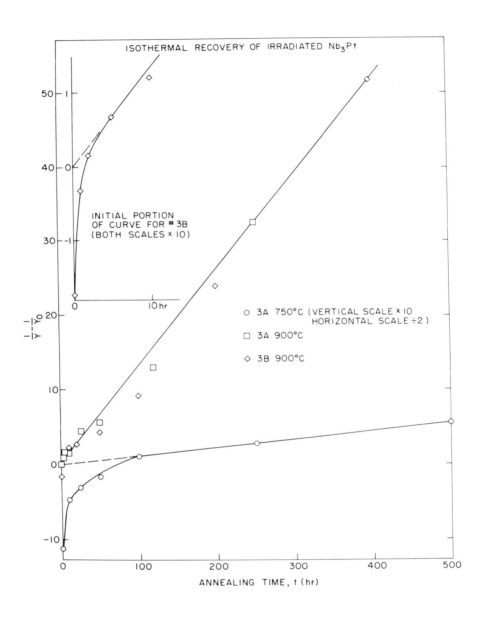

FIGURE 3. $1/y - 1/y_o$ versus t for irradiated Nb_3Pt annealed at $750^\circ C$ and $900^\circ C$. $1/y_o$ is defined by the intercept of the straight line portion of the curve with the t = 0 axis.

with the results on the thermally disordered material. The ratio of the initial slopes gives an activation energy (U_R) ~2.8 eV. Taking an average value for (U_R+U_f) of 3.7 eV, this suggests that U_f ~0.9 eV. Fitting exponentials to the initial portions of the curves yields values for τ of 30 h at 750°C and 0.75 h at 900°C. An activation energy for vacancy motion U_m, of ~2.5 eV, is deduced from this change in τ with temperature. These activation energies cannot be regarded as being very accurate as they are derived from only two temperatures. Isothermal anneals carried out at least five different temperatures are necessary to give reasonably accurate values.

The similarity of recovery behavior, at long times, for both thermally disordered and irradiated samples reinforces the belief that the degradation in T_c by radiation is mainly due to anti-site disorder. Reordering requires that both atom species move to their correct sites, and the controlling step is that of the slowest movement. That this is the motion of A atoms from B sites into A sites is indicated by consideration of the detailed kinetic theory (Dew-Hughes et al. 1978) and the fact that reordering follows second order kinetics. Diffusion on the other hand, will be controlled by motion on the sub-lattice on which vacancies move the fastest and thus U_m is expected, as indeed it is found to be, less than U_R. The activation energy for self-diffusion $=(U_m+U_f)$ ~3.4 eV compared to 2.5-3.0 eV for Nb_3Sn and V_3Ga. Moehlecke et al (1978) state that the recovery behavior of Nb_3Pt differs from that of other Al5 compounds. It is, in fact, different only in that activation energies in Nb_3Pt are somewhat larger than in the other compounds so far studied.

REFERENCES

Dew-Hughes, D., Moehlecke, S., and Welch, D. O. (1978). J. Nucl. Mater. 72, 225.
Moehlecke, S., Cox, D. E., and Sweedler, A. R. (1977). Solid State Commun. 23, 703 (1978). J. Less Common Met. 62, 111.

RADIATION DAMAGE IN A-15 MATERIALS: EXAFS STUDIES*

G. S. Knapp, R. T. Kampwirth
P. Georgopoulos and B. S. Brown

Argonne National Laboratory
Argonne, IL   60439

EXAFS measurements are useful in determining the local atomic environment of a particular element in a solid. Since there has been some controversy about the nature of the defects produced in A-15 materials by radiation damage, we have carried out such studies on some A-15 compounds. We have studied V_3Ga which was damaged by neutrons, as well as Nb_3Ge damaged by 2.5 MeV α particles. In the V_3Ga sample, site exchange disorder seems to be the most important result of the neutron damage with less than 20% of the vanadium atoms on wrong sites. However, in the Nb_3Ge sample we find in addition to site exchange disorder, an unusual splitting of the first near-neighbor distance between the Ge and Nb. This splitting, approximately 0.2 Å, may explain the large Debye Waller factors observed by Burbank et al. (1979).

The high T_c, A-15 compounds, have been shown to be very sensitive to radiation damage (Sweedler and Cox, 1978; Sweedler et al., 1974; Poate et al., 1976).   Sharp reductions in the critical temperature, T_c, and in the resistance ratios are observed. When suitably scaled, it is possible to make universal plots of T_c and resistance ratio versus damage for a number of A-15 compounds. However, the nature of the defects introduced is still unclear.

Sweedler et al. (1974) concluded that anti-site disorder was the main result of the damage, based on neutron and x-ray diffraction measurements on certain A-15 materials damaged by

---

*
Work supported by the U.S. Department of Energy.

neutrons. Burbank et al. (1979), however, using x-ray diffraction techniques on thin films of Nb_3Sn concluded that
large static displacements (0.2 Å) were produced in Nb_3Sn
when subjected to α particle irradiation. In addition, they
observed large amounts of anti-site disorder. We have attempted to obtain a different view of the local order by carrying
out extended x-ray absorption fine structure (EXAFS) measurements, which are sensitive to the local atomic environment of
the absorbing atom. We have initiated a series of experiments
on a number of A-15 compounds. The studies of neutron damage
of V_3Ga are nearly complete, and a number of measurements
have been made on Nb_3Ge damaged by 2.5 MeV α particles.

The equation describing the EXAFS oscillations can be
written as (Stern, 1974)

$$\chi(k) = \sum_j \frac{N_j A_j(k)}{k \, r_j^2} \sin\left[2kR_j + \phi_j(k)\right] e^{-2k^2 \sigma_j^2} e^{-R_j/\lambda} ,$$

where k is the momentum of the photoemitted electron, N_j the
number of backscattering atoms from shell j, $A_j(k)$ the backscattering amplitude, $\phi_j(k)$ a phase shift term, R_j the distance to shell j, $\sigma_j(k)$ the root-mean-square relative displacement, and λ the elastic mean-free path of the electrons. The
parameters $A_j(k)$ and $\phi_j(k)$ have been calculated (Teo and Lee,
1979), so that applying Eq. (1) to the EXAFS data permits the
determination of N, R_j, σ, (k) and λ. The EXAFS pattern for
the A-15 crystal structure is particularly simple if the
scattering atom is on a B site (formula written as A_3B).
Only the first near neighbor shell makes a significant contribution to the EXAFS oscillations (12 A atoms at $\sqrt{5/16} \, a_o$,
where a_o is the lattice constant). An atom on an A site,
however, has neighbors at $a_o/2$, $\sqrt{5/16} \, a_o$ and $\sqrt{3/8} \, a_o$. These
differences allow an estimation of the amount of anti-site
disorder.

Figure 1 shows the experimental $\chi(k)$ for V_3Ga, measured
at 85 and 402 K. We present this data to illustrate the
effect of disorder induced by the thermal motion. Fits to
this data indicate that σ increased from 0.04 to 0.07 Å.
In Fig. 2 we show k $\chi(k)$ for both undamaged and neutron damaged samples of V_3Ga. The neutron fluence depressed T_c from
14 to 4.5 K. Little difference is seen between the two
patterns, except for a decrease in the total amplitude. Increases in the root-mean-square relative displacements, σ,
larger than 0.02 Å are inconsistent with the data. Using
Eq. (1) and a simple model of anti-site defects, we can show
that 20% or less of the Ga atoms are on V sites. We have made
additional measurements of another sample of V_3Ga, irradiated

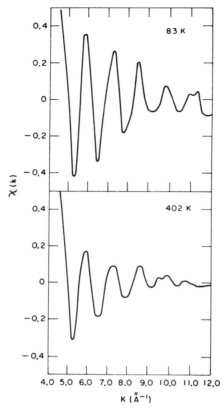

FIGURE 1.   Ga-EXAFS spectra, $\chi(k)$, for V_3Ga at 402 and 83K.

such that T_c = 8.8 K, and compared results at 77 K where even
smaller changes in σ could be detected. Again, negligible
increases in displacements were found.

EXAFS measurements were also made on a one micron thick
sample of Nb_3Ge, resistance ratio 2.4. X-ray diffractometer
scans reveal no extra lines, allowing us to conclude that our
sample was at least 95% A-15 phase. The lattice parameter
before irradiation was 5.138 Å. In Fig. 3 we show k $\chi(k)$
versus k for this sample of Nb_3Ge as a function of α particle
damage. The transition temperature after each irradiation
is also shown in this figure.

A striking feature of these patterns is the noticeable
frequency skewing at large wavevectors as a function of
radiation damage. Note in particular the position of the
oscillation peaking near 7.5 A^{-1}. We have made extensive

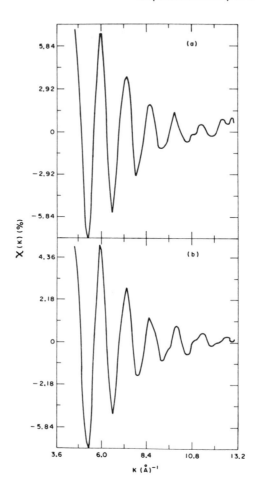

FIGURE 2.   Ga-EXAFS spectra, χ(k), for undamaged and
damaged V₃Ga at 295 K.

computer modeling studies in an attempt to understand these
frequency shifts. Models involving simple site exchange
disorder, as well as a large range of RMS relative displacements simply failed to reproduce the observed changes in
these patterns. The only model, which results in a reasonable fit, is one in which the main 2.87 Å ($\sqrt{5/16}\ a_o$) near
neighbor distance splits symmetrically in two distances,
2.77 Å and 2.97 Å, respectively. This bimodal distribution
with ΔR ≃ 0.2 Å is consistent with the large static displacements in α particle irradiated Nb₃Sn reported by Burbank
et al. (1979). They fit their data using a model of random,

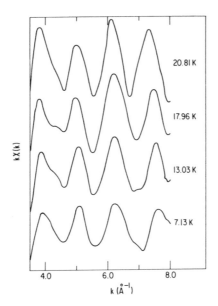

FIGURE 3. Ge-EXAFS spectra, kχ(k) for Nb₃Ge subjected to increasing amounts of α particle damage. The resulting T_c are also shown.

uncorrelated displacements. However, as they point out, strongly correlated motions, such as the bimodal distribution observed here, would be more consistent with the crystal structure. Computer fits to our data clearly rule out large random displacements of even 0.1 Å. It should be noted that x-ray diffraction measurements are much less sensitive than our EXAFS studies since the effective k range of the EXAFS experiments are much larger.

REFERENCES

Burbank, R.D., Dynes, R.C. and Poate, J.M. (1979). To be published.

Meyer, O. and Seeber, B., (1977). *Sol. St. Comm. 22*, 603.

Poate, J.M., Dynes, R.C., Testardi, L.R., and Hammond, R.H., (1976). *Phys. Rev. Lett. 37*, 1308.

Stern, E.A., (1974). *Phys. Rev. B 10*, 3027.

Sweedler, A.R., Schweitzer, D.G., and Webb, G., (1974). *Phys. Rev. Lett. 33*, 168.

Sweedler, A.R. and Cox, D.E., (1978). *J. Nucl. Materls. 72,*
    50.
Teo, B.K. and Lee, P.A. (1979). To be published.
Testardi, L.R., Poate, J.M., Weber, W., Augustyniak, W.M., and
    Barrett, J.H., (1977). *Phys. Rev. Lett. 39,* 716.

TRANSPORT PROPERTIES, ELECTRONIC DENSITY OF STATES
AND T_c IN DISORDERED A15 COMPOUNDS

P. Müller, G. Ischenko, H. Adrian, J. Bieger, M. Lehmann

Physikalisches Institut der Universität
Erlangen-Nürnberg
D 8520 Erlangen

E. L. Haase

Institut fur Angewandte Kernphysik der GFK
D 7500 Karlsruhe

INTRODUCTION

The emphasis of this work is the investigation of the
effects of defects in order to check the idea of universal behavior of A15 compounds (J. M. Poate et al., 1975, A. R.
Sweedler, 1978, G. Ischenko et al., 1978) and their transition
into the metallic glass phase at high degrees of disorder.
The main point of interest is not the study of defect structures, but the well defined variation of sample parameters by
the irradiation induced defects as a continuous sample preparation procedure of the same sample. Because of the high
Rutherford dislocation cross section, fast heavy ions are the
best tool to achieve this within a short irradiation time.

EXPERIMENTS

Nb_3Ge-, Nb_3Sn-, V_3Si-films on Al_2O_3 substrates were prepared by cosputtering (H. F. Braun et al., 1978) or electron
beam codeposition and characterized by X-ray diffraction with
Bragg-Brentano or Seemann-Bohlin geometry, backscattering
spectrometry (P. Müller et al., 1976,1978), elastic recoil
detection analysis (V. Brückner et al., 1979), and scanning

electron microscopy. Amorphous layers of Nb_3Si and Nb_3Ge *
were obtained by cold condensation.

T_c, residual resistance ρ_o, ρ vs T, dH_{c2}/dT were measured
in situ at the low temperature irradiation facility of the
Erlangen tandem accelerator. Irradiations were performed with
12.5 MeV ^{16}O, 20 MeV ^{32}S at low ($T_{Irr} \leqslant 30$ K) and at room temperature. Homogeneous damage of the films was guaranteed by
the fact that the mean range of projectiles (3-5 μm) exceeded
the sample thickness (500-2000 Å) by a factor of 10 or more.

RESULTS

T_c and ρ_o

Figure 1 shows results of low temperature irradiation for
some of the Nb_3Ge samples (P. Müller et al., 1978,2):
- No T_c-threshold effect was observed (H. Wiesman et al.,
  1978).
- After a continuous decrease (increase) (J. M. Poate
  et al., 1976) T_c and ρ_o show a flat minimum (maxi-
  mum) before reaching the saturated state
  ($T_{c,min} \approx 4$ K, $T_{c,sat} \approx 4.4$ K, $\rho_{o,sat} \approx 140$ μΩcm)
  (Ischenko et al., 1977).
- The dose curves at high doses are independent of the
  starting values of T_c and ρ_o.
- The T_c and ρ_o saturation values agree well with those of
  cold condensed, amorphous layers (C. C. Tsuei et al.,
  1977).
The Nb_3Sn and V_3Si samples in general show similar behavior
(H. Adrian et al., 1978). Low temperature irradiation of
amorphous (cold condensed) samples shows a T_c increase with
dose of a similar magnitude (Bieger et al., 1979).

Effects of Isochronal Annealing after Low Temperature
Irradiation

Compared with the large changes after irradiation, the
effect of annealing up to room temperature is only a few percent. Room temperature irradiations with 20 MeV ^{32}S show a
behavior similar to the low temperature experiments. A surprising result is the inverse annealing behavior:

**Prepared by C. C. Tsuei, IBM Thomas J. Watson Research
Center, Yorktown Heights, New York.*

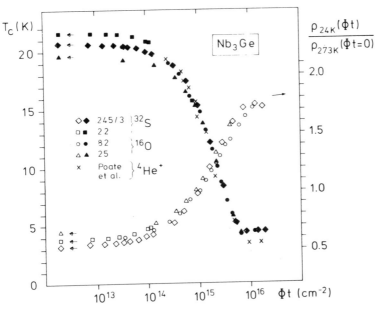

FIGURE 1. Low temperature irradiation results. The dose axis is in equivalents of 12.5 MeV ^{16}O. ^{32}S-values (this work) and room temperature $^4He^+$-data are normalized to this abscissa by the Rutherford dislocation cross section. T_c: full symbols, left scale, ρ: open symbols right scale.

- "Before" the T_c-minimum T_c increases with annealing temperature T_A (ρ_o decreases): the T_c-ρ_o correlation is conserved (Fig. 2, upper part).
- "After" the minimum in the T_c vs dose curves T_c decreases with T_A (ρ_o increases), Fig. 2, lower part. Again - this behavior was observed for all Nb_3Ge, Nb_3Sn and V_3Si samples.

ρ vs T-Curves after Annealing to Room Temperature

Representative for all three materials, Fig. 3 shows ρ vs T curves of several Nb_3Ge samples of different states of damage:
- At low defect concentrations $\rho(T)$ is proportional to T^2 at low temperatures and saturates with $\sim 1/T$.
- For large degrees of damage only small positive or negative temperature coefficients (Nb_3Sn and Nb_3Ge) or a flat resistance minimum (V_3Si) were observed for ρ_o values around 130 $\mu\Omega cm$. This corresponds well with the behavior of metallic glasses (F.Y.Ohkawa, 1978).

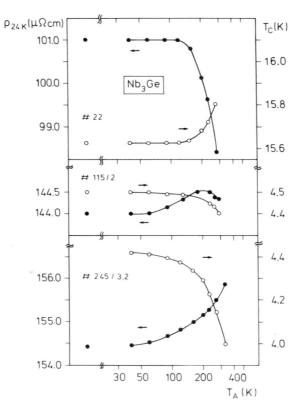

FIGURE 2.   *Annealing behavior of T_c and ρ_0 of Nb₃Ge samples at different degrees of damage; further explanations see text.   Note the weak effects on sample #115/2, which was irradiated just before the T_c-minimum.*

- The thermal resistance ρ_{th} decreases strongly,
        Matthiessen's rule is not conserved (L. R. Testardi
        et al., 1977).
The curves were fitted with the Woodward and Cody formula
(S. J. Williamson et al., 1976).   At high fluences the characteristic temperature T_0 increases strongly.   Figure 4 displays
the decrease of the exponential term with damage.

Measurements of $dH_{c2}/dT(T=T_c)$

        Figure 5 shows a representative result for H'_{c2}: $= dH_{c2}/dT$
$(T=T_c)$ measured in situ after low temperature irradiation of a
Nb₃Ge sample.   The other materials show similar behavior.   The
parameter $\eta\gamma(\eta \approx 1 \ldots 1.2$, strong coupling correction

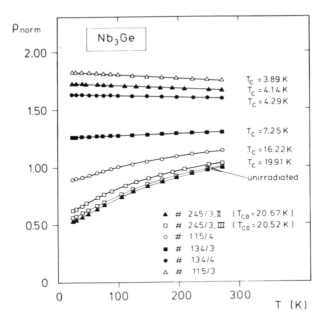

FIGURE 3. ρ *(normalized with* ρ*(273 K;* φ*t=0) vs tempera-ture for several Nb₃Ge samples.*

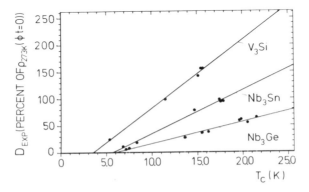

FIGURE 4. *Normalized "strength" of the exponential term of the resistivity vs* T_c *for V₃Si, Nb₃Sn and Nb₃Ge.*

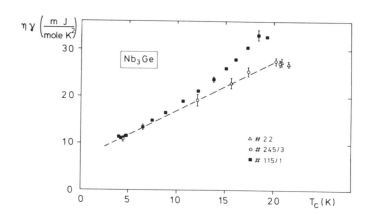

FIGURE 5. $dH_{c2}/dT(T=T_c$ vs T_c of a Nb_3Ge sample after low temperature irradiation with 20 MeV ^{32}S ions.

parameter; γ: coefficient of the electronic specific heat) was extracted from the measured data of H'_{c2} and ρ_0 via the "coherence length-formula" (H. Wiesman et al. 1978). The validity of this formula has been tested recently by several authors (A. K. Ghosh et al., 1979, T. P. Orlando et al., 1979). Figure 6 shows results for Nb_3Ge. Specific heat data ($\gamma(T_c \approx 21$ K) = 24–30 mJ/mole K^2) (G. R. Stewart et al., 1978) agree well

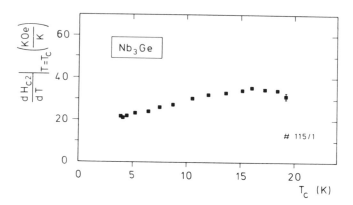

FIGURE 6. $\eta\gamma$ vs T_c for Nb_3Ge; further explanations see text.

with our H'_{c2} values of the unirradiated samples ($\eta \approx 1.2$)
(H. Wiesman et al., 1978). Note the $\eta\gamma$-data (11.4-12.8 ±
0.6 mJ/mole K^2) after T_c-saturation (T_c = 3.9-4.4 K):
(C.C. Tsuei et al., 1978) obtained γ = 12 ± 2 mJ/mole K^2 from
specific heat data of amorphous Nb_3Ge indicating $\eta \approx 1$ and
therefore weak coupling behavior. Δ/kT-data show the same
result.

Discussion of ρ_{th}, T_c and $\eta\gamma$

In the framework of the s-d-interband-scattering model
(N. F. Mott, 1964, I. Nakayama et al. 1978) the thermal part
ρ_{th} of the specific resistivity is correlated with the d-band
density of states. The similar decrease of ρ_{th} and $\eta\gamma$ with the
degree of disorder ($\sim\Delta\rho_{Res}$) supports strongly this assumption
(Fig. 7). Furthermore, even at different states of damage the
good agreement of the ρ_{sd}-formula of I. Nakayama et al. (1978)
with fitting of the ρ vs T curves with the Woodward and Cody
formula using the *same* T_0's shows strong evidence for a selec-
tive electron phonon scattering mechanism (A. I. Golovashkin,
1978).
According to the model of (L. R. Testardi, L. F. Mattheis,
1978) smearing out of sharp structures in N(E) near the Fermi
level due to electron lifetime effects causes a strong decrease

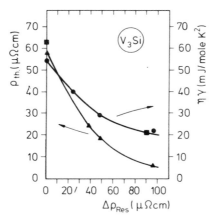

FIGURE 7. ρ_{th} and $\eta\gamma$ for V_3Si as a function of the irradiation induced increase $\Delta\rho_{Res}$ of ρ_0. The two squares are γ-
values obtained from the specific heat data of a neutron irrad-
iated V_3Si single crystal (R. Viswanathan, R. Caton, Phys. Rev.
B18, 15 (1978)).

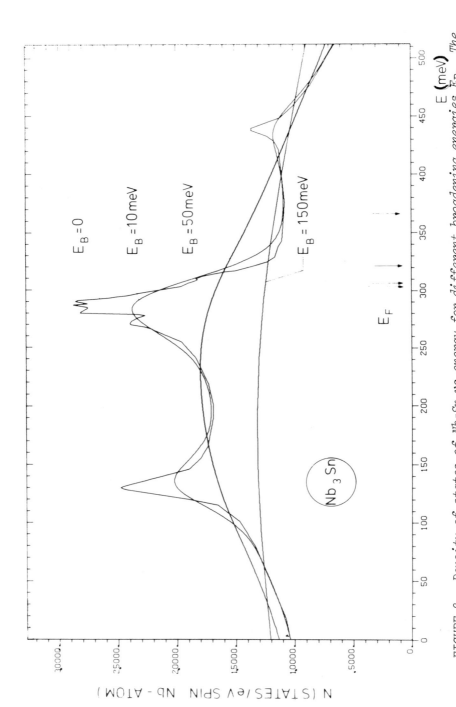

FIGURE 8. Density of states of Nb₃Sn vs energy for different broadening energies E_B. The arrows indicate the increase of E_F upon broadening. $E_B = 0$ data were taken from B.M. Klein et al. (1978).

of $N(E_F)$ with increasing ρ_o. Our $N_{\eta\gamma}$ values obtained from the data of V_3Si after renormalization agree well with this theory. We therefore used this procedure in order to calculate life-time broadened densities of states of Nb_3Sn using the data of B. M. Klein et al. (1978). Figure 8 shows the results. We then used these data as input for T_c-calculations. The most remarkable support for the dominance of the electronic-density of states in determining T_c of A15 compounds is obtained by a calculation of T_c with the Eliashberg formalism of P. Horsch and H. Rietschel (1977) and S. G. Lie et al. (1978). Based on the T_c-equations for the dirty limit (G. Gladstone et al. 1969) these theories include explicitly the energy dependence of $N(E)$ near the Fermi level. We calculated T_c of Nb_3Sn as function of damage. Different amounts of damage were taken into account by different broadening energies of $N(E)$ (see above). Figure 9 shows a comparison of input $N(E_F)$ (renormalized with $\lambda = 1.62$ and in units of mJ mole^{-1} K^{-2}) with our measured data from Nb_3Sn as a function of T_c (calculated and measured). Bearing in mind that the shape of $\alpha^2F(\omega)$ (input

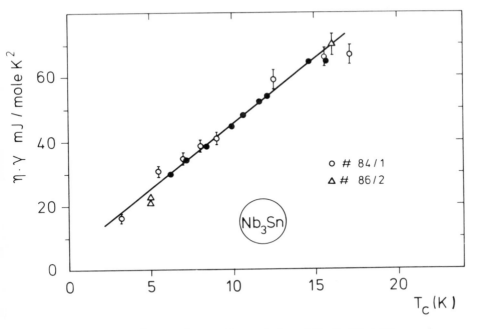

FIGURE 9. Comparison of $\eta\gamma$-data with $N_\gamma(E_F)$(theory) vs T_c (calculated or measured resp.); for further explanations see text.

data taken from tunneling results of L. Y. L. Shen (1972) was
held constant, the T_c degradation is mainly governed by N(E)
degradation. Recent evaluations (B. M. Klein et al., 1979) of
self-consistent APW band structure calculations (B. M. Klein
et al., 1978) show a similar behavior of η_H (Hopfield parameter) vs $N(E_F)$ for the Nb_3B and V_3B family of A15 materials.
Together with the constancy of $M\langle\omega\rangle^2$ (transition metal)
(B. P. Schweiss, 1976) this leads to a similar result. It is
expected that the strong increase of T_c and $N(E_F)$ of Mo_3Ge with
irradiation dose (M. Gurvitch et al., 1978) is also explainable
within the framework of these arguments.

REFERENCES

Adrian, H., Ischenko, G., Lehmann, M., Müller, P., Braun, H.,
    and Linker, G. (1978). *J. Less-Common Metals 62*, 99.
Bieger, J., Ischenko, G., Adrian, H., Lehmann, M., Müller, P.,
    Söldner, L., Haase, E. L., and Tsuei, C. C. (1979). NATO
    Advanced Study Institute on Amorphous Metals, Zwiesel
    (W. Germany).
Braun, H. F. and Saur, E. J. (1978). *J. Low Temp. Physics 33*,
    87.
Brückner, V., Ischenko, G., Löffler, F., Müller, P., Schmidt,
    W., and Schubert, W. (1979). *Verhandl.DPG (VI) 14*, 290.
Ghosh, A. K., Gurvitch, M., Weisman, H., and Strongin, M.
    (1978), *Phys. Rev. B18*, 6116.
Gladstone, G., Jensen, M. A., Schrieffer, J. R. (1969). "Super-
    conductivity, R. D. Parks, ed., vol. II, p. 865, Marcel
    Dekker, New York.
Golovashkin, A. I. (1978). *JETP Lett.26*, 74.
Gurvitch, M., Ghosh, A. K., Gyorffy, B. L., Lutz, H.,
    Kammerer, O. F., Rosner, J. S., and Strongin, M. (1978).
    Phys. Rev. Lett. 41, 1616.
Horsch, P. and Rietschel, H. (1977). *Z. Physik B27*, 153.
Klein, B. M., Boyer, L. L., Papaconstantopoulos, D. A., and
    Mattheis, L. F. (1978). *Phys. Rev. B18*, 6411.
Klein, B. M., Boyer, L. L., and Papaconstantopoulos, D. A.
    (1979). *Phys. Rev. Lett. 42*, 530.
Lie, S. G., Daams, J. M., and Carbotte, J. P. (1978).
    J. Phys. (Paris) Colloq. 39, C6-468.
Ischenko, G., Adrian, H., Klaumünzer, S., Lehmann, M.,
    Müller, P., Neumüller, H., and Szymczak, W. (1977). *Phys.
    Rev. Lett. 39*, 43.
Ischenko, G., Klaumünzer, S., Neumüller, H., Adrian, H., and
    Müller, P. (1978). *J. Nucl. Mat. 72*, 212 and references
    therein.

Mott, N. F. (1964). *Advances in Physics 13*, 325.

Müller, P. and Ischenko, G. (1976). *J. Appl. Phys. 47*, 2811.

Müller, P., Szymczak, W., Ischenko, G. (1978). *Nucl. Instr. and Methods 149*, 271.

Muller, P., Adrian, H., Ischenko, G., and Braun, H. (1978). *J. Phys. (Paris) Colloq.39*, C6-387.

Nakayama, I., and Tsuneto, T. (1978). *Progr. Theor. Phys. 59*, 1418.

Ohkawa, F. J. (1978). *J. Phys. Soc. Jap.44*, 1105, 1112.

Orlando, T. P., McNiff, E. J., Foner, S., and Beasley, M. R. (1979). Stanford University, preprint.

Poate, J. M., Testardi, L. R., Storm, A. R., Augustyniak, W.M. (1975). *Phys. Rev. Lett. 35*, 1290.

Poate, J. M., Dynes, R. C., Testardi, L. R., Hammond, R. H. (1976). *Phys. Rev. Lett.37*, 1308.

Schweiss, B. P., Renker, B., Schneider, E., and Reichardt, W. (1976). Proceedings of the 2nd Rochester Conference on Superconductivity in d- and f-Band Metals, Douglass, D.N., ed., p. 189, Plenum Press, New York.

Sweedler, A. R. (1978). *J. Nucl. Mat. 72*, 50.

Stewart, G. R., Newkirk, L. R., and Valencia, F. A. (1978). *Solid State Commun. 26*, 417.

Shen, L. Y. L. (1972). *Phys. Rev. Lett. 29*, 1082.

Testardi, L.R., Poate, J. M., and Levinstein, H. J. (1977). *Phys. Rev. B15*, 2570.

Testardi, L. R., and Mattheis, L. F. (1978). *Phys. Rev. Lett. 41*, 1612.

Tsuei, C. C., Johnson, W. L., Laibowitz, R. B., and Viggiano, J. M. (1977). *Solid State Commun. 24*, 615.

Tsuei, C. C., Molnar von, S., and Coey, J. M. (1978). *Phys. Rev. Lett. 41*, 664.

Weisman, H., Gurvitch, M., Ghosh, A. K., and Strongin, M. (1978). *J. Low Temp. Phys. 30*, 503.

Weisman, H., Gurvitch, M., Ghosh, A. K., Lutz, H., Kammerer, O. P., and Strongin, M. (1978). *Phys. Rev. B17*, 122.

Williamson, S. J., Milewitz, M. (1976). Proceedings of the 2nd Rochester Conference on Superconductivity in d- and f-Band Metals, Douglass, D. N., ed., p. 551, Plenum Press, New York.

MAGNETIC ORDERING IN RARE-EARTH
TERNARY SUPERCONDUCTORS

G. Shirane
W. Thomlinson

Brookhaven National Laboratory[1]
Upton, New York

D. E. Moncton

Bell Laboratories
Murray Hill, New Jersey

A review is given of current neutron scattering studies of rare-earth(R) ternary superconductors: $RMo_6X_8(X = S \ or \ Se)$ and $ErRh_4B_4$. Most of these compounds develop antiferromagnetic long-range order with coexists microscopically with superconductivity. Two compounds, $HoMo_6S_8$ and $ErRh_4B_4$, become ferromagnetic causing the destruction of superconductivity. In $ErRh_4B_4$, both superconducting and ferromagnetic regions are simultaneously present between 0.9 and 1.2 K but microscopic coexistence is not indicated. However, in this temperature range, magnetic fluctuations occurring in the superconducting regions take the form of a magnetic spiral with a wavelength of ~ 100 Å in order to accommodate superconductivity. This observation is in good agreement with the theoretical predictions of Blount and Varma.

[1]*Research supported by the Division of Basic Energy Sciences, DOE, under Contract No. EY-76-C-0016.*

381

Extensive studies have been carried out in recent years on two new classes of rare-earth (R) ternary compounds: RMo_6X_8 (X = S or Se) (Fischer et al.,1977). Some of these compounds develop both superconductivity and magnetism and they exhibit fascinating physical properties as a result of the interaction between these two competing order parameters. In two cases, $HoMo_6S_8$ (Ishikawa and Fischer, 1977a) and $ErRh_4B_4$ (Fertig et al., 1977), the superconducting state is destroyed at a lower transition temperature T_{c2}. The other compounds remain superconducting (Ishikawa and Fischer, 1977b) below the magnetic transition temperature T_M. In this review, we will discuss recent neutron scattering studies of magnetic ordering in these superconductors.

The results are summarized in Table I together with the appropriate references. This work has been carried out in collaboration with D. B. McWhan and P. H. Schmidt (Bell Laboratories); J. Eckert and C. F. Majkrzak (Brookhaven); M. Ishikawa and Ø. Fischer (Univ. Genève); M. B. Maple, H. B. MacKay, L. D. Woolf, Z. Fisk, D. C. Johnston, and R. N. Shelton (U.C.S.D.); J. W. Lynn (Univ. Maryland). The magnetic transition temperatures T_M in Table I were determined by our neutron scattering measurements and they are in excellent accord with other physical measurements. It was absolutely essential that all of our neutron measurements

TABLE I. Magnetic Ordering in Superconducting Ternary Compounds

Compound	T_c (°K)	T_M		Reference	
$GdMo_6S_8$	1.4	0.84	AF	Majkrzak	(1979)
Tb "	2.05	1.05	"	Thomlinson	(1979a)
Dy "	2.05	0.40	"	Moncton	(1977b)
Ho "	1.2	0.67	F	Lynn	(1978b)
Er "	2.2	0.2	AF	Thomlinson	(1979b)
$GdMo_6Se_8$	5.5	0.8	AF	Maple	(1979)
Er "	6.0	1.1	" (complex)	Lynn	(1978a)
$ErRh_4B_4$	8.5	1.2	F	Moncton	(1977a) (1979)

were done on samples which had been properly characterized by
the electrical, magnetic, and thermal measurements. Only by
this type of systematic study could we positively identify the
correspondence between the magnetic and superconducting
characteristics.

As shown in Table I, most of these compounds develop long-
range antiferromagnetic (AF) order. Rare-earth ions in these
Chevrel phase crystals form simple cubic lattices with slight
rhombohedral distortion. Except for the complex magnetic
structure of $ErMo_6Se_8$ and the ferromagnetic order in $HoMo_6S_8$),
all others order in a simple antiferromagnetic scheme of alternating ferromagnetic sheets along the cubic [100]
direction. As shown in Fig. 1 for $TbMo_6S_8$, some rhombohedral
peaks are clearly split. This permits us to determine the
spin direction uniquely as the rhombohedral [111] direction.

Experimental details were given in our original publica-
tions; we would like to emphasize here only a few specific
problems unique to the current studies of ternary superconductors. Large magnetic moments on rare-earth ions permit
routine neutron diffraction experiments <u>if</u> one can (1) achieve
low temperatures and (2) obtain pure samples. We have utilized two cryogenic systems: a He^3 cryostat (0.3 K) and a
dilution refrigerator (0.05 K) with an option of the application of magnetic fields up to 70 kOe. A small amount of
magnetic impurity poses a special problem in these ternary
compounds because of a low density of the host magnetic ions.
We recall that the magnetic Bragg intensities are proportional
to

$$v \, N_M^{\,2}$$

where v is the volume fraction of each component and N_M is
the number of magnetic atoms per cubic centimeter. Suppose we
have a magnetic impurity such as DyS in which N_M is several
times larger than the host $DyMo_6S_8$. Then this impurity of a
few per cent in volume can give a magnetic peak intensity comparable to those of the host. We have been fortunate, in most
cases, to obtain samples with sufficient purity not to cause
any difficulty. Moreover, most of the magnetic structures
turned out to be very simple as is the case with antiferromagnetic $TbMo_6S_8$ and ferromagnetic $ErRh_4B_4$. Only in one case,
$ErMo_6Se_8$, did we encounter basic difficulties of resolving the
magnetic structure because of impurity phases.

It is now clearly established that antiferromagnetic long-
range order can coexist with superconductivity. It was shown
by Ishikawa and Fischer (1977b) that antiferromagnetic interactions influence the upper critical field H_{c2}. Fig. 2 de-
picts the temperature dependence of sublattice magnetization

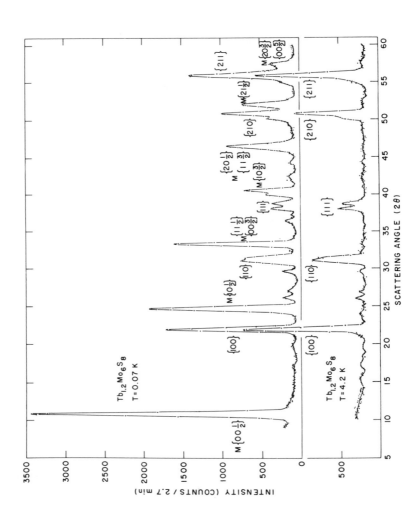

Figure 1. Neutron diffraction data for TbMo₆S₈ above (T = 4.2 K) and below (T = 0.07 K) the antiferromagnetic ordering temperature at T_M = 1.05 K. After Thomlinson et al. (1979a).

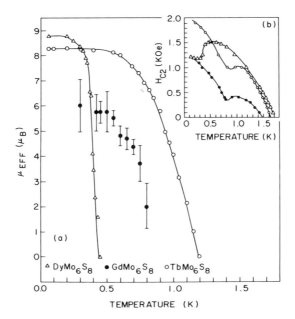

FIGURE 2. *Temperature dependence of the magnetization*
for RMo$_6$S$_8$ (R = Dy, Gd, Tb) as determined by neutron diffraction,
together with the H$_{c2}$ for these compounds. After Majkrzak
et al. (1979).

for three sulfides. A sharp rise was noted for the Dy compound compared with the much slower increases observed for Tb
and Gd. There is a striking correlation between the magnetic
order parameter and H_{c2}, shown in the inset. Further theoretical study is needed along this line.

The most recent neutron scattering activity has been centered around ferromagnetic ErRh$_4$B$_4$. This reentrant superconductor was first reported by Fertig et al.(1977) to undergo a
sudden destruction of superconductivity at T_{c2} = 0.92 K. Subsequently, Moncton et al. (1977) demonstrated, by neutron
scattering, that the compound develops long-range ferromagnetic order below T_M = 1.0 K. This point is demonstrated in
Fig. 3 by additional magnetic components superimposed on
nuclear peaks at low temperatures. Assuming T_{c2} = 0.92 K for
this sample, we concluded that the temperature region in which
the sample had long-range magnetic order and was also superconducting was small if it existed at all. Subsequent measurements of ac susceptibility for this sample (Moncton, 1979)
showed that T_{c2} was 0.74 K, and a clear overlap region is
indicated. We believe, however, that microscopic coexistence
does not obtain; the sample simply contains both ferromagnetic

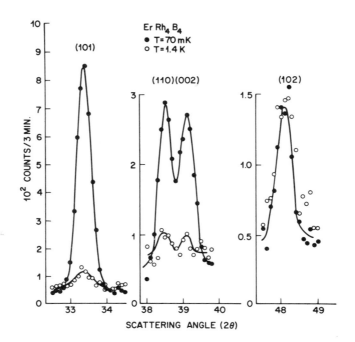

FIGURE 3. Four powder peaks for ErRh₄B₄ at temperatures above (T = 1.4 K) and below (T = 0.07 K). The (102) peak has no magnetic contribution. After Moncton et al. (1977a).

and superconducting domains. Apparently, T_{c2} is quite sensitive to the annealing and other conditions of sample preparation (Rowell et al., 1979). This two-phase behavior is consistent with the observed smearing of the magnetic intensity versus temperature (Moncton et al., 1977). Furthermore, convincing theoretical arguments by Blount and Varma (1979) preclude the coexistence of ferromagnetism and superconductivity due to electromagnetic interactions.

A new and well-annealed sample of $ErRh_4B_4$ (with B^{11} isotope) was prepared by M. B. Maple and collaborators; neutron and specific heat measurements were carried out with the results shown in Fig. 4. These data positively establish the relation between the magnetic order and the reentrant temperature T_{c2}. At the temperature $T_{c2} = 0.92$ K, where the superconductivity appears on heating, the magnetization still retains approximately 70% of its saturation value ($I_M \propto M^2$). Observed magnetic peaks remain sharp until their disappearence around 1.2 K, indicating that a long-range magnetic order exists in the domains which are present with superconducting regions above 0.92 K.

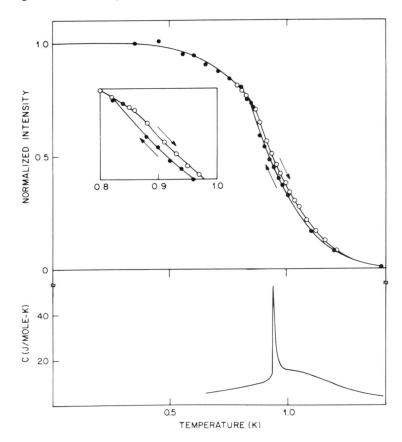

FIGURE 4. Ferromagnetic Bragg intensities for $ErRh_4B_4$, compared with the specific heat data of the identical sample (on heating). After Moncton et al. (1979).

One may note a unique hysteresis covering a temperature range of 0.8 - 1.2 K. This hysteresis is <u>not</u> connected with the first order superconducting normal transition at 0.92 K. We actually see no effect on the magnetic intensity connected with the sharp spike in the specific heat at 0.92 K. The hysteresis of neutron intensity indicates some sort of acti-vation process in this temperature range. This probably cor-responds to the creation of superconducting "domains" out of uniform magnetization. Similar hysteresis was observed in the temperature dependence of the transverse periodicity of charge density waves (Ellenson et al., 1978).

The most striking feature of the mixed phase region above 0.9 K is the nature of the magnetic fluctuations. In ordinary ferromagnetic phase transitions, the magnetic critical scattering is centered at q=0 and peaks at T_c. In $ErRh_4B_4$, the magnetic scattering cross section shows a broad maximum around $2°$ in 2θ for $\lambda = 4$ A. This scattering indicates spiral magnetization with a wavelength of approximately 100 A, in agreement with recent theoretical predictions of Blount and Varma. They show that this wavelength results from a minimization of the magnetic free-energy in the presence of superconductivity. It depends upon both a magnetic length and a superconducting penetration depth. We note that this.is the first experimental observation of the influence of superconductivity on magnetic fluctuations. The details of this theoretical model will be discussed by the next speaker, Chandra Varma.

We have now reached the stage which requires a single crystal for further study. What is the direction of the propagation vector of spiral fluctuations? What are the spin-wave excitations in this unusual ferromagnet? And, finally, is the mixed phase region above 0.92 K due to sample inhomogeneity or does it, as we believe, reflect another aspect of the interaction between magnetism and superconductivity?

We are grateful for stimulating discussions with many of our colleagues, in particular Ø. Fischer, M. Ishikawa, J. W. Lynn, C. F. Majkrzak, M. B. Maple, D. B. McWhan, and C. M. Varma.

REFERENCES

Blount, E. I., and Varma, C. M. (1979). *Phys. Rev. Lett. 42,* 1079.

Ellenson, W. D., Shapiro, S. M., Shirane, G., and Garito, A.F. (1977). *Phys. Rev. 16,*3244.

Fertig, W. A., Johnston, D. C., DeLong, L. E., McCallum, R.W., Maple, M. B., and Matthias, B. T. (1977). *Phys. Rev. Lett. 38,* 987.

Fischer, Ø., Treyvand, A., Chevrel, R., and Sergent, M. (1975). *Solid State Commun. 17,* 21.

Ishikawa, M., and Fischer, Ø. (1977a). *Solid State Commun.23,* 37.

Ishikawa, M., and Fischer, Ø. (1977b). *Solid State Commun. 24,* 747.

Lynn, J. W., Moncton, D. E., Shirane, G., Thomlinson, W., Eckert, J., and Shelton, R. N. (1978a). *J. Appl. Phys. 49,* 1389.

Lynn, J. W., Moncton, D. E., Thomlinson, W., Shirane, G., and Shelton, R. N. (1978b). *Solid State Commun. 26,* 493.

Majkrzak, C. F., Shirane, G., Thomlinson, W., Ishikawa, M., Fischer, Ø., and Moncton, D. E. (1979). *Solid State Commun.* to be published.

Maple, M. B., Woolf, L. D., Majkrzak, C. F., Shirane, G., Thomlinson, W., and Moncton, D. E. (1979). to be published.

Matthias, B. T., Corenzwit, E., Vandenberg, J. M., and Barz, H. E. (1977). *Proc. Nat. Acad. Sci. U. S. A.*, *74*, 1334.

Moncton, D. E., McWhan, D. B., Eckert, J., Shirane, G., and Thomlinson, W. (1977a). *Phys. Rev. Lett. 39*, 1164.

Moncton, D. E., Shirane, G., Thomlinson, W., Ishikawa, M., and Fischer, Ø. (1977b). *Phys. Rev. Lett. 41*, 1133.

Moncton, D. E. (1979). *J. Appl. Phys. 50*, 1880.

Moncton, D. E., McWhan, D. B., Schmidt, P. H., Shirane, G., Thomlinson, W., Maple, M. B., McKay, H. B., Woolf, L. D., Fisk, Z., and Johnston, D. E. (1979). to be published.

Rowell, J. M., Dynes, R. C., and Schmidt, P. H. (1979). Reported at this conference.

Thomlinson, W., Shirane, G., Moncton, D. E., Ishikawa, M., and Fischer, Ø. (1979a). *J. Appl. Phys. 50*, 1981.

Thomlinson, W., Shirane, G., Moncton, D. E., Ishikawa, M., and Fischer, Ø. (1979b). to be published.

TRANSITIONS OF SUPERCONDUCTORS INTO
FERRO- AND ANTIFERRO- MAGNETS

C. M. Varma

Bell Laboratories
Murray Hill, New Jersey

INTRODUCTION

Recent discoveries (Fischer et al., 1975; Matthias et
al., 1977) of superconducting compounds that contain a
lattice of (magnetic) rare-earth ions have reopened the
questions of interaction of superconductivity and magnetic
order that were a topic (Jensen and Suhl, 1966) of considerable discussion in the 1960's. At that time it was not
possible to come to clear conclusions because the materials
investigated contained magnetic ions that were disordered in
the lattice and possibly existed in a clustered form. The
principal conclusions of the new experimental work, for the
points of interest in this talk, are:

(1) The ferromagnetic transition below the superconducting transition is of first-order and superconductivity
disappears below this transition (Maple, 1978).

(2) Antiferromagnetism often co-exists with superconductivity. However there is evidence that the pair-
breaking effects due to the magnetic ions are enhanced near
and in the antiferromagnetic phase (Ishikawa et al., 1978).

The work I will report on has been done in collaboration
with E. I. Blount (Blount and Varma, 1979) and with T. V.
Ramakrishnan (1979). Related work for superconductors undergoing a transition to the ferromagnetic state has been done
by Suhl (1979) and by Matsumoto, Tachiki and Umezawa (1979).

PRINCIPAL INTERACTIONS

There are two principal interaction effects of the magnetization   M with the superconducting order parameter ψ. One
is through the coupling of ψ to the electromagnetic field B,
(Meissner effect) and the coupling of the electromagnetic
field to the magnetization M. The important point is that
the lowest order coupling of the electromagnetic field is
only to the uniform ($q=0$) magnetization. Thus this coupling
may be important for the consideration of ferromagnetic
instabilities, but not for the consideration of antiferromagnetic instabilities.

The other effect is the direct interaction of M with ψ.
As is well known from experience with dilute concentrations of
magnetic impurities in superconductors, the conduction electron spin-magnetic moment interaction is a time reversal non-
invariant perturbation and tends strongly to suppress superconductivity. The theory of this effect has been worked out
by Abrikosov and Gor'kov (1961) for the case of dilute impurities with long-lived local moments to second order in the exchange interaction parameter J. For the case of a lattice of
rare-earth ions, one must adopt a more general point of view
and consider the interaction of spin-fluctuations with the
superconducting order parameter. As in the case of phonons,
one has effects due to both the exchange of spin-fluctuations
between the particles of the Cooper pair and due to the self-
energy of electrons. If one is interested in the lowest
order coupling between M_q and ψ, the term in the free-energy
will be of the form

$$F_{M-\psi} = \int d^3q \int d\omega g(q,\omega) M_q(\omega) M_{-q}(\omega) |\psi|^2 \quad . \tag{1}$$

But generally a $|M_q|^2$ term in the free-energy can be written
in terms of the susceptibility $\chi(q)$

$$F_{M-(q,\omega)} \underset{\sim}{\sim} \tfrac{1}{2}\chi^{-1}(q,\omega) M_q M_{-q}$$

$$\underset{\sim}{\sim} [\tfrac{1}{2}\chi_0^{-1}(q,\omega) - \delta\chi^{-1}(q,\omega)|\psi|^2] M_q(\omega) M_{-q}(\omega) \tag{2}$$

Thus $g(q,\omega)$ can be identified as arising from the change in
the susceptibility due to superconductivity. This quantity
can readily be calculated. In Fig. (1) the q-dependence of
the zero frequency susceptibility of a normal metal and of a
superconductor (at finite temperature and/or with spin-orbit
coupling so that $\chi(0) \neq 0$) is represented schematically.

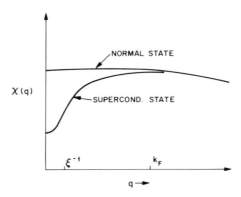

FIGURE 1.   The wave-vector dependence of the normal-state
and the superconducting-state magnetic susceptibility for a
metal with a spherical Fermi-surface.

For small q, $\delta\chi(q,0) \sim (q\xi)^2$, where ξ is the coherence length
of the superconductor, whereas for $q \sim k_F$, $\delta\chi(q) \sim k_F/q$.
Correspondingly, $g(q)$ can be shown to be

$$g(q,0) = \beta J^2 \rho(\varepsilon_f)(q\xi)^2 \ , \ q < \xi^{-1}$$

$$\sim \beta J^2 \rho(\varepsilon_f)\left(\frac{k_F}{q}\right) \ , \ q \sim k_F \qquad (3)$$

where ξ is the coherence length and J is the exchange energy
between the local moment and the conduction electrons. For the
problems of present interest J is only about 10 meV (Maple et
al., 1977; Fischer et al., 1979), whereas it may be as large as
1 eV for transition metal impurities in superconductors.

Above the ferromagnetic transition temperature the q of
importance is around 0, whereas above the antiferromagnetic
transition temperature it is around the reciprocal-vector of
the Neél phase. From (3), the effect of the direct coupling
being considered is of the same order near a ferro- or an
anti-ferromagnetic transition, because of the large phase-
space involved in the latter. I believe that these interactions determine the enhancement of pair-breaking observed
in the antiferromagnetic phase and that in the materials
investigated that show a ferromagnetic transition, the phenomenon is mostly controlled by electromagnetic effects. This
will be elaborated as we go along.

PAIR-BREAKING DUE TO SPIN-FLUCTUATIONS

In a mean-field approximation, we can write

$$F_{M-\psi} \approx a'|\psi|^2 \quad , \tag{4}$$

$$a' \approx \int d^3q \int d\omega \, g(q,\omega) <M_q(\omega) M_{-q}(\omega)> \tag{5}$$

The effect of antiferromagnetism can therefore be investigated in this approximation by calculating a'.

The Abrikosov-Gorkov theory would have

$$a' \approx J^2 \rho^2 (\varepsilon_f) S(S+1)/k_B T_C \quad . \tag{6}$$

The $a'|\psi|^2$ renormalizes the usual

$$a|\psi|^2 \quad , \quad \left(a = \rho(\varepsilon_f) \frac{T-T_C}{T_C} \right)$$

term so that the coefficient of the $|\psi|^2$ term is

$$\tilde{a} = a + a' \quad .$$

The coherence length ξ is given by

$$\xi^2 \approx \frac{\hbar^2}{2m\tilde{a}} \quad ,$$

and in type II superconductors H_{C2} is given by

$$H_{C2} \approx \phi_0/\xi^2$$

where ϕ_0 is the unit of flux.

The evaluation of $g(q,\omega)$ to lowest order in perturbation theory may be done by calculating the two vertices shown in Fig. (2). Fig. (2a) is the leading diagram of the ladder type while (2b) is the leading self-energy correction. The vertices correspond as shown to the exchange interaction of the conduction electrons with the localized spins S, and to the BCS pair interaction parameter λ. Note that if there were phonon interactions instead of the s·S interaction the

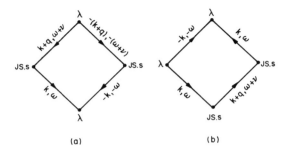

FIGURE 2.   *Leading diagrams for the vertex for the local*
moment conduction electron spin interactions in a
superconductor.

zero frequency contribution of the two diagrams would cancel
identically. However, for the s·S interaction they are
identical and add. Moreover, Fig. (1b) is negligible for
finite frequencies. For q's of interest here, the contribution of the two diagrams can be evaluated to be

$$g(q,m) \approx \frac{\pi^4}{2} J^2 S(S+1) \lambda^2 \frac{\rho(\epsilon_f)}{\epsilon_f} \frac{k_F}{q}$$

$$\cdot \beta \left[1 + \frac{4}{\pi^2} \frac{1}{m} \left(1 + \frac{1}{3} + \ldots + \frac{1}{(2m+1)} \right) \right] \qquad (7)$$

where the frequency $\omega_m = 2m\pi k_B T$, and m is an integer (not
zero).

Next we approximate $<S(q,\omega)S(-q,\omega)>$ by an Ornstein-Zernike
(O.Z.) form near T_N and get $F_{M-\psi}$ in the form (4), with

$$a'(T) \simeq \alpha [\{ 1 - \frac{\pi}{2} (\Delta T/T_N)^{\frac{1}{2}} \}$$

$$+ \ln^2(\omega_0/k_B T)] \quad , \quad \text{for } \frac{\Delta T}{T_N} > 0 \text{ but } << 1 \quad . \qquad (8)$$

In Eq. (8)

$$\alpha \stackrel{\sim}{=} J^2 S(S+1) \lambda^2 \frac{\rho(\epsilon_f)}{\epsilon_f} \cdot \frac{1}{\hbar \omega_0} \quad ,$$

where $\omega_0 \simeq DR^2$, where D is the diffusion constant in the O.Z. expression and R is the lattice constant. The contribution in the curly bracket is due to the static fluctuations and the rest is due to the finite frequency fluctuations. For $T \gg T_N$, $\hbar\omega_0 \stackrel{\sim}{\sim} k_\beta T$, the finite frequency contribution vanishes and the result of Abrikosov-Gorkov to the same order as here is obtained.

Below the magnetic transition temperature,

$$<M_q M_{-q}> = <M_Q>^2 + <\delta M_q \delta M_{-q}> \quad , \tag{8a}$$

where Q are the antiferromagnetic Bragg points. The first term represents a new periodic potential, which will alter the band structure and weaken superconductivity. This is not really a magnetic effect on superconductivity, but an altered electron-phonon effect due to the new band structure. Note that for the ferromagnetic case (Q = 0), the first term in (8a) acts as a polarization term, whose effect apart from a factor $<S>^2/S(S+1)$ would be of the same nature below and above the transition.

Below T_N, the longitudinal fluctuations are again of a diffusive nature, and Eq. (8) is again obtained for them with $\Delta T \rightarrow |\Delta T|$. The transverse fluctuations, however, develop into spin waves. Taking a simple mean field form for the spin waves in $<\delta M(q,\omega) \delta M(-q,\omega)>$ gives

$$a'(T) \simeq \alpha \left[1 + \ln^2 \frac{\omega_s}{k_\beta T} \right] ; \quad T < T_N \tag{9}$$

where ω_s is a typical spin-wave frequency.

The result of these calculations is a small increase in $a'(T)$ and therefore of pair-breaking as T_N is approached, Eq. (8), and a further increase in pair-breaking for $T < T_N$ due to the spin-waves as given by Eq. (9). The theory is valid only for $a' \ll a$ and thus breaks down at low temperatures. The whole class of pair-breaking diagrams must then be summed. This will be reported later.

The theory near T_N suffers from use of the O.Z. approximation. In that regime, the physics is analogous to that of electrical resistivity of metals near the anti-ferromagnetic transition points (Hohenberg and Halperin, 1979), where improved approximations for the spin-spin

correlations have been used. Another fault of the theory is
that $\chi(q)$ for the actual case will probably not correspond to
the spherical Fermi-surface but will peak, both for χ_N and χ_S
near the antiferromagnetic Bragg points. Remedying both
these faults again leads to an enhancement of pair-breaking
as T_N is approached, but with the exponent $\frac{1}{2}$ in (8) replaced
by the much smaller value $1-\alpha$, where α is the specific heat
exponent.

PAIR-BREAKING IN FERROMAGNETS

Before we part with the effects due to the direct $M-\psi$
interactions, let me mention that for the ferromagnet the
free-energy due to these can be written (see Fig. (1) and
Eq. (2)).

$$\eta_1 M^2 \psi^2 + \eta_2 |\nabla M|^2 \psi^2 \ , \tag{10}$$

where $\eta_1 \approx \frac{1}{2}\chi_0^{-1}(0)J^2$, and $\eta_2 \approx \xi^2 \chi_0^{-1}J^2$. Since the susceptibility begins to recover towards the normal value at large q,
superconductors may prefer to have magnetic order only at finite
q as was realized from these considerations by Anderson and
Suhl (1959). But as we shall soon see, the Anderson-Suhl
effect is overshadowed for the present cases of interest
(type II materials and small J) by electromagnetic effects.

ELECTROMAGNETIC EFFECTS NEAR THE FERROMAGNETIC TRANSITION

Our starting point is the free-energy functional:

$$F\{\psi,\vec{M},\vec{A}\} = \int d^3r \{ \tfrac{1}{2}a|\psi|^2 + \tfrac{1}{4}b|\psi|^4 + \tfrac{1}{2}p_0|(\nabla - ir_0\vec{A})\psi|^2$$

$$+ (\vec{B}^2/8\pi) + \tfrac{1}{2}\alpha|\vec{M}|^2 + \tfrac{1}{4}\beta|\vec{M}|^4 + \tfrac{1}{2}\gamma^2|\nabla\vec{M}|^2$$

$$- \vec{B}\cdot\vec{M} + \tfrac{1}{2}[\eta_1\vec{M}^2 + \eta_2|\nabla\vec{M}|^2]|\psi|^2 \} \ . \tag{11}$$

In Eq. (11) $\vec{B} = \nabla \times \vec{A}$. Also $a = a_0(T-T_c)/T_c$, where T_c is
the upper superconducting transition temperature and
$r_0 = 2e/\hbar c$. The London penetration depth $\lambda(T)$ is given by
$\lambda^{-2} = 4\pi p_0 r_0^2 |\psi|^2$, $\alpha = \alpha_0(T-T_m')/T_m'$.

The uniform superconducting state has the free-energy
density

$$F_c = -a^2/4b \quad \text{for} \quad T > T_c \ . \tag{12}$$

If $|\psi| \neq 0$, then $B = 0$ in the bulk of the sample, and the question of magnetic order does not arise until $T < T_m{}'$. If, however, $|\psi| = 0$, then $B = 4\pi M$, and the free-energy density for the uniform magnetic state (with $H=0$) is

$$F_M = -(\alpha-4\pi)^2/4\beta \quad , \quad T < T_m^0 \quad , \tag{13}$$

and

$$T_m^0 = (1+4\pi/\alpha_0) T_m{}' \quad . \tag{14}$$

The latter state is locally stable with respect to fluctuations in $|\psi|$ because A increases with the size of the system when B is uniform. For situations of present interest, viz., $T_m^0 \sim 1$ K, α_0 is of order unity, so that $T_m^0 \gg T_m{}'$.

We now consider states of uniform $|\psi|$ with variable M(r) and A(r). It turns out A(r) can again be eliminated leaving an extra magnetic self-energy term at finite q given by

$$\left(-2\pi + \frac{2\pi}{(1+\lambda^2 q^2)}\right) M_q^2 \quad . \tag{15}$$

At $q = 0$, this is zero in accord with our discussion above, while for $q \gg \lambda^{-1}$ the full benefit of the electromagnetic field is felt. Thus superconductivity expels electromagnetic fields effectively only for $q \ll \lambda^{-1}$. The term (15) together with $\hat{\gamma}^2 q^2 M_q^2$, $(\hat{\gamma}^2 = \gamma^2 + \eta_2|\psi|^2)$ the magnetic stiffness energy, leads to magnetization being favored at a finite q. This is illustrated in Fig. (3). It is easy to convince oneself that the magnetic structure would be of a spiral type. The inverse susceptibility at $q_0 \sim (\hat{\gamma}\lambda)^{-\frac{1}{2}}$ vanishes at a temperature $T_S(|\psi|)$ given by

$$[T_m^0 - T_S(|\psi|)] \sim 4\pi^{\frac{1}{2}}(\hat{\gamma}/\lambda)/(\alpha_0+4\pi) \quad . \tag{16}$$

If we compare the free-energy of the three states discussed above, we find that, since $a^2/4b$ is of order $k^2 T_c^2/E_F V$ as $T \to 0$ whereas $\alpha^2/4\beta$ is of order $k_B T_m^0/V$ as $T \to 0$ (V is the volume per atom), the uniform magnetic state is always favored over the superconducting state at some temperature below T_m^0. The uniform magnetic state is also favored over the state of spiral magnetism over the entire temperature range below T_m^0 if

$$\frac{(\alpha-4\pi)^2}{4\beta}\left(\frac{T_S-T_m^C}{T_m^0}\right)^2 > \frac{a^2}{4b} \ . \tag{17}$$

If the inequality is not satisfied, the spiral state is favored in a small temperature region below T_S; at lower temperatures the uniform magnetic state is again favored. A schematic comparison of the free energies of the three states is given in Fig. (3).

To sum up, the mean-field results give a first-order transition from the superconducting state to the uniform magnetic state if the inequality (17) is satisfied. For smaller $|T_S-T_m^0|$, a second-order transition to the spiral state is indicated followed by a first-order transition to the uniform magnetic state. For a reasonable choice of parameters the temperature
state is indicated followed by a first-order transition to the uniform magnetic state. For a reasonable choice of parameters the temperature region in which the spiral state is favored is very small, if it exists at all.

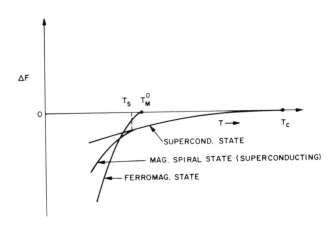

FIGURE 3.   Coefficients of M_q^2 term in the free-energy illustrating why the susceptibility peaks at a finite q due to electromagnetic effects and magnetic stiffness. The figure also illustrates why for type II superconductors with moderate values of J, spin-spin scattering effects are not as important as electromagnetic effects to determine q at which susceptibility peaks.

Using (11) it is also possible to discuss, to a first approximation, the fluctuations about the mean-field result. $\chi_{BB'}$ which is measured by neutron diffraction, peaks at

$$q_B = q_0 + \frac{\alpha - [(4\pi)^{\frac{1}{2}} - \hat{\gamma}/\lambda]^2}{\hat{\gamma}^2 \lambda^2 q_0^2}$$

$$\hat{\alpha} = \alpha + \eta_1 |\psi|^2 \tag{18}$$

Thus the above first order transition to a uniform ferromagnetic state χ_{BB} peaks at a finite wave-vector which gradually deviates from q_0. The prediction of such unusual critical scattering has been verified in small-angle neutron scattering (Moncton, 1978).

The effect of spin-spin scattering in the above discussion has been taken into account by $\alpha \to \hat{\alpha}$ and $\gamma^2 \to \hat{\gamma}^2$. From (10), we conclude that if J is small and $\lambda \gg \xi$ (type II materials), the minimum as a function of the coefficient of the M_q^2 term is primarily determined by the electromagnetic effect (and the ordinary spin-stiffness). This is illustrated in Fig. (4). Anderson and Suhl had visualized a 'Cryptoferromagnetic' state by consideration of the spin-spin interaction

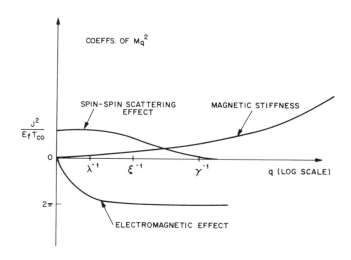

FIGURE 4.   *Schematic comparison of the free-energies for the states discussed in the text.*

terms alone. Our conclusion is that for type II materials, electromagnetic effects dominate over the mechanism considered by Anderson and Suhl.

Ferrel, et al. (1979), have recently considered modifications of the theory proposed by E. I. Blount and me to include non-local effects of electromagnetic screening. These effects are however unimportant for $\lambda \gg \xi$.

In the discussion above, we have neglected questions of magnetic anisotropy. It would seem, however, that if the anisotropy is of the Ising type, spiral fluctuations will not be favored. This is probably the reason why in $Er_{0.6}Ho_{0.4}Rh_4B_4$ (Mook, 1979), where magnetization points along the c-axis, spiral fluctuations are not observed, whereas they are observed in $ErRh_4B_4$ (Moncton, 1978), where the moments lie in the basal plane.

Anisotropy probably plays a major role in the ferromagnetic transition temperature. Neutron irradiation damage sharply reduces T_m (Rowell et al., 1979) and the most plausible reason we can think for this is that damage destroys the coherence of the anisotropy axis.

An interesting possibility for superconductors that become ferromagnetic might have been the realization of a spontaneous vortex structure. After all, if the B field produced due to M_{sat} is small compared to H_{C2} and if $\lambda \gg \xi$, a phase with spontaneous net vorticity may be imagined. The reason this cannot happen is that the correlation length associated with M is small and that the effective penetration depth becomes of the order $(\lambda\gamma)^{\frac{1}{2}}$, so that the material begins to behave like a type I material.

ACKNOWLEDGEMENTS

As mentioned earlier, the work reported here was done in collaboration with E. I. Blount and with T. V. Ramakrishnan. I also wish to thank Ø. Fischer, B. Maple and D. Moncton for discussions about the experimental results.

REFERENCES

Abrikosov, A. A. and Gorkov, L. P. (1961), Sov. Phys. JETP12, 1243

Anderson, P. W. and Suhl, H. (1959), Phys. Rev. 116, 898.

Blount, E. I. and Varma, C. M. (1979), Phys. Rev. Lett. 42, 1079.

Ferrel, R. A., Bhattacharjee, J. K. and Bagchi, A. (1979), Phys. Rev. Lett. 43, 154.

Fischer, Ø., Treyvaud, A., Chevrel, R. and Sergent, M. (1975), Solid State Comm. 17, 721.

Fischer, Ø., Ishikawa, M., Pelizzone, M. and Treyvaud, A. (1979), J. Physique 40, C5-89.

Hohenberg, P. C. and Halperin, B. I. (1979), Rev. Mod Phys. 49, 435.

Ishikawa, M., Fischer, Ø., and Müller, J. (1978), J. Physique Supplement 3, C6-1379, and references cited therein.

Jensen, M. A. and Suhl, H. (1966), in Magnetism, Vol 11 B, ed. G. T. Rado and H. Suhl (Academic Press, New York).

Maple, M. B., DeLong, L. E., Fertig, W. A., Johnston, D. C., McCallum, R. W. and Shelton, R. N. (1977), in Valence Instabilities and Related Narrow Band Phenomenon, ed. R. D. Parks (Plenum, New York), 17.

Maple, M. B. (1978), J. Physique Supplement 3, C6-1374, and references cited therein.

Matsumoto, H., Tachiki, M. and Umezawa, H. (1979), preprint.

Matthias, B. T., Cerenzwit, E., Vandenberg, J. M. and Barz, H. E. (1977), Proc. Natl. Acad. Sci. USA, 74, 1334.

Moncton, D. E. (1978), Proc. 24th Conference on Magnetism and Magnetic Materials, J. Appl. Phys. (to be published).

Mook, H. A., Koehler, W. C., Maple, M. B., Fisk, Z. and Johnston, D. C. (1979), private communication and Proc. of this conference.

Ramakrishnan, T. V. and Varma, C. M. (1979), unpublished.

Rowell, J. M., Dynes, R. C. and Schmidt, P. H. (1979), Solid State Comm. 30, 191, and Proc. of this conference.

Suhl, H. (1979), preprint.

LOW TEMPERATURE THERMAL CONDUCTIVITY OF $ErRh_4B_4$

H.R. Ott
W. Odoni

Laboratorium für Festkörperphysik
ETH-Hönggerberg
Zürich, Switzerland

The thermal conductivity of polycrystalline $ErRh_4B_4$ was measured between 0.05 K and 4 K, and in magnetic fields up to 3 kOe with directions parallel and perpendicular to the heat flow. The transition from the superconducting to the magnetically ordered normal state is hysteretic in temperature as observed before in resistivity measurements.

I. INTRODUCTION

The phenomenon of reentrant superconductivity upon long range magnetic ordering has recently been studied mainly in two types of materials, namely in rare-earth Rh_4B_4-compounds and in rare-earth Chevrel phases (Maple, 1978; Ishikawa *et al.*, 1978). Of special interest is, of course, the temperature range where with decreasing temperature the compounds undergo a transition from the superconducting to a normal but magneti-

Work supported by the Schweizerische Nationalfonds

403

cally ordered state at a critical temperature T_c. It seems
likely that the fundamental problem of the coexistence of
superconductivity and long range magnetic order can best be
studied in such cases and it has indeed been found that antiferromagnetic order and superconductivity apparently coexist
in some materials (Fischer *et al.*, 1979). Ferromagnetic order,
however, seems to destroy superconductivity but several experiments have revealed that the states of superconductivity and
net magnetization may coexist in a very limited temperature
range around T_c (Moncton *et al.*, 1978; Maple, 1978; Ott *et al.*,
1978). Investigating such problems it is essential to probe
real bulk properties in order to make sure that the observed
effects are not merely of spurious nature.

A material in which superconductivity is destroyed by the
onset of ferromagnetic order is $ErRh_4B_4$ (Fertig *et al.*, 1977)
and we report on measurements of the thermal conductivity λ of
$ErRh_4B_4$ between 50 mK and 4 K. Among the transport properties
the thermal conductivity is certainly better suited to sample
bulk properties than e.g. the electrical resistivity, because
in our case superconductivity is involved. Furthermore $\lambda(T)$
can also give some information on microscopic properties of
$ErRh_4B_4$ as a superconductor.

II. SAMPLES AND EXPERIMENT

The sample was a prism of polycrystalline material with
dimensions of $7 \times 1.5 \times 1.5$ mm^3 and in previous work it was denoted as sample A (see Ott *et al.*,1978). The thermal conductivity λ was determined by measuring the temperature difference
ΔT across the specimen caused by a steady heat flow $\dot{Q}$. The
whole temperature range was covered in a dilution refrigerator.
The accuracy of the absolute temperature measurement was of the
order of 2 to 3% and ΔT was always chosen to be less than 3%
of the absolute temperature. We also did measurements of $\lambda(T)$
in external magnetic fields up to 3 kOe with the field directions parallel or perpendicular to the heat flow.

III. RESULTS AND DISCUSSION

In Figure 1 we show our results for λ in zero magnetic
field (earth field not compensated) as a function of T over the
entire temperature range investigated. Within the covered temperature range many contributions are likely to influence the
thermal conductivity but it appears that at least two familiar
features may readily be identified. Between 0.05 K and 0.9 K
we observe a linear T-dependence of λ, as expected for a domi-
nantly electronic thermal conductivity limited by impurity and
imperfection scattering at low temperatures. This fits well
with the temperature independent electrical resistivity observed below the phase transition to the normal state in previous
work (Fertig et al., 1977; Ott et al., 1978). In the superconducting state and provided T is considerably below T$_c$ one expects a T^3-dependence of λ due to the lattice conductivity.
Between 1.3 K and 2.5 K our measurements of the temperature
dependence of λ indeed reveal a trend towards a T-exponent of
about 3.
    In order to investigate a possible temperature hysteresis
of the phase transition at 1 K as was found before in the electrical resisitivity (Fertig et al., 1977; Ott et al., 1978), λ
was measured for both decreasing and increasing temperature
through the transition. It may be seen from Figure 1 that also

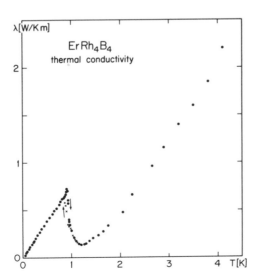

FIGURE 1.   Low temperature thermal conductivity of
ErRh$_4$B$_4$ in zero magnetic field.

for the thermal conductivity the warming and the cooling curves
are not identical in a limited temperature range and this indicates that this thermal hysteresis is a bulk phenomenon.

An analysis of the zero field temperature dependence of λ
has to take into account that $ErRh_4B_4$ is a superconductor with
a high concentration of magnetic impurities and even magnetic
order at low enough temperatures. The magnetic ordering induces
a transition from the superconducting to the normal state. Our
data are not yet sufficient to allow for a thorough analysis
in the superconducting region, especially for $t = T/T_c > 0.5$.
From the results below 4 K it is difficult to separate the electronic and the lattice contribution but they indicate that
the electronic thermal conductivity ratio $\lambda_{es}/\lambda_{en}$ varies much
less with decreasing temperature than what is expected from the
theory of Bardeen, Rickayzen and Tewordt (Bardeen *et al.*, 1959).
This agrees in general with calculations of Ambegaokar and
Griffin (Ambegaokar and Griffin, 1965) where for $t < 0.5$
$\lambda_{es}/\lambda_{en}$ tends to increase above the BRT values with increasing
number of magnetic impurities. Of special interest is, of
course, the temperature range between 0.85 K and 1.3 K where
we interpret the temperature dependence of λ to be due to a decrease of the energy gap with decreasing temperature originating from an additional pair-breaking due to the onset of magnetic ordering (Fulde and Maki, 1966) and hence a reactivation
of the depaired conduction electrons to take part in the heat
transport process.

Neutron diffraction experiments show an increasing intensity of magnetic Bragg-peaks below 1.2 K (Moncton *et al.*, 1978).
At this temperature we observe a minimum for λ. At 0.85 K where
with decreasing temperature $\lambda(T)$ shows a maximum and then decreases linearly with T, the spontaneous magnetization of the
Er ions has reached about 65% of their saturation value. It is
therefore justified to claim that superconductivity and a magnetically ordered state with a net spontaneous magnetization
in some way coexist between 0.85 K and 1.2 K. We note that at
0.95 K, according to the neutron experiments the magnetization
has reached about half of its saturation value but only about
40% of the potential electronic thermal conductivity is observed. This "coexistence" is probably possible because $ErRh_4B_4$
is a type II superconductor with a negative surface energy between the normal and superconducting regions. Taking this into
account and considering the problem of nucleation will probably
also explain the thermal hysteresis. Comparing the results of
different experiments like neutron diffraction (Moncton *et al.*,
1978), specific heat (Maple, 1978), resistivity (Ott *et al.*,
1978) and the present work (see also Odoni and Ott, 1979), it

appears that the hysteretic first order phase transition is related to the superconducting to normal transition influenced
by the magnetic ordering but not to the magnetic ordering itself.

In order to check that the observed anomaly between 0.85 K
and 1.3 K is due to a transition from a normal to a superconducting state we also measured λ(T) in external magnetic
fields. Figure 2 shows the difference of how λ is affected by
a parallel or a perpendicular magnetic field. In the parallel
case we expect that the normal cores in the mixed state immediately lead to an increase of λ at constant temperature. In
perpendicular fields the mixed state leaves the sample in a
completely different situation and the increase of λ(T) in external fields should be much less pronounced than in the parallel case. This is exactly what we find and it is displayed
in Figure 2.

In perpendicular fields one might even expect a decrease
of λ(T) in small fields due to additional scattering on vorti-
ces. We do not, however, observe such a behaviour at any temperature up to 1.5 K. It is rather round that at a fixed temperature λ increases linearly with external field, a feature already found in previous work on type II superconductors
(Parks et al., 1967). It could also indicate that H$_{c1}$ of
ErRh$_4$B$_4$ is always less than 250 Oe in the temperature range
covered so far.

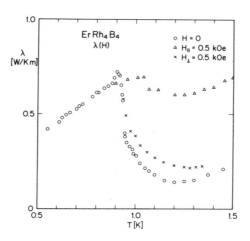

FIGURE 2. Influence of magnetic fields on the thermal
conductivity of ErRh$_4$B$_4$.

REFERENCES

Ambegaokar, V., and Griffin, A. (1965). *Phys.Rev.* *137*, 1151.
Bardeen, J., Rickayzen, G., and Tewordt, L. (1959).
    Phys.Rev. *113*, 982.
Fertig, W.A., Johnston, D.C., DeLong, L.E., McCallum, R.W.,
    Maple, M.B., and Matthias, B.T. (1977). *Phys. Rev. Letters*
    38, 987.
Fischer, Ø., Ishikawa, M., Pelizzone, M., and Treyvaud, A.
    (1979). *J. Physique C5*, 89.
Fulde, P., and Maki, K. (1966). *Phys. Rev. 141*, 275.
Ishikawa, M., Fischer, Ø., and Müller, J. (1978). *J. Physique*
    C6, 1379.
Maple, M.B. (1978). *J. Physique C6*, 1374.
Moncton, D.E., McWhan, D.B., Eckert, J., Shirane, G., and
    Thomlinson, W. (1978). *Phys. Rev. Letters 39*, 1164.
Odoni, W., and Ott, H.R. (1979). *Phys. Letters 70A*, 480.
Ott, H.R., Fertig, W.A., Johnston, D.C., Maple, M.B., and
    Matthias, B.T. (1978). *J. Low Temp. Phys. 33*, 159.
Parks, R.D., Zumsteg, F.C., and Mochel, J.M. (1967). *Phys. Rev.*
    Letters 18, 47.

ION DAMAGE, CRITICAL CURRENT AND
TUNNELING STUDIES OF $ErRh_4B_4$ FILMS

J. M. Rowell
R. C. Dynes
P. H. Schmidt

Bell Laboratories
Murray Hill, NJ

We recently speculated (Rowell et al., 1979) that the
unusual sensitivity of the A15 superconductors to damage by
neutrons, α particles or electrons will be a common behavior
in all cluster compound superconductors (Vandenberg and
Matthias, 1977). This was demonstrated, at least for
$ErRh_4B_4$, by showing the rapid depression of the superconducting and magnetic transition temperatures (T_c and T_m) with α
particle damage, and that T_c depended on the resulting
resistance ratio in a fashion exactly similar to that in the
high T_c A15s.

We report here some further details of these damage
studies of $ErRh_4B_4$ films, and also some preliminary critical
current and tunneling measurements. Details of the film
preparation and references to similar work on the A15s will
not be repeated here.

In an extension of the data shown in Fig. 2 of our
earlier publication, resistive measurements of two samples
were made down to .035 K. At a damage level of 3.75×10^{16}
α/cm^2, no re-entry to the normal state was observed, thus
allowing us to extend the superconducting region of the
phase diagram to low T. At the higher damage level of
$1 \times 10^{17} \alpha/cm^2$, no indication of a superconducting transition
was observed to .035 K. Thus there appears to be no saturation of either T_c or T_m, at least down to .035 K. This is
in contrast to A15s, where T_c generally saturates between
1 and 4 K.

A direct comparison of the sensitivities of $ErRh_4B_4$,
Nb_3Ge and Nb_3Sn to ion-beam damage is shown in Fig. 1, where

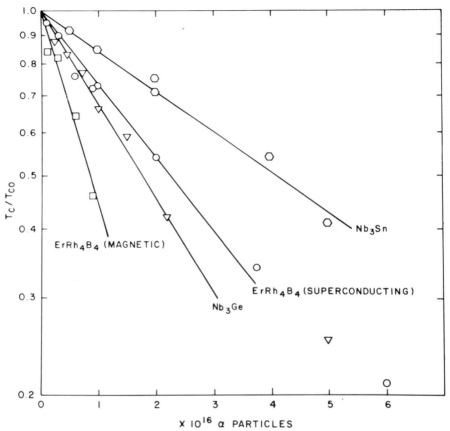

FIGURE 1. *Sensitivity of the superconducting and magnetic transitions of* $ErRh_4B_4$ *to 1.8 MeV* α *particle irradiation. For comparison similar results for* Nb_3Sn *and* Nb_3Ge *are shown.*

a plot of $\log(T_C/T_{CO})$ [T_{CO} being the transition temperature of the undamaged film] versus α particle dose (D) yields reasonable straight lines, at least down to $T_C/T_{CO} \sim 0.4$. Such a depression of T_C, given by $T_C = T_{CO}\exp(-D/D_O)$, has been noted earlier for both A15s and Chevrel compounds. From similar plots of the published data we have summarized the sensitivity to damage for A15s, Chevrels and $ErRh_4B_4$ in Table I. A clear difference between neutron and α particle damage is the comparatively small value of D_O for neutron damage of vanadium based A15s, not surprising in view of the large neutron cross section of vanadium.

The resistivity versus temperature of $ErRh_4B_4$ as a function of α particle dose was measured on a well-defined

TABLE I

| | Neutron Damage | | | α Particle Damage | |
	Ref. A	Ref. B	Ref. C	Ref. D	Ref. E
$PbMo_6S_8$	4.5×10^{18}		4.3×10^{18}		
$NbSe_2$	4.5×10^{18}				
$SnMo_6S_8$			5.6×10^{18}		
V_3Si	1.2×10^{19}	1.1×10^{19}		1.4×10^{17}	
Nb_3Pt		1.3×10^{19}			
Nb_3Ga	1.3×10^{19}				
Nb_3Ge	1.6×10^{19}	1.5×10^{19}		2.5×10^{16}	
$ErRh_4B_4$ (T_m)					1.2×10^{16}
(T_c)					3.4×10^{16}
Nb_3Al	2.0×10^{19}	1.8×10^{19}	1.5×10^{19}		
Nb_3Sn	2.2×10^{19}	1.5×10^{19}		6.2×10^{16}	
V_3Ge				9.3×10^{16}	

[A] Sweedler et al., 1976.

[B] Sweedler et al., 1978.

[C] Brown et al., 1977.

[D] Dynes et al., 1977.

[E] Rowell et al., 1979.

bridge-shaped sample prepared by photolithography and ion milling (Fig. 2). However the film had a rough surface which limited the accuracy with which its thickness could be determined. Thus the absolute resistivities of Fig. 2 are uncertain to ∿ ±20%. Again, the similarity of the data of Fig. 2 to that for many A15s is striking. However, a careful analysis of the resistivities between 10 and 30 K, using the method described by Gurvitch (Gurvitch, 1979), shows $\rho \propto T^3$ whereas $\rho \propto T^2$ has been reported for A15s.

From Fig. 2 it can be seen that ρ_o is in the range 30 to 300 μΩcm for these films. A plot of T_c versus ρ_o is shown in Fig. 3 and again compared with compounds reported by others previously. Except in the case of Nb_3Sn, T_c appears to vary linearly with ρ_o. In all cases, for $\rho_o = 130 \pm 20$ μΩcm, $T_c \sim 5$ K.

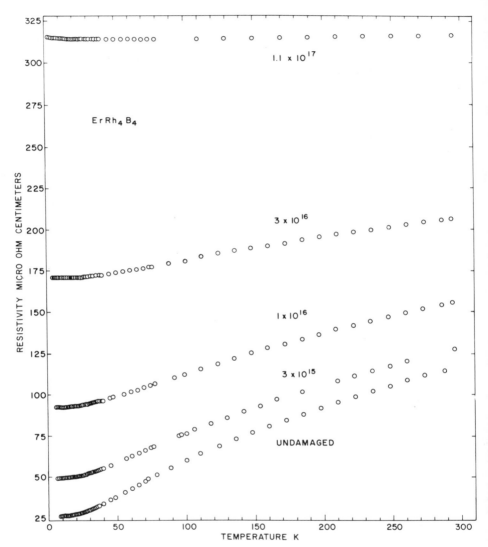

FIGURE 2. Resistivity of ErRh₄B₄ as a function of
temperature for various damage levels. 1.8 MeV α particle
fluences are in units of α/cm².

Further analysis of the data of Fig. 2, using the circuit
model of a parallel saturation resistivity $\rho_s [1/\rho = 1/\rho_s +
1/[\rho_o + \rho(T)]]$, gives a value for ρ_s of 400 $\mu\Omega$cm and $\rho(T)$ as

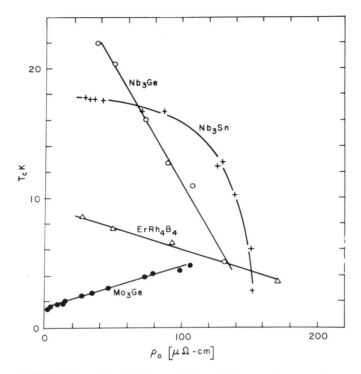

FIGURE 3. Variation of T_c with ρ_o for various A15
compounds and $ErRh_4B_4$.

shown in Fig. 4. It is clear that $\rho(T)$, due to phonon
scattering, is unchanged with damage. As it is generally
believed that the coefficient of the temperature dependent
term in the resistivity reflects the electron-phonon coupling
strength λ, this suggests that this quantity is not changing
drastically over this range. At the highest level of damage,
1.1×10^{17} α/cm^2, the resistivity has an additional activated
term.

Preliminary studies of the transition from the
superconducting to the magnetic state in $ErRh_4B_4$ have been
made by measurement of the critical current and by tunneling.
This temperature region is of particular interest in view of
double structure observed in specific heat measurements
(MacKay et al., 1979), the broad transition to ferromagnetism
observed by neutron scattering (D. E. Moncton et al., 1979)
and the prediction of a spiral state by Blount and Varma
(Blount and Varma, 1979).

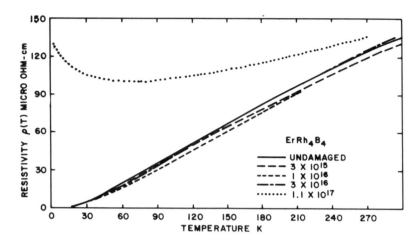

FIGURE 4. The temperature dependent portion of the
resistivity of ErRh$_4$B$_4$ for various fluences of 1.8 MeV α
particles. These data are extracted from the data of Fig. 2
assuming the parallel resistance model and a ρ$_s$ of 400 μΩcm.

The critical currents of an ErRh$_4$B$_4$ bridge, of cross
section 8×0.3 μ^2 and one of composition Er$_{.43}$Ho$_{.57}$Rh$_4$B$_4$, of
cross section 100×0.39 μ^2, are shown in Fig. 5. These
critical currents are low, not exceeding 3.5×10^6 amps/cm^2, and
in Er$_{.43}$Ho$_{.57}$Rh$_4$B$_4$ decrease very suddenly at the magnetic
transition, within a temperature range of ∿ 0.1 K. However in
ErRh$_4$B$_4$ the critical current reaches a broad maximum at 1.6 K
and then decreases slowly down to 1 K.

We have not been able to prepare tunnel junctions on the
freshly sputtered surface of these compounds, presumably
because rhodium oxide is conducting. Junctions were fabricated by sputtering ≤ 100 Å of Al over the boride film before
any exposure to air. The I-V characteristics shown in Fig. 6
indicate that the energy gap in this junction is surprisingly
small, Δ = 0.7 meV, which is half the BCS value. It is
possible that the gap is depressed by the Al layer but there
is no indication of such a proximity effect from the shape of
I versus V at higher voltages. Junctions using a Si artificial
barrier had much poorer characteristics with a broad gap
extending from 0.7 to 1.4 meV, the latter being much more
reasonable for material with a T$_c$ of 8.5 K. As seen in Fig.
6, with decreasing temperature from 4.2 to 1.5 K the gap
increases, but from 1.4 K to 1 K the I-V characteristic
indicates a progressive smearing of the ErRh$_4$B$_4$ gap with only
a small decrease in the energy of the peak in the density of
states. Below 0.9 K the characteristic is that of tunneling

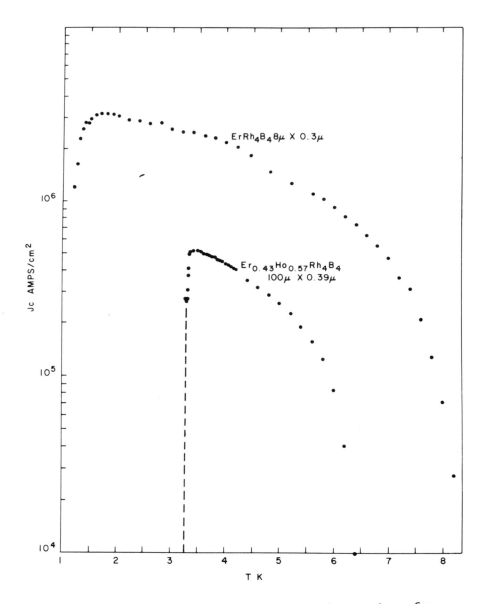

FIGURE 5. *Critical current versus temperature for undamaged bridges of* $ErRh_4B_4$ *and* $Er_{.43}Ho_{.57}Rh_4B_4$.

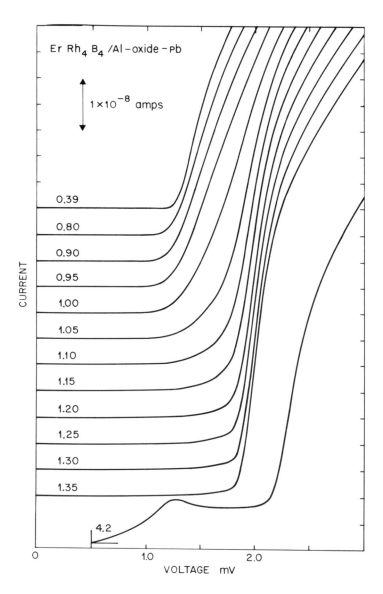

FIGURE 6. The I-V characteristics of an ErRh$_4$B$_4$/Al-oxide-Pb junction at the temperatures indicated. Between 1.35 and 0.39 K the plots have been offset vertically. The 4.2 K trace has also been offset horizontally by 0.5 mV.

from a normal metal into Pb. More detailed measurements and extraction of the superconducting density of states from this and other data still remains to be done.

These tunneling and critical current measurements, like the specific heat and neutron scattering data, indicate that the transition to ferromagnetism, or the spiral state, begins about 1.4-1.5 K. There is no clear indication of a first order transition in our results for ErRh$_4$B$_4$, but the decrease in critical current is essentially abrupt in Er$_{.43}$Ho$_{.57}$Rh$_4$B$_4$. It is interesting to note that in the Ho doped compounds the specific heat transition is also a simple step structure once the Ho concentration exceeds 30%. This result leads us to suggest that the spiral state, if it exists at all in the higher Ho concentrations, occurs in an extremely narrow temperature range.

In summary, we have shown that many of the aspects of α particle damage in the A15 compounds occur also in the ternary borides, at least ErRh$_4$B$_4$ and Er$_{1-x}$Ho$_x$Rh$_4$B$_4$. The resistivity demonstrates a saturation effect with $\rho_s \simeq 400$ $\mu\Omega$cm. In ErRh$_4$B$_4$ the re-entrant transition to the normal state signalling a ferromagnetic state is also strongly suppressed by α particle damage. Preliminary critical current and tunneling measurements indicate that fluctuations into the spiral state of Blount and Varma begin at 1.5-1.6 K in ErRh$_4$B$_4$ and strongly affect the superconducting properties.

ACKNOWLEDGMENTS

We thank D. J. Bishop for measurements down to .035 K, and G. J. Dolan for definition of the bridge-shaped samples.

REFERENCES

Blount, E. I. and Varma, C. M. (1979). *Phys. Rev. Lett. 42,* 1079; also this volume.

Brown, B. S., Hafstrom, J. W. and Klippert, T. E. (1977). *J. Appl. Phys. 48,* 1759.

Dynes, R. C., Poate, J. M., Testardi, L. R., Storm, A. R. and Hammond, R. H. (1977). *Trans. on Magnetics MAG13,* 640.

Gurvitch, M., this volume.

MacKay, H. B., Woolf, L. D., Maple, M. B. and Johnson, D. C. (1979). *Phys. Rev. Lett. 42,* 918.

Moncton, D. E., McWhan, D. B., Schmidt, P. H., Shirane, G. and Thomlinson, W. (1979). *Bull. Am. Phys. Soc. 24,* 391; see also Moncton, D. E. (1979). *J. Appl. Phys. 50,* 1880.

Rowell, J. M., Dynes, R. C. and Schmidt, P. H. (1979). *Sol.
    St. Comm. 30*, 191.
Sweedler, A. R., Snead, C. L., Newkirk, C., Valencia, F.,
    Geballe, T. H., Schwall, R. H., Matthias, B. T. and
    Corenzwit, E. (1976). *Proc. Intl. Conf. on Radiation
    Effects and Tritium Technology* (Gatlinburg, TN) Conf.
    #750989, Vol. III, p. 422.
Sweedler, A. R., Cox, D. E. and Moehlecke, S. (1978).
    Radiation Effects on Superconductivity, eds. B. S. Brown,
    H. C. Freyhardt and T. H. Blewitt [North-Holland Pub.
    Co.].
Vandenberg, J. M. and Matthias, B. T. (1977). *Science 198,*
    194.

MICROPROBE FIELD MEASUREMENTS
IN $ErRh_4B_4$

R. Dean Taylor
J. O. Willis

Los Alamos Scientific Laboratory
Los Alamos, New Mexico

I. INTRODUCTION

Over two decades ago Matthias, Suhl, and Corenzwit (1958) proposed the coexistence of superconductivity and magnetic order in certain Laves-phase alloys of the system $Ce_{1-x}R_xRu_2$, where R is a rare earth element such as Gd. Verification of coexistence in this and other similar systems was hampered because the usual techniques for characterizing magnetic order, such as bulk magnetic susceptibility, failed because of the superconductor's diamagnetic shielding. Techniques which require external magnetic fields are also suspect because of the possible influence of the field on the superconductivity.

One of the earliest attempts to overcome the difficulties encountered in the conventional measurements of the bulk properties was to use the Mossbauer Effect (ME) to measure local internal magnetic fields (Erickson *et al.*, 1973; Taylor *et al.*, 1974; Steiner *et al.*, 1973). The ME is unaffected by the superconductivity and does not require an applied field to obtain hyperfine spectra signaling the possible presence of magnetic order. Only a limited number of useful ME nuclides exist, but the number includes several rare earth elements common to many of the suggested coexistence systems. Magnetic studies have been reported on $La_{1-x}Eu_x$ (Steiner *et al.*, 1973; 1977), $Ce_{1-x}Gd_x Ru_2$ (Ruebenbauer *et al.*, 1977), $Eu_xSn_{1-x}Mo_6S_y$ (Bolz *et al.*, 1977; Fradin *et al.*, 1977) and $ErRh_4B_4$ (Shenoy *et al.*, 1979) using the ME of the magnetic rare earth component. Alternatively, Erickson *et al.* (1973) used the well-known ME in ^{57}Fe as a

very dilute impurity probe to detect the presence of hyperfine magnetic order transferred to the ^{57}Fe in samples of
$Ce_{1-x}Gd_xRu_2$. Samples dilute in Gd (0.08 < x < 0.13) become
superconducting at T_c; ME probe measurements (Erickson et al.,
1973) show magnetic ordering occurs below T_c. Samples more
concentrated in Gd (0.13 < x) are magnetically ordered at T_m.
In a narrow composition range just above x = 0.13 these samples become superconductors at a T_c below T_m; ME probe measurements show the magnetic ordering persists in the superconducting regime. Magnetic Fe sites gave a saturation field
of 7.6 T; this large internal field enhanced the capability
of the ME impurity method. The concentration of the ^{57}Co
parent was too low to produce significant impurity-impurity
interactions and too low to affect any of the measured bulk
properties of the host.

Although the ME is useful to detect magnetic ordering on
a microscopic scale, it is generally not possible to determine whether the interaction is ferromagnetic or antiferromagnetic. However, from a concentration-dependence study of
the ME hyperfine spectra of $La_{1-x}Eu_x$ Steiner and Gumprecht
(1977) concluded that the ordering was of a spin-glass type
in the coexistence region. A further complication in the ME
technique sometimes arises from relaxation-time effects,
which tend to mask any magnetic hyperfine effects due to ordering. Nevertheless the ME technique provides a tool simple
in concept and application which provides unique complemen-
tary information on magnetic ordering in magnetic superconductors.

The recent neutron scattering experiments (Moncton, 1979)
have been extremely valuable in developing a better picture
of the nature of the magnetic interaction in superconductors.
However, this powerful technique cannot be used with metallurgical systems which have nuclei with a high neutron absorption cross-section.

In this introduction we have emphasized techniques which
sense microscopic spin order in the absence of an applied
field. Recent more general reviews of the subject of coexistence include Roth (1978) and Fischer (1978).

II. $ErRh_4B_4$

The discovery of the superconducting and magnetic properties of the ternary compounds RRh_4B_4 (Matthias et al.,
1977) and more particularly the reentrant behavior of $ErRh_4B_4$

(Fertig *et al.*, 1977) provide the experimentalist an excellent metallurgical system to study the coexistence of superconductivity and magnetism. These ternary compounds have a well-defined crystal structure and can be prepared relatively free of voids and second phases. Pseudoternary compounds obtained by systematic substitution of one or more rare earth elements for Er provide an added dimension for establishing trends in an effort to develop an understanding of the competition of magnetism and superconductivity in this system.

Shenoy *et al.* (1979) measured the ME of ^{166}Er in ErRh$_4$B$_4$ at temperatures from 4.2 to 0.1 K. Resolved hyperfine spectra, mostly attributed to crystal-field and electronic spin relaxation-time effects, were obtained at all temperatures. The differences in the ME spectra between 0.1 K and 1.5 K were subtle but were surely a result of the presence of magnetic order below T_{c2}, the temperature at which superconductivity disappears and below which ferromagnetism is present.

Our motivation in using a microscopic magnetic ME probe in ErRh$_4$B$_4$ was to try to determine whether the ferromagnetic order established below T_{c2} persists above T_{c2}, into the superconducting region. We chose the ME of ^{57}Fe impurities because the applicability of the technique was already established and because relaxation time effects were not likely to interfere. In ErRh$_4$B$_4$ one expects the ^{57}Co to subsitute for Rh. The sensitivity of the method depends on the electronic configuration of Fe in ErRh$_4$B$_4$.

III. EXPERIMENTAL

Two different ErRh$_4$B$_4$ samples were used in this study. Sample A was made by D.C. Johnston (Fertig *et al.*, 1977) and Sample B was made by F. E. Wang. Sections about 1 mm thick of the annealed samples were doped with ^{57}Co, the parent of ^{57}Fe, for the ME studies. About 0.5 mCi of carrier-free ^{57}Co was electroplated onto one side of Sample A giving a source whose active area was about 0.1 cm^2; the cobalt was diffused into the host by heating at 900°C first in a H$_2$-Ar atmosphere and then in a high vacuum for a total time of 1.3 hours. Source B was similarly prepared except the heat treatment was entirely in a vacuum of 10^{-6}-10^{-8} torr. A deeper diffusion was achieved by treating for 26 hours at 960°C followed by 17 hours at 1080°C. By observing the relative intensities of the 122- and 14.4-keV γ rays of the ^{57}Fe as a function of the heat treatment we could determine the average depth of diffusion and the average impurity composition. For Source B we found

that the average concentration of the doped region was roughly 100 ppm ^{57}Co. For Source A we only know that the average
^{57}Co concentration was much higher. T_{c1} (superconducting
onset), determined from ac magnetic susceptibility measurements for both samples before and after doping agreed with
Fertig *et al.*, (1977). T_{c2}, determined for the undoped samples, also agreed; T_{c2} was not measured for the doped samples.

The source was cooled in a ^{3}He cryostat and analyzed by an
external room-temperature single-line absorber in conjunction
with a conventional constant-acceleration mode ME spectrometer. A superconducting solenoid provided fields up to 6 T
on the source.

IV. DATA DISCUSSION

The ME spectra of ^{57}Fe impurities in Sample B at 0.38
and 1.1 K are shown in Fig. 1. The splitting of the quadrupole doublet was essentially independent of temperature from
0.38 to 300 K and the same for both samples. Equal intensities suggest that the ^{57}Fe impurities were in equivalent
sites. We have analyzed the low-temperature data assuming a
symmetrical Lorentzian doublet. We have summarized the low
temperature data for both sources in Table I. Much of the
excessive line width quoted is attributable to the rather
thick potassium ferrocyanide absorber used. (A thin absorber
gave component line widths of 0.33 mms^{-1}).

TABLE I. *Measured Field at the Impurity Site*

Source	Temperature K	Quadrupole Splitting mms^{-1}	Line width mms^{-1}	Effective field[a] T
A	4.0	0.523 (15)	0.589 (26)	–
	1.2	0.510 (8)	0.616 (7)	Ref. A
	0.54	0.499 (15)	0.627 (13)	0.06 (4)
B	4.0	0.491 (5)	0.591 (7)	–
	1.1	0.488 (4)	0.584 (6)	Ref. B
	0.38	0.489 (4)	0.659 (6)	0.49 (6)

[a]Entries calculated with respect to Ref. A or B data.

   In Sample A the spectra at 1.2 and 0.54 K were virtually
indistinguishable. For $T/T_{c2} \simeq 0.6$ we assumed nearly complete
ferromagnetic ordering of the sample. The absence of any
broadening at 0.54 K implied the absence (to within 0.1 T)
of any internal field at the Fe sites. To obtain this upper
limit to the field we assumed that the electronic state of
the Fe was nonmagnetic. Such a nonmagnetic state has
been observed for Fe in the superconductors V, Nb, and Ta
(Kitchens et al., 1965). In $ErRh_4B_4$ Shenoy et al., (1979)
measured the field at the Er nucleus to be 770 T at 0.1 K.
Although an internal field at the Er sites does not neces-
sitate a field at the Rh and B neighbors, it does offer the
possibility of a hyperfine field at the Fe.

   The general features of the ME spectra of Source B confirm those of Source A except statistically significant
broadening was found at 0.38 K. We presume this broadening
is associated with the lower measuring temperature rather
than differences in the sample-source preparation, but this
point could not be fully checked because the source was
ruined in a subsequent heat treatment. The field at 0.38 K
at the Fe site, calculated from the fitted ME spectra shown
in Fig. 1, was 0.49 T. To obtain this value we assumed that
the principal quadrupole and magnetic axes were co-linear,
but the value was rather insensitive to this particular
choice.

   We measured the response of the ^{57}Fe ME impurity to an
applied magnetic field in order to confirm the supposition
that Fe in $ErRh_4B_4$ was in a nonmagnetic electronic state.
The measurements showed that the hyperfine field at the Fe
site in Sample A at 1.3 and 0.53 K was 5.98 $\pm$0.11 and 6.01
$\pm$0.09 T, respectively, in an applied field of 6.00 T. Measurements at a lower field also indicated that the field at
the ^{57}Fe nucleus was within a few percent of the value of
the applied field. We have taken this as direct evidence
that the Fe impurity in $ErRh_4B_4$ is indeed nonmagnetic. The
absence of a moment on Fe precludes relaxation effects as the
source of broadening at 0.38 K. It seems safe to speculate
that the field observed (Table I) arises from the ordered Er
moments.

V. CONCLUSIONS

   We have observed a small hyperfine field at Fe impurity
sites in $ErRh_4B_4$ at 0.38 K. The magnitude of the field at
the nonmagnetic impurity sites was 0.5 T. We presume this
hyperfine field reflects spin order at the Er sites but the

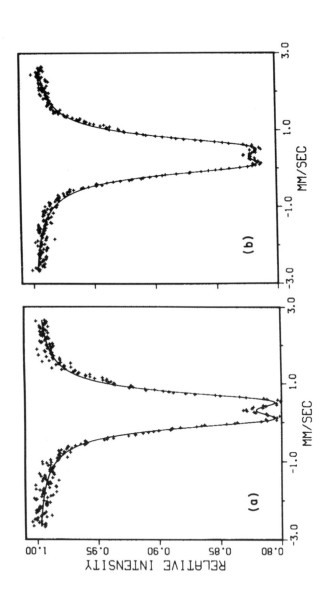

FIGURE 1. Mossbauer effect spectra of ^{57}Fe in $ErRh_4B_4$ at 1.1 (a) and 0.38 K (b).

sensitivity of the probe technique employed turned out to be marginal for investigating quantitatively the shape of the ordering curve with temperature near and below T_{c2}.

REFERENCES

Bolz, J., Crecelius, G., Maletta, H., and Pobell, F. (1977). *J. Low Temp. Phys.*, *28*, 61.

Erickson, D. J., Olsen, C. E., and Taylor, R. D. (1973). "Mossbauer Effect Methodology," Vol. 8, p. 73. Plenum Publishing Corp., New York.

Fertig, W. A., Johnson, D. C., DeLong, L. E., McCallum, R. W., Maple, M. B., and Matthias, B. T. (1977). *Phys. Rev. Lett. 38*, 987.

Fischer, Ø. (1978). *Appl. Phys. 16*, 1.

Fradin, F. Y., Shenoy, G. K., Dunlap, B. D., Aldred, A. T., and Kimball, C. W. (1977). *Phys. Rev. Lett. 38*, 719.

Kitchens, T. A., Steyert, W. A., and Taylor, R. D. (1965). *Phys. Rev. 138A*, 467.

Matthias, B. T., Corenzwit, E., Vandenberg, J. M., and Barz, H. E. (1977). *Proc. Natl. Acad. Sci. 74*, 1334.

Matthias, B. T., Suhl, H., and Corenzwit, E. (1958). *Phys. Rev. Lett. 1*, 449.

Moncton, D. E. (1979). *J. Appl. Phys. 50*, 1880 and related references therein.

Roth, S. (1978). *Appl. Phys. 15*, 1.

Ruebenbauer, K., Fink, J., Schmidt, H., Czizek, G., Tomala, K. (1977). *Phys. Stat. Sol. 84*, 611.

Shenoy, G. K., Dunlap, B. D., Fradin, F. Y., Kimball, C. W., Potzel, W., Probst, F., and Kalvius, G. M. (1979). *J. Appl. Phys. 50*, 1872.

Steiner, P., Gumprecht, G. (1977). *Solid State Commun. 22*, 501.

Steiner, P., Gumprecht, D., and Hufner, S. (1973). *Phys. Rev. Lett. 30*, 1132.

Taylor, R. D., Decker, W. R., Erickson, D. J., Giorgi, A. L., Matthias, B. T., Olsen, C. E., and Szklarz, E. G. (1977). "Low Temperatures Physics—LT 13," Vol. 2, p. 605. Plenum Publishing Corp., New York.

NEUTRON SCATTERING STUDY OF THE MAGNETIC TRANSITION IN THE REENTRANT SUPERCONDUCTOR $Ho_{0.6}Er_{0.4}Rh_4B_4$

H. A. Mook[1]
W. C. Koehler[1]

Solid State Division
Oak Ridge National Laboratory
Oak Ridge, Tennessee

M. B. Maple[2]
Z. Fisk[3]
D. C. Johnston[4]

Institute for Pure and Applied Physical Sciences
University of California, San Diego
La Jolla, California

Since the discovery of reentrant superconductivity in $ErRh_4B_4$ (Fertig et al., 1977) and $Ho_{1.2}Mo_6S_8$ (Ishikawa and Fischer, 1977), there has been a great deal of interest in the nature of the superconducting-magnetic transition at the lower critical temperature T_{c2}. The pseudoternary system $(Er_{1-x}Ho_x)Rh_4B_4$ is a particularly good testing ground for

[1]*Research sponsored by the Division of Materials Sciences, U.S. Department of Energy under contract W-7405-eng-26 with the Union Carbide Corporation.*
[2]*Research sponsored by the U.S. Department of Energy under Contract Ey-76-S-03-0034-PA227-3.*
[3]*Research sponsored by the National Science Foundation under Grant NSF/DMR77-08469.*
[4]*Present address: Corporate Research Laboratories, Exxon Research and Engineering Co., Linden, New Jersey.*

427

experimental and theoretical studies of reentrant superconducting behavior. The specific heat anomalies (MacKay et al.,
1979) and neutron scattering results (Lander et al., 1978;
Moncton et al., 1977) obtained for this system are of particular interest. $HoRh_4B_4$ becomes a ferromagnet at low temperatures, ordering directly from the paramagnetic normal
state (Matthias et al., 1977). The heat capacity has a saw
tooth-shaped feature at the magnetic ordering temperature, the
sharp rise occurring at about 6.7 K. Neutron scattering measurements show that the material orders ferromagnetically
along the c-axis with nearly the full free ion Ho moment. The
transition is found to be second order with considerable magnetic precursor or critical scattering observed above the
ordering temperature. $ErRh_4B_4$ is a reentrant superconductor
with an upper and lower transition temperature T_{c1} and T_{c2} of
about 8.7 and 1.0 K. The specific heat in the neighborhood of
T_{c2} has unusual behavior (MacKay et al., 1979; Maple, 1978)
with a large spike at about 0.93 K but a pronounced high temperature tail. Neutron scattering measurements have shown
that $ErRh_4B_4$ orders ferromagnetically in the basal plane with
a moment of about 5.6 μ_B, a value considerably below the free
ion value. The temperature dependence of the magnetic intensity was obtained from the intensity of the (101) reflection.
In the initial measurements smooth second-order behavior was
found with no apparent hysteresis. Recent measurements have
suggested hysteresis in the temperature region between 1.2 and
0.8 K (Moncton, 1979). However, considerable precursor
scattering was found above the transition temperature T_{c2}. It
appears that there may be some overlap between the magnetic
and superconducting state in this material. The measurements
we have made on $Ho_{0.6}Er_{0.4}Rh_4B_4$ show that the magnetic behavior at T_{c2} is altogether different than that found in either
the pure Ho or Er based materials.

The $Ho_{0.6}Er_{0.4}Rh_4B_4$ compound was prepared by arc melting
the rare earth tetraborides with Rh, followed by annealing.
The [11]B isotope was used to decrease the absorption cross section for the slow neutrons. The rare earth tetraborides were
prepared as small crystals by precipitation from molten Al
which appeared to remove any small traces of C found in the
boron isotope. $Ho_{0.6}Er_{0.4}Rh_4B_4$ is a reentrant superconductor
with T_{c1} and T_{c2} being 7.2 and 3.6 K (Johnston et al., 1978).
The sample was in the form of small particles and neutron
powder scans showed no preferred crystallographic orientations. A number of different diffraction patterns were
obtained with various resolutions. Most of the measurements
were made using a pyrolytic graphite monochromator and an
incident wavelength of 2.348 Å. Pyrolytic graphite filters
were used to remove order contamination. Diffraction patterns

were also obtained with a Be monochromator which made it possible to measure additional reflections. Figure 1 shows a high resolution powder diffraction scan of the (101), (110) and (002) peaks at 1.6 and 4.2 K. These peaks all have very small nuclear components as observed in the 4.2 K data. Large magnetic (101) and (110) peaks are found at 1.6 K showing long range ferromagnetic order, however no (002) peak is observed. This shows that the moment is parallel to the c-axis. High resolution scans on some of the higher angle powder peaks show that the ferromagnetic order has a correlation length of at least 200 Å at 1.6 K. Analysis of the nuclear and magnetic diffraction peaks shows that the average ordered moment is about 5.0 ± .5 μ_B at 1.6 K. This value is considerably below the free ion moment for Ho or Er but a very similar moment to that found for ErRh$_4$B$_4$ (Moncton et al., 1977). The temperature dependence of the intensity of the (101) powder peak is shown in Fig. 2. Upon cooling no long-range magnetic scattering is found above 3.60 K. At 3.60 K the magnetic intensity jumps suddenly to a finite value then increases nearly linearly as the temperature is lowered to about 2.5 K. Below 2.5 K the intensity rounds off and appears to saturate at

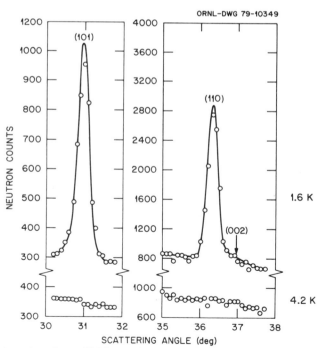

FIGURE 1. Powder diffraction peaks at temperatures of 1.6 and 4.2 K.

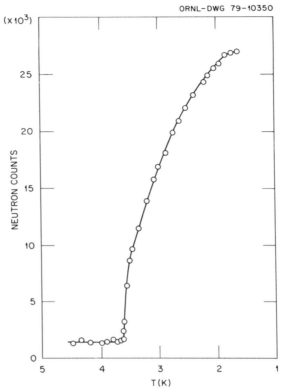

ORNL-DWG 79-10350

FIGURE 2. Temperature dependence of (101) peak intensity. Only temperature independent nuclear scattering is found above T_{c2}.

about 1.6 K which was the lowest temperature presently achievable in our cryostat. It is possible that the magnetic intensity would start to rise again at some lower temperature or that only the Ho moments order at 3.6 K while the Er moments order at some much lower temperature. If the magnetic intensity is assumed to be associated with ordering of the Ho moments alone, the resultant Ho ordered moment at 1.6 K would be equal to about 8.3 μ_B, in reasonable agreement with the value of 8.7 $\pm$ 0.3 μ_B determined from neutron scattering experiments on HoRh$_4$B$_4$ (Lander et al., 1978). However, we note that there are no specific heat anomalies between 0.5 K, the low temperature limit of the heat capacity experiments, and 1.6 K.

Figure 3 shows a more detailed plot of the magnetic intensity in the neighborhood of the transition. The error

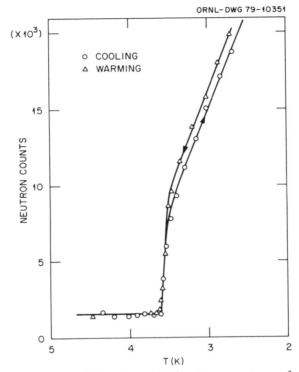

ORNL-DWG 79-10351

FIGURE 3. Expanded view of the temperature dependence of the (101) peak intensity in the neighborhood of the transition.

bars are about the size of the points and the temperature stability better than 0.005 K. The temperature was measured with a Ge resistor mounted directly on the sample holder. We see clearly the jump in the magnetic intensity at 3.60 K and the change in slope of the intensity vs temperature curve in the neighborhood of the transition. Since the magnetic scattering intensity is proportional to the square of the magnetic moment the fast rise in the magnetic intensity signifies a rapid increase in the ordered moment. In fact, the system develops almost one half its saturated moment in a temperature interval of about 0.05 K. Little hysteresis is found in the transition temperature but hysteresis is found in the magnetization developed at a given temperature. This hysteresis disappears around 2 K as saturation is approached. The fast rise in magnetic order at 3.60 K and the magnetization hysteresis suggests a first order transition. The specific heat of $Ho_{0.6}Er_{0.4}Rh_4B_4$ shows a large spike at 3.60 K in excellent agreement with our results. No high temperature tail is found on the side of the spike as is found for $ErRh_4B_4$. We observe

no region of coexistence of superconductivity and long range
ferromagnetic order. These results differ dramatically from
those found in $ErRh_4B_4$ or the Chevrel phase material
$HoMo_6S_8$ (Lynn et al., 1978) where the magnetic intensity
changes smoothly with temperature and precursor scattering is
observed above T_{c2}.

The fact that the magnetic transition appears first
order is important in understanding the interactions responsible for producing the magnetic state. MacKay et al. (1979)
discuss mechanisms that could cause a first-order superconducting-magnetic transition. It seems clear that different
types of transitions occur within the $Ho_xEr_{1-x}Rh_4B_4$ system and
plans are underway to make more detailed measurements on this
system. It has been suggested that structure in the small
angle scattering pattern for $ErRh_4B_4$ near the transition
T_{c2} indicates magnetic fluctuations with a wavelength of 100 Å
(Moncton et al., 1979). We find no evidence of this type of
small-angle scattering in $Ho_{0.6}Er_{0.4}Rh_4B_4$ although we plan to
make higher resolution measurements. This is further evidence
of the different nature of the transitions in these materials.

REFERENCES

Fertig, W. A., Johnston, D. C., Delong, L. E., McCallum,
    R. W., Maple, M. B., Matthias, B. T. (1977). Phys. Rev.
    Lett. 38, 987.
Ishikawa, M. and Fischer, O. (1977). Solid State Commun. 23,
    37.
Johnston, D. C., Fertig, W. A., Maple, M. B. and Matthias,
    B. T. (1978). Solid State Commun. 26, 141.
Lander, G. H., Sinha, S. K., and Fradin, F. Y. (1979). Pro-
    ceedings of the 24th Conference on Magnetism and Magnetic
    Materials, Cleveland, Ohio, November 14-17, 1978, in press.
Lynn, J. W., Moncton, D. E., Thomlinson, W., Shirane, G. and
    Shelton, R. N. (1978). Solid State Commun. 26, 493.
MacKay, H. B., Woolf, L. D., Maple, M. B., and Johnston, D. C.
    (1979). Phys. Rev. Lett. 42, 918.
Maple, M. B. (1978). Proceedings of the 15th International
    Conference on Low Temperature Physics, Grenoble, France,
    August 23-29, 1978.
Matthias, B. T., Corenzwit, E., Vandenberg, J. M., and Barz,
    H. E. (1977). Proc. Natl. Acad. Sci. 74, 1334.
Moncton, D. E., 1979. Private communication.
Moncton, D. E., McWhan, D. B., Eckert, J., Shirane, G., and
    Thomlinson, W. (1977). Phys. Rev. Lett. 390 1164.
Moncton, D. E., McWhan, D. B., Schmidt, P. H., Shirane, G.,
    and Thomlinson, W. (1979). AIP Conf. Proc. 24, 391.

EQUILIBRIUM PROPERTIES OF THE FLUXOID LATTICE IN SINGLE-CRYSTAL NIOBIUM[1]

H. R. Kerchner
D. K. Christen
S. T. Sekula
P. Thorel[2]

Solid State Division
Oak Ridge National Laboratory
Oak Ridge, Tennessee

I. INTRODUCTION

The dimensions and symmetry of the fluxoid lattice in a single-crystal sphere of niobium have been measured by using a double-perfect-crystal small-angle neutron-scattering technique (DCSANS). The bulk magnetization of the same sample has been measured by a field-sweep technique. In addition, the misalignment between the fluxoids and the applied magnetic field was observed by DCSANS. The experimental methods and most of the results are reported elsewhere (Kerchner *et al.*, 1979; Christen *et al.*, 1979). Here we summarize our findings and compare the measurements with realistic microscopic theory where it is available.

The DCSANS technique gives directly the symmetry, orientation, and dimensions of the fluxoid-lattice unit cell. The flux density per unit cell, deduced by using the flux quantization condition, is plotted in Fig. 1. In the intermediate mixed state (IMS) of the sphere the flux density per unit cell is independent of the applied field H_a while the bulk flux

[1]*Research sponsored by the Division of Materials Science, U. S. Department of Energy, under contract W-7405-eng-26 with the Union Carbide Corporation.*
[2]*Guest Scientist from Départemente de la Recherche Fondementale, Centre d'Etudes Nucléaires, Grenoble, France.*

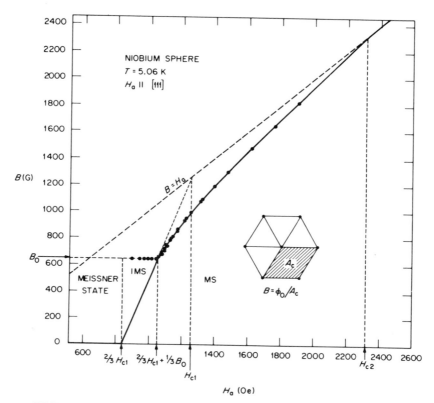

FIGURE 1. The bulk flux density (solid curve) vs applied field H_a, compared to the average flux density/unit cell obtained from DCSANS. The inset shows how the flux density/ unit cell is found from the flux quantum condition.

density (solid curve) rises linearly with H_a. In the mixed state (MS) the sample is filled with fluxoid lattice and the two quantities are identical.

The fluxoid lattice observed was hexagonal only when the applied field was parallel to a <111> crystal direction. Usually the lattice was constructed of isosceles triangles whose orientation sometimes reflected the symmetry of the crystal. At some orientations the crystal symmetry was reflected only in the random occurrence of two different, symmetrically disposed fluxoid lattices. For fields in a <100> direction, we observed a structural phase transition near 2 K between a square lattice and a scalene-triangle lattice.

The field-sweep technique gives directly the field derivative of the magnetization, dM/dH_a. The magnetization $M(H_a)$ was obtained by numerical integration. The Meissner state,

the IMS, the MS, and the normal state show up as distinct
regions of the $M(H_\alpha)$ and dM/dH_α plots shown in Fig. 2. Since
the Meissner state usually superheats and the magnetization is
hysteretic in the IMS, only MS data were used to determine the
equilibrium properties.

Detailed comparisons between our magnetization measurements and various results of Gorkov theory are described elsewhere (Kerchner *et al.*, 1979; Christen *et al.*, 1979). We have
little new to add to the well-known discrepancies between

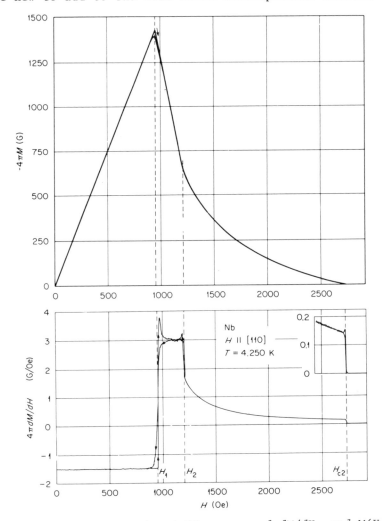

FIGURE 2. *Experimentally measured dM/dH_α and $M(H_\alpha)$
obtained by numerical integration.*

Gorkov theory and the magnetization of niobium. Some recent theoretical effort has been made to relate the relative anisotropy of the lower and upper critical fields H_{c1} and H_{c2} and the limiting value of dM/dH_a at H_{c2} to the anisotropy of the Fermi surface. In addition, the effects of a strong, retarded electron–phonon interaction on H_{c2} and the thermodynamic critical field H_c have recently been investigated theoretically. Those theoretical results are compared with our data below. Since our measurements of the anisotropy of H_{c2} are substantially identical to earlier results (Seidl *et al.*, 1978), they will not be discussed here.

II. ANISOTROPY RESULTS

In Fig. 3 are plotted detailed measurements of the orientation dependence of the lower critical field H_{c1} and the equilibrium flux density B_o at H_{c1}. The anisotropy of H_{c1} was measured by three independent techniques. The circles in Fig. 3 were derived from DCSANS measurements of B_o and the field $H_2 = \frac{2}{3} H_{c1} + \frac{1}{3} B_o$ bounding the IMS and the MS. The solid curve in the upper part of the figure was found by fitting a cubic harmonic expansion of the thermodynamic expression (Christen *et al.*, 1979; Takanaka and Nagashima, 1979),

$$H_a \Delta\psi = -\frac{2}{3} \, dH_{c1}/d\alpha,\tag{1}$$

to the measured misalignment angle $\Delta\psi$ between $\vec{B}_o$ and $\vec{H}_a$ in the IMS. Here α is the angle between $\vec{H}_a$ and the [001] direction in a symmetry plane of the crystal. The bulk magnetization was measured at many temperatures for fields in the three high-symmetry directions, and the deduced values of H_{c1}/H_c were fitted to an expansion in cubic harmonics and in powers of $(1 - T/T_c)$. The temperature dependences of the cubic harmonic coefficients deduced are shown in Fig. 4, and the corresponding values of $H_{c1}/\bar{H}_{c1}$ at 4.3 K for the three high-symmetry directions are shown in Fig. 3. The agreement among the three measurements demonstrates the reliability of our determination of the anisotropy of H_{c1}. The anisotropy of B_o could be resolved only by DCSANS measurements of the fluxoid lattice in the IMS.

The limiting temperature dependence of $\Delta H_{c1}/H_{c1}$ near T_c has been related to the Fermi surface average,

$$\langle v_x^4 \rangle / \langle v^4 \rangle - .2 \simeq .058 \, [2\langle v^2 H_4(\hat{v}) \rangle / \langle v^2 \rangle - \langle H_4(\hat{v}) \rangle]\tag{2}$$

(Takanaka and Hubert, 1975), where v_x is the projection on [001] of the Fermi velocity $\vec{v}$ and $H_4(\hat{v})$ is the $\ell = 4$ cubic

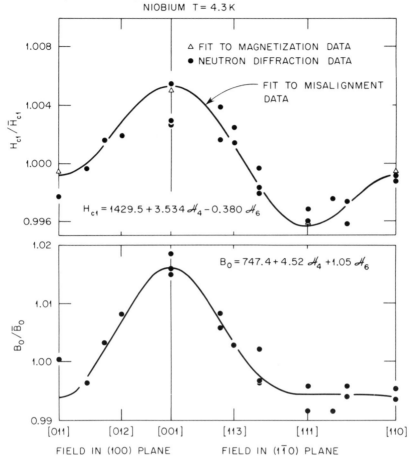

NIOBIUM T = 4.3 K

△ FIT TO MAGNETIZATION DATA
● NEUTRON DIFFRACTION DATA

FIT TO MISALIGNMENT DATA

$H_{c1} = 1429.5 + 3.534\ \mathcal{H}_4 - 0.380\ \mathcal{H}_6$

$B_0 = 747.4 + 4.52\ \mathcal{H}_4 + 1.05\ \mathcal{H}_6$

FIELD IN (100) PLANE   FIELD IN (1Ī0) PLANE

*FIGURE 3.   Experimental data at T = 4.30 K for the anisot-
ropy of the low-field critical parameters B_0 and H_{c1} for
applied fields in the (1Ī0) and (100) niobium crystal planes.*

harmonic evaluated in the direction of $\vec{v}$. The right-hand side
of Eq. (2) has been calculated by using band-theory (Butler,
1979) and has been deduced from the measured anisotropy of H_{c2}
(Seidl *et al.*, 1978). The two values are in good agreement,
and the predicted anisotropy of H_{c1} agrees well with the
observations as shown in Fig. 4.

The limiting value of dM/dH_a at H_{c2} was read off the
field-sweep plots and the parameter κ_2 deduced by using the
definition,

$$4\pi [dM/dH_a]_{H_{c2}} = [\beta(2\kappa_2^2 - 1) + \tfrac{1}{3}]^{-1}. \tag{3}$$

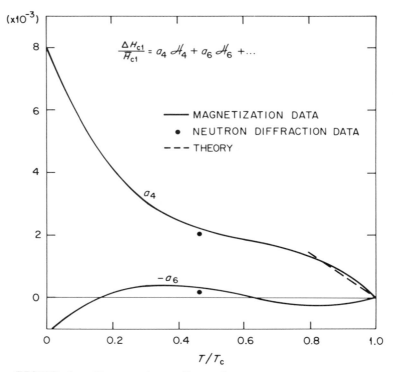

FIGURE 4. *Temperature dependences of the cubic harmonic coefficients for H_{c1}.*

The experimental values of κ_2 were fitted as for the H_{c1}/H_c data. The best-fit temperature dependences of the cubic harmonic coefficients are plotted in Fig. 5. The limiting temperature dependence of $\Delta\kappa_2/\kappa$ depends on the parameter

$$\langle v_x^4 - .2v^4\rangle/\langle v^2\rangle^2 \simeq .058[2\langle v^2 H_4(\vartheta)\rangle/\langle v^2\rangle - \langle H_4(\vartheta)\rangle] \quad (4)$$

(Berthel and Pietrass, 1978). Again, the theoretical result agrees well with the observations.

III. H_c RESULTS

The temperature dependence of H_c is sensitive to the strength of the electron-phonon interaction in strong-coupling superconductors. Our measurements are compared with recent theoretical calculations (Daams and Carbotte, 1978) based on two different tunneling-derived phonon spectra in the

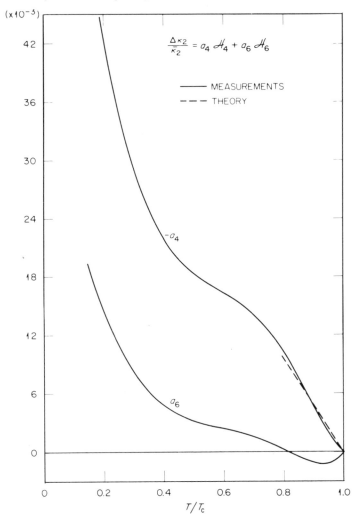

FIGURE 5. Temperature dependences of the cubic harmonic coefficients for κ_2.

deviation-function plot of Fig. 6. The MIT spectrum ($\lambda = .33$) is too weak. The Stanford spectrum ($\lambda = .98$) is slightly too strong. Daams and Carbotte adjusted the Stanford spectrum ($\lambda = .89$) to make the deviation function fall close to earlier calorimetric determinations. This latter curve falls on our measurements at low temperatures but below them near T_c. Using functional derivatives evaluated for the Stanford spectrum by Daams and Carbotte, we have determined that adding

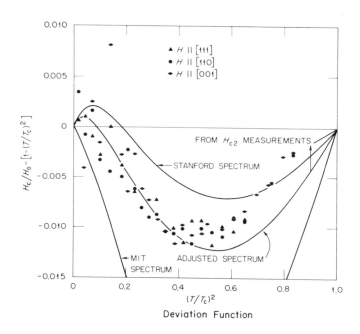

FIGURE 6. The deviation function for the temperature dependence of the thermodynamic critical field H_c. The solid curves represent calculations based on tunneling-derived phonon spectra.

weight to the phonon spectrum at high energy (subject to the constraint of constant T_c) raises the high-temperature portion of the predicted deviation function as indicated by the vertical arrow without appreciably affecting the low-temperature portion.

Our H_c measurements indicate that the electron-phonon enhancement parameter $\lambda = .91$. This result is too small to account for the experimental enhancement of the band velocity deduced by Crabtree *et al.* (1979). The mean-square (renormalized) Fermi velocity of Crabtree *et al.* gives nearly the observed H_{c2} near T_c (Hohenberg and Werthamer, 1967) when the strong-coupling enhancement (Rainer and Bergmann, 1974) is taken into account. Thus the discrepancy is clearly a theoretical problem, probably due either to incorrect band velocities or to the neglect of a significant electron-electron component of the many-body-theory mass enhancement.

IV. CONCLUSIONS

To conclude, we want to emphasize that in favorable cases mixed-state measurements on type-II superconductors can be made very precisely. A number of features of the mixed state are sensitive to the microscopic electronic structure of the material; therefore, they are not described accurately by Gorkov theory, and they can serve as sensitive tests of realistic microscopic theory. The theoretical connection between the mixed state and microscopic material properties is difficult and correspondingly incomplete, but where the connection exists it works well, as demonstrated by our results.

ACKNOWLEDGMENTS

The authors are grateful to H. Harmon and Y. K. Chang for preparing the samples. W. H. Butler and J. P. Carbotte explained much of the microscopic theory to us and supplied results of their theoretical calculations before publication.

REFERENCES

Berthel, K.-H, and Pietrass, B. (1978). *J. Low Temp. Phys.* *33*, 127.

Butler, W. H. (1979). Private communication and this volume.

Christen, D. K., Kerchner, H. R., Sekula, S. T., and Thorel, P. (1979). Submitted to Phys. Rev.

Crabtree, G. W., Dye, D. H., Karim, D. P., Koelling, D. D., and Ketterson, J. B. (1979). *Phys. Rev. Letters 42*, 390.

Daams, J. M., and Carbotte, J. P. (1978). *In* "Inst. Phys. Conf. Ser. No. 39," p. 715.

Hohenberg, P. C., and Werthamer, N. R. (1967). *Phys. Rev.* *153*, 493.

Kerchner, H. R., Christen, D. K., and Sekula, S. T. (1979). Submitted to Phys. Rev.

Rainer, D., and Bergmann, G. (1974). *J. Low Temp. Phys. 14*, 501.

Seidl, E., Weber, H. W., and Teichler, H. (1978). *J. Low Temp. Phys. 30*, 273.

Takanaka, K., and Hubert, A. (1975). *In* "Proceedings of the Fourteenth International Conference on Low Temperature Physics" (M. Krusius and M. Vuorio, ed.), p. 309. American Elsevier, New York.

Takanaka, K., and Nagashima, T. (1979). To be published in J. Phys. Soc. Jap.

PROGRESS IN CALCULATIONS OF THE
SUPERCONDUCTING PROPERTIES OF
TRANSITION METALS[1]

W. H. Butler

Metals and Ceramics Division
Oak Ridge National Laboratory
Oak Ridge, Tennessee

First principles calculations of the electron-phonon parameters of d-band metals can now be performed to an accuracy of about 10% for "averaged" quantities such as the mass enhancement or the room temperature resistivity. Quantities such as the spectral function $\alpha^2 F(\omega)$ or the phonon linewidth which describe the electron-phonon interaction in more detail can also be calculated. Agreement between calculated and experimental phonon linewidths is generally good but there are differences between the experimental and calculated versions of $\alpha^2 F(\omega)$. Calculations of the thermodynamic critical field and the upper critical field for Nb agree well with experiment.

I. TECHNIQUES

The two basic techniques which have been used in calculating electron-phonon parameters are the rigid-muffin-tin approximation (RMTA) and the modified tight-binding approximation (MTBA). The RMTA is based upon the Bloch formulation of the electron-phonon interaction (Bloch, 1928) in which the electron-phonon matrix element $I_{kk'}^{\alpha} = \langle \psi_k | \hat{x}_\alpha \cdot \delta V | \psi_{k'} \rangle$ is calculated using the periodic crystal wave functions ψ_k and the change in crystal potential due to an atomic displacement δV. The RMTA simply consists in approximating this change δV by

[1]*Research sponsored by the Materials Sciences Division, U.S. Department of Energy under contract W-7405-eng-26 with the Union Carbide Corporation.*

$\vec{u} \cdot \delta V = \vec{u} \cdot \nabla V$ where $\vec{u}$ is a small displacement and ∇V is the gradient of the crystal potential of the perfect lattice. The name derives from the fact that the crystal potential of transition metals is usually cast into "muffin-tin" form and for a potential of this type the above approximation is equivalent to assuming that when an atom is displaced its muffin-tin potential shifts with it rigidly. One nice feature of the RMTA is that for energy conserving transitions a very simple expression for the matrix elements has been found (Gaspari and Gyorffy, 1972). Given the assumption of a rigid muffin-tin, the remainder of an RMTA calculation for electron-phonon parameters can be performed without further approximation (other than the usual approximations of band theory). We will present evidence that the RMTA is accurate to about 10% for quantities such as the mass enhancement, λ.

The MTBA is based upon the Fröhlich (Fröhlich, 1966) formulation of the electron-phonon interaction in which the crystal wave function is expanded in terms of atomic orbitals $\phi_m(\vec{r}-\vec{R}_i)$ which move with the nucleus as it is displaced. The electron-phonon matrix element is obtained by considering the change in the tight-binding Hamiltonian matrix elements due to an atomic displacement,

$$I^{\alpha}_{kk'} = \sum_{\substack{ij \\ mn}} \left[e^{i\vec{k}\cdot\vec{R}_{ij}} - e^{i\vec{k}\cdot\vec{R}_{ij}} \right] \hat{x}_{\alpha} \cdot \nabla_{\vec{R}_{ij}} \langle \phi_m(\vec{r}-\vec{R}_i) | H | \phi_n(\vec{r}-\vec{R}_j) \rangle \quad . \qquad (1)$$

where $\vec{R}_{ij} = \vec{R}_i - \vec{R}_j$. A helpful way of remembering the difference between the two approaches is to note that in the Bloch formulation one calculates the *matrix element* of a *potential gradient* while in the Fröhlich formulation one calculates the *gradient* of the *potential matrix elements*. The MTBA electron-phonon matrix elements are of course never evaluated rigorously using a true tight-binding basis set, although this would be an interesting calculation, instead, the usual procedure is to set up a tight-binding interpolation Hamiltonian with parameters which can be adjusted until agreement is achieved between the tight-binding energy bands and those obtained from a first principles calculation. If additional assumptions are made concerning the dependence of the Hamiltonian matrix elements on intersite separation the electron-phonon matrix elements can be evaluated. There are several variations on the MTBA technique having to do with precisely how one deduces the Hamiltonian matrix elements and their derivatives (Birnboim and Gutfreund, 1975, Peter et al., 1974, and Varma et al., 1979).

Recently, it has been shown that the Bloch and Fröhlich formulations are equivalent (Ashkenazi et al., 1979, Varma, et al., 1979), and that for energy conserving transitions ($E_k = E_{k'}$) with which we are concerned here it makes no difference whether one calculates the matrix element of the potential gradient or the gradient of the potential matrix element. If the potentials are the same and if both calculations are carried through without further approximation the results should be the same.

II.   RESULTS

Table I displays results of calculations of the Fermi surface average of the square of the electron-phonon matrix element $\langle I^2 \rangle$,

$$\langle I^2 \rangle = \int \frac{dS_k}{v_k} \int \frac{dS_{k'}}{v_{k'}} \sum_\alpha (I^\alpha_{kk'})^2 \bigg/ \left(\int \frac{dS_k}{v_k} \right)^2 . \tag{2}$$

The agreement between the RMTA and empirical values is very gratifying. $\langle I^2 \rangle$ varies greatly in the transition metals being largest for those materials which show strong d-d intersite bonding (Nb, Mo, Tc). This variation in $\langle I^2 \rangle$ when coupled with the known variations in phonon frequencies and in Fermi energy density of states explains the Matthias e/a rules (Matthias, 1957).

TABLE I.   *Calculated and Empirical Values*
of $\langle I^2 \rangle$ for the 4-d Metals

		$\langle I^2 \rangle$		
	e/a	Empirical[a]	RMTA[a]	MTBA[b]	MTBA[c]
Y	3	0.0008	0.0021		
Zr	4	0.008	0.007		
Nb	5	0.011	0.0148	0.0144	0.023
Mo	6	0.030	0.0235	0.0242	
Tc	7	0.024	0.0189		
Ru	8	0.017	0.014		
Rh	9	0.0054	0.009		
Pd	10	0.0037[d]	0.0029		

[a]*Butler, 1977.*
[b]*Varma et al., 1979.*
[c]*Peter et al., 1977.*
[d]*Estimated from resistivity.*

Most MTBA calculations of $\langle I^2 \rangle$ yield values which are
much higher than the empirical ones, but the very recent MTBA
results of Varma et al., who use a non-orthogonal basis set
and include "p-type" orbitals as well as "s" and "d" are in
excellent agreement with the RMTA results. Papaconstantopoulos
et al., (1977) have obtained RMTA results very similar to those
shown in Table I. Birnboim and Gutfreund, (1975) obtained a
value for $\langle I^2 \rangle$ for Nb of 0.024 using an MTBA approach.

Substantial uncertainty is associated with the empirical
estimates of $\langle I^2 \rangle$ in Table I because they depend on the poorly
known values of the Coulomb pseudopotential μ^* and upon estimates of the effective mean square phonon frequencies $\langle \omega^2 \rangle$ defined below. Details of the precedure used in estimating these
quantities are given by Butler, (1977). Uncertainties in
$\langle \omega^2 \rangle$ can be avoided by calculating the mass enhancement λ
which is defined by

$$\lambda = \frac{\Omega}{(2\pi)^3} \int \frac{dS_k}{v_k} \int \frac{dS_{k'}}{v_{k'}} \sum_{\alpha\beta j} \frac{I_{kk'}^{\alpha} \cdot I_{kk'}^{\beta} \cdot \varepsilon_j^{\alpha}(k-k') \varepsilon_j^{\beta}(k-k')}{M\omega_j^2(k-k') \int \frac{dS_k}{v_k}} \quad , \quad (3)$$

and is related to $\langle I^2 \rangle$ and $\langle \omega^2 \rangle$ by $\lambda = N\langle I^2 \rangle / M\langle \omega^2 \rangle$ where N is
the Fermi energy density of states and $\hat{\varepsilon}_j$ is a phonon polari-
zation vector. Table II compares experimental and calculated
values of λ for Nb, Mo, and Pd. The RMTA values seem to be
about 10% too high for Nb and Pd and essentially exact for Mo
(taking μ^* to be 0.11).

The uncertainties in μ^* can be avoided by calculating the
room temperature resistivity which is given by Chakraborty
et al., (1976) $\rho = (3\pi k_B / \hbar e^2 N \langle v^2 \rangle) \lambda_{tr} T$ where $\langle v^2 \rangle$ is the mean
square Fermi velocity and λ_{tr} is a quantity very similar to λ

$$\lambda_{tr} = \frac{\Omega}{(2\pi)^3} \int \frac{dS_k}{v_k} \int \frac{dS_{k'}}{v_{k'}} \sum_{\alpha\beta j} \frac{I_{kk'}^{\alpha} \cdot I_{kk'}^{\beta} \cdot \varepsilon_j^{\alpha}(k-k') \varepsilon_j^{\beta}(k-k')}{M\omega_j^2(k-k') \int \frac{dS_k}{v_k} 2v_k^2} (\vec{v}_k - \vec{v}_{k'})^2 \quad . \quad (4)$$

RMTA calculations yield $\lambda_{tr} = 1.07$ for Nb and $\lambda_{tr} = 0.46$ for
Pd (Pinski et al., 1978a). Both values appear to be about 10%
larger than experiment.

TABLE II. Calculated and Empirical Values of λ for Nb, Mo, and Pd

	Empirical[a]	Empirical[b]	Tunneling	RMTA
Nb	0.96	1.04	$1.01,^c 0.98^d$	1.12^e
Mo	0.44			0.40^f
Pd		0.38		0.41^f

[a]*From Tc using $\mu^* = 0.13$*
[b]*From Resistivity: $\lambda = (\lambda_{tr})emp \times (\lambda/\lambda_{tr})calc.$*
[c]*Arnold et al., (1979).*
[d]*Robinson and Rowell, (1977).*
[e]*Butler et al., (1979).*
[f]*Pinski et al., (1978b).*

More detailed investigations of the strength of the electron-phonon interaction have also been performed. The most detailed information can be obtained from measurements of the phonon linewidth (Allen, 1972). The phonon linewidth $\gamma_j(q)$ tells the amount of electron-phonon coupling contributed by each particular phonon and is related to λ by

$$\lambda = \sum_{qj} \gamma_j(q)/\pi N\omega_j^2(q) \ . \tag{5}$$

RMTA calculations (Butler et al., 1977 and Pinski and Butler, 1979a), have successfully predicted two regions of strong electron-phonon coupling before they were observed experimentally (Fig. 1). The peak in the calculated Pd [100] LA linewidth appeared at $\zeta = 0.34$, but by identifying the Fermi surface transitions responsible for the peak and using experimental Fermi surface data it was possible to *predict* that the true peak position would be at $\zeta = 0.4$ in precise agreement with experiment (Youngblood et al., 1979).

Another detailed measure of the electron-phonon interaction is the spectral function $\alpha^2 F(\omega)$ which can be defined as an average over the linewidth

$$\alpha^2 F(\omega) = \sum_{qi} \gamma_j(q)\delta[\omega-\omega_j(q)]/2\pi\omega N \ . \tag{6}$$

RMTA calculations of $\alpha^2 F(\omega)$ for Nb (Butler et al., 1977, Harmon and Sinha, 1977, and Butler et al., 1979) yield a spectral function which is quite similar in shape to the phonon density

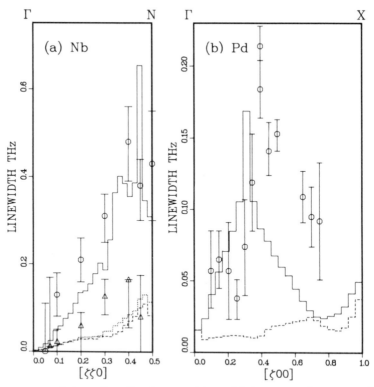

*FIGURE 1(a).   [100] phonon linewidths for Nb.   (b) [100]
phonon linewidths for Pd.   Solid (dotted or dashed) histograms
are calculated longitudinal (transverse) linewidths.   Circles
(triangles) with error bars are experimental longitudinal
(transverse) linewidths.*

of states (Fig. 2). Similar results have been obtained by the
Varma-Weber version of the MTBA (Weber, 1977).

The experimental spectral function of Nb is somewhat controversial. Three experimental spectra are shown in Fig. 2.
The differences arise partly from different experimental techniques and partly from different ways of analyzing the tunneling data. The most recent of the experimental spectra (Arnold
et al., 1979) shows excellent agreement with the calculation
except for the high frequency peak. The phonon linewidth
measurements seem to favor the calculated spectrum because
they show strong coupling for the longitudinal phonons. It
has recently been shown (Otschik, 1978) that any mechanism
(e.g., disorder, dissolved oxygen) which decreases the strength
of the electron-phonon coupling within ≈50 Å of the surface
will appear in the deduced spectral function as an enhancement
at low frequency and a reduction at high frequency.

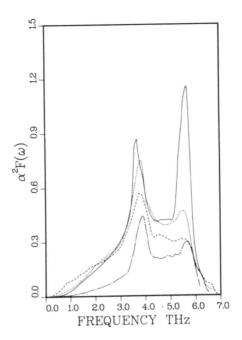

FIGURE 2.   Calculated and experimental spectral functions for Nb.   Solid curve is RMTA calculation (Butler et al., 1979), dotted, dashed and chain-dotted curves are experimental results of Arnold et al. (1979), Robinson and Rowell (1977), and Bostock et al. (1976), respectively.

Since the shape and strength of the spectral function determine the thermodynamic critical field $H_c(T/T_c)$, it is possible in principle to test a proposed spectral function by using it to calculate a deviation function $D(t) = [H_c(t)/H_c(0)] - (1-t^2)$ which may be compared with experiments. In Fig. 3, we compare the deviation function obtained from the calculated spectral function of Fig. 2 with three sets of experimental data. The agreement is quite good especially with the calorimeteric data of Ferreira et al. Unfortunately, the deviation function is insufficiently sensitive to the shape of $\alpha^2 F(\omega)$ and the different sets of experimental data show too much variation to conclusively rule out either the Arnold–Wolf spectrum or the Robinson–Rowell spectrum (Daams and Carbotte, 1978). The deviation function does indicate however that λ is approximately 1.0. It should be noted that the Arnold–Wolf data has λ of about the right magnitude but the deduced T_c is

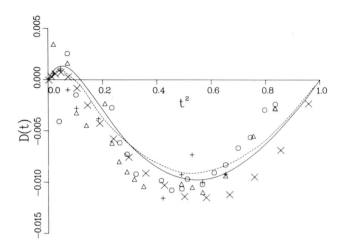

FIGURE 3.   Calculated and experimental deviation functions for Nb. Circles, triangles, and +'s are magnetization data (Kerchner et al., 1979) taken with applied field parallel to [100], [110] and [111] directions, respectively. X's and dashed line are calorimetric data of Leupold and Boorse (1964) and Ferreira et al. (1969), respectively. Solid line is calculated using the RMTA α^2F of Fig. 2.

only 7.9 K and the Robinson-Rowell data although it yields the correct T_c gives a value for the zero frequency gap which is too low.

The upper critical field H_{c2} of a cubic type II superconductor is sensitive to anisotropy in the Fermi velocity and in the gap function. Figures 4 and 5 show the results of recent calculations of $H_{c2}(T,\hat{B})$ for pure Nb. The calculations are based on an extension of the theory of Hohenberg and Werthammer, (1967) and amount to an essentially exact solution of their equations (5) and (26). Fermi velocities were calculated at 1057 points in the irreducible Brillouin zone using Korringa-Kohn-Rostocker band theory and a potential adjusted to give a good fit to Fermi surface data (Butler et al., 1979). Small strong coupling corrections were applied using the factors $\eta_{H_{c2}}(0)$ and $\eta_{H_{c2}}(1)$ calculated by Rainer and Bergmann, (1974). It is primarily the anisotropy of the Fermi velocity which is responsible for the anisotropy of H_{c2} and for the deviation of its temperature dependence from that predicted by isotropic theory. The calculated anisotropy in Fig. 4 has the same shape as the experiment but it is somewhat

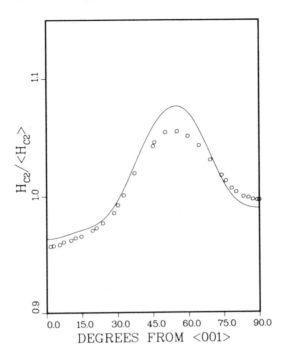

FIGURE 4. *Anisotropy of H_{c2} for Nb in [110] plane. Solid curve is calculated for $T = 0$. Circles are experimental results (Williamson, 1970) for $T/T_c = 0.04$.*

larger. The calculations do not include gap anisotropy because RMTA calculations (Butler et al., 1979 and Pinski and
Butler, 1979b) indicate it is ignorably small. MTBA calculations of Peter et al., (1977) yield a larger anisotropy but it
is of such character as to increase the observed anisotropy
and worsen the agreement with experiment.

ACKNOWLEDGMENTS

　　J. Carbotte, J. M. Daams and B. Mitrovic graciously provided the programs which were used to calculate $H_c(T)$.
G. Arnold, E. L. Wolf and R. Kerchner made their data available
to us prior to publication. Helpful conversations with
P. B. Allen, F. J. Pinski, J. S. Faulkner, J. F. Harris, and
G. M. Stocks are gratefully acknowledged. We also extend thanks
to G. Golliher for rapid and accurate typing of the camera-
ready manuscript.

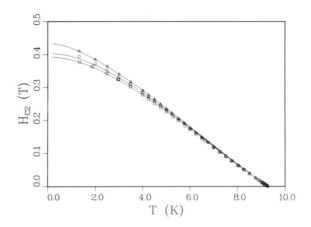

FIGURE 5.   Temperature dependence of H$_{c2}$ for Nb.   Solid curves are calculated results for an applied field in the [111] (upper), [110] (middle) and [100] (lower) directions.   Experimental results (Kerchner et al., 1979) are shown by the triangles [111] direction, circles [110] direction, and rectangles [100] direction.

REFERENCES

Allen, P.B. (1972). *Phys. Rev. B 6*, 2577.
Arnold, G. B., Zasadzinski, J., Osmun, J. W., and Wolf, E. L. (1979). *Phys. Rev. B* (to be published).
Ashkenazi, J., Dacorogna, M., and Peter, M. (1979). *Sol. State Comm. 29*, 181.
Birnboim, A. and Gutfreund, H. (1975). *Phys. Rev. B 12*, 2682.
Bloch, F. (1928). *Z. Physik 52*, 555.
Bostock, J., Lo, K. H., Cheung, W. N., Diadiuk, V., and MacVicar, M.L.A. (1976). In "Superconductivity in d- and f-band Metals," (ed. D. H. Douglass), Plenum, New York.
Butler, W. H. (1977). *Phys. Rev. B 15*, 5267.
Butler, W. H., Olson, J. J., Faulkner, J. S., and Gyorffy, B. L. (1976). *Phys. Rev. B 14*, 3823.
Butler, W. H., Pinski, F. J., and Allen, P. B. (1979). *Phys. Rev. B 19*, 3708.
Butler, W. H., Smith, H. G., and Wakabayashi, N. (1977). *Phys. Rev. Lett. 39*, 1004.
Chakraborty, B., Pickett, W. E., and Allen, P. B. (1976). *Phys. Rev. B 14*, 3227.
Daams, J. M. and Carbotte, J.P. (1978). In "Transition Metals-1977," (eds. M.J.G. Lee, J. M. Perz, E. Fawcett), the Institute of Physics, London.

Ferreira, J., Burgemeister, E. A., and Dokoupil, A. (1969).
    Physica 41, 409.
Fröhlich,H. (1966). In "Perspectives in Modern Physics,"
    (R. E. Marshak, ed.) p. 539, Interscience, New York
Gaspari, G. D. And Gyorffy, B. L. (1972). *Phys. Rev. Lett. 28*,
    801.
Harmon, B. M. and Sinha, S. K. (1977). *Phys. Rev. B 16*, 3919.
Hohenberg, P. C. and Werthamer, N. R. (1966). *Phys. Rev. 153*,
    493.
Kerchner, H. R., Christen, D. K., and Sekula, S. T. (1979).
    Phys. Rev. B (to be published) and this volume.
Leupold, H. A. and Boorse, H. A. (1964). *Phys. Rev. 134*, A1322.
Matthias, B. T. (1957). In "Progress in Low Temperature
    Physics," (C. J. Gorter, ed.) p. 138, North Holland
    Amsterdam.
Otschik, P. (1978). *Phys. Stat. Sol. (b) 88*, 563.
Papaconstantopoulos, D. A., Boyer, L. L., Klein, B. M.,
    Williams, A. R., Moruzzi, V. L., and Janak, J. R. (1977).
    Phys. Rev. B 15, 4221.
Peter, M., Klose, W., Adam, G., and Entel, P. (1974). *Helv.
    Phys. Act. 47*,807.
Peter, M. Ashkenazi, J., Dacorogna, M. (1977). *Helv. Phys. Act.
    50*, 267.
Pinski, F. J., Allen, P. B., and Butler, W. H. (1978a). *Phys.
    Rev. Lett. 41*, 431.
Pinski, F. J., Allen, P. B., and Butler, W. H. (1978b). *J.
    Phys. (Paris) 39*, C6-472.
Pinski, F. J. and Butler, W. H. (1979a). *Phys. Rev. B* (in press).
Pinski, F. J. and Butler, W. H. (1979b). Unpublished.
Rainer, D. and Bergmann, G. (1974). *J. Low Temp. Phys. 14*, 1974.
Robinson, B. and Rowell, J. M. (1977). Private communication.
Shapiro, S. M., Shirane, G., and Axe, J. D. (1975). *Phys. Rev.
    B 12*, 4899.
Varma, C. M., Blount, E. I., Vashishta, P., and Weber, W. (1979).
    Phys. Rev. B (in press).
Weber, W. (1977). Private communication.
Williamson, S. J. (1970). *Phys. Rev. B 2*, 3545.
Youngblood, R., Noda, Y., and Shirane, G. (1979). *Phys. Rev. B*
    (in press).

CALCULATIONS OF THE SUPERCONDUCTING AND
TRANSPORT PROPERTIES OF
A15 STRUCTURE COMPOUNDS

Barry M. Klein
Dimitrios A. Papaconstantopoulos
Larry L. Boyer

Naval Research Laboratory
Washington, D.C.

I. INTRODUCTION

Klein, et. al., (1978) have performed electronic band structure calculations for a series of 10 different A15 structure compounds V_3X and Nb_3X, X = Al, Ga, Si, Ge, and Sn having both high and low superconducting transition temperatures, T_c. These results have been used (Klein, et. al., 1979) to study the superconducting properties of these materials along with a discussion of the thermal anomalies present in some of them. In this paper we make use of the aforesaid results, and also some new results for Nb_3Sb, to present calculations of Fermi velocities, plasma frequencies, mean free paths, and resistivities for all of these materials. In addition we put forth arguments on how to use these results to predict even higher T_c materials as recently formulated by Papaconstantopoulos, et. al. (1979). Specifically we predict that $T_c > 15K$ for the pseudobinary A15 alloy $(V_{1-x} Ti_x)_3$ Ge, $x \sim 0.13$.

II. BAND STRUCTURE AND DENSITY OF STATES RESULTS

The electronic band-structure calculations used as a basis for our studies have been performed using the self-consistent augmented-plane-wave method (APW). These calculations include non-spin-dependent relativistic corrections, "warped muffin-tin" corrections to the usual APW approach, and make use of the local density approximation to the exchange-correlation potential. Full details of these calculational approaches are given by Klein, et. al., (1978). The density-of-states (N(E)) calculations have been performed using a new fourier series interpolation procedure (Boyer, 1979) with the final integrations being done using the tetrahedral method (Lehmann and Taut 1972). Our calculations of the Fermi velocities and other related transport quantities to be described have been performed using appropriate modifications of these density-of-states techniques. We also note that several other groups have done A15 band structure calculations and applications (Mattheiss, 1965, 1975; Ho, et. al., 1978; Arbman and Jarlborg, 1978; and van Kessel, et. al., 1978).

In the paper by Klein, et. al., (1978), band structure and N(E) results for the A15 compounds V_3X and Nb_3X, $X = $ Al, Ga, Si, Ge, and Sn have been presented. In Fig. 1 we show new results for Nb_3Sb, one of the low-T_c ($=0.2$ K) A15 compounds. We see from Fig. 1 that N(E_F) (E_F is the Fermi energy) is quite small for Nb_3Sb, having a value $\sim \frac{1}{3}$ that of high-T_c Nb_3Sn. The fact that Nb_3Sb has a low-T_c can be anticipated from this result if one assumes that T_c correlates with N(E_F), which was shown to be approximately true by Klein, et. al., (1979) for the A15 compounds. In fact on a per atom basis, N(E) for Nb_3Sb is considerably smaller than that of metallic bcc Nb. Our detailed calculations give $\lambda = 0.22$ for Nb_3Sb and $T_c \cong 0.0$ K.

We have previously discussed (Klein, et. al., 1979) the fine structure in N(E) near E_F, and its correlation with the anomalous thermal behavior of the electrical, magnetic and elastic properties of several A15 compounds. We showed that for Nb_3Sn, a very good fit to the observed magnetic susceptibility as a function of temperature could be obtained using the Pauli susceptibility calculated from our N(E) results. We claimed that those A15 compounds showing strong thermal variations in quantities such as the magnetic susceptibility, Knight shift and elastic constants are those with strong variations in N(E) near E_F.

In Figs. 2 and 3, we show plots of N(E) in a range $\sim \pm 3$mRy ($\sim \pm 500°$K) around E_F for 11 different A15 materials. We see that V_3Si, Nb_3Sn and V_3Ga, the materials exhibiting the most pronounced anomalous thermal behavior in the aforesaid properties have the biggest variations in N(E). It is our contention that this sharp structure is primarily responsible for the observed anomalous thermal variations because of the resultant unusual temperature dependencies introduced via the Fermi functions (chemical potential, Pauli susceptibility, etc.). In addition, we point out that this argument does not imply that there exists a strong correlation between high-T_c and fine structure in N(E) near E_F. Indeed the high T_c compounds Nb_3Al and Nb_3Ga have high values of N(E_F), but much weaker *structure* in N(E) compared with Nb_3Sn, V_3Si and V_3Ga. It is satisfying to note that the observed thermal variations are much smaller in Nb_3Al and Nb_3Ga than in Nb_3Sn, V_3Si, and V_3Ga.

III. CALCULATIONS OF TRANSPORT PROPERTIES

We now wish to turn to a discussion of some new calculations of several of the transport parameters of the A15 compounds — Fermi velocities (v), plasma frequencies (Ω_p), electron-phonon mean free paths (l), and resistivities ($\rho_1 T$).

$$v^n(E) = <|\vec{\nabla}_{\vec{k}} E(\vec{k})|^n> \tag{1}$$

$$\Omega_p^2(E) = \frac{8\pi}{3} e^2 N(E_F) v^2(E) \tag{2}$$

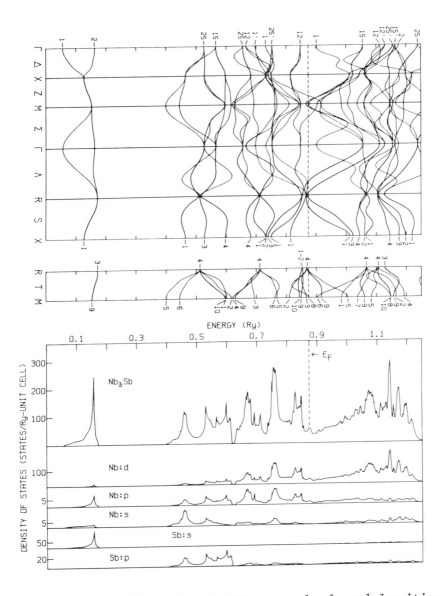

FIGURE 1. Self-consistent APW energy bands and densities-of-states for Nb₃Sb. The calculational procedure follows Klein et al. (1978).

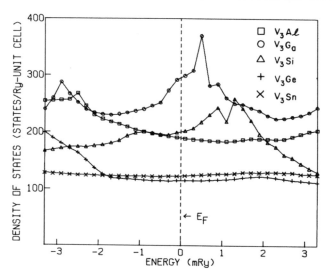

FIGURE 2.   *Densities-of-states near E_F for a series of V_3X*
A15 compounds.

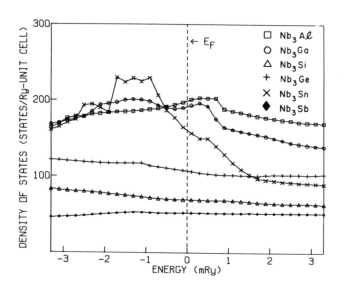

FIGURE 3.   *Densities-of-states near E_F for a series of Nb_3X*
A15 compounds.

$$\hbar \tau_{ep}^{-1} = 2\pi \lambda_{tr} K_B T \tag{3}$$

$$l = \hbar v^2 (E_F)/2\pi \lambda_{tr} K_B T \tag{4}$$

$$\rho_1 T = 4\pi/\Omega_p^2 \tau_{ep} \tag{5}$$

In Eqs. 1-5, E is an energy, $\vec{k}$ a Brillouin zone wave-vector, τ_{ep} is the electron-phonon collision time, λ_{tr} is the "transport" equivalent to the electron-phonon mass enhancement factor (see Allen 1971; and Chakraborty, et. al., 1976), K_B is Boltzmann's constant, and T the temperature. We note that in our calculations, λ_{tr} has been replaced by the λ used in superconductivity theory (see the articles by W. Butler and P.B. Allen in this volume) in our calculations.

There are three band-structure derived quantities entering into Eqs. 1-5; $v^2(E)$, $N(E)$, and λ (used in place of λ_{tr}), with the former two calculated using Boyer's (1979) interpolation method. λ has been determined using the method of Gaspari and Gyorffy (1972) in conjunction with the phonon spectra data of Schweiss, et. al., (1976), as discussed by Klein, et. al., (1979).

Figures 4 and 5 show plots of $N(E)$, $v_F(E) \equiv (v^2(E))^{1/2}$ and $\hbar\Omega_p$ for Nb_3Sn and Nb_3Sb respectively; and Table I gives a tabulation of all of our transport parameter calculations. Referring first to Figs. 4 and 5 we see that the magnitude of the

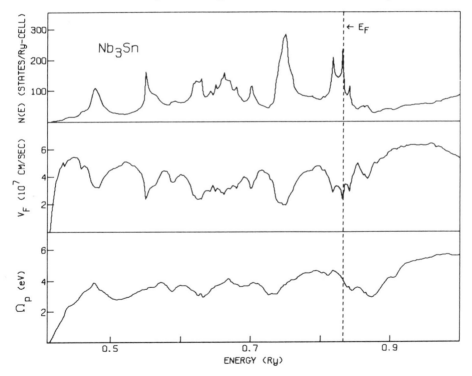

FIGURE 4. Density-of-states, $N(E)$; Fermi velocity, $v_F(E)$; and plasma frequency, $\Omega_p(E)$ for Nb_3Sn.

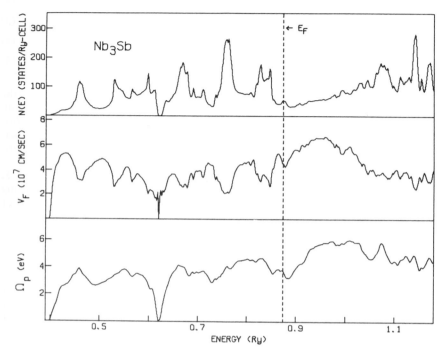

FIGURE 5. *Density-of-states, N(E); Fermi velocity, vF(E);
and plasma frequency, $\Omega_p(E)$ for Nb₃Sb.*

variations of $v_F(E)$ and N(E) with energy are similar. Energy variations of $\Omega_p(E)$ are also considerable, but we note that we have examined $\Omega_p(E)$ on an energy scale of several mRy around E_F (as in Figs. 2 and 3) and found very small variations in those materials where both N(E) and $v^2(E)$ have very big variations (viz. Nb₃Sn, V₃Si, V₃Ga). The fact that near E_F, $\Omega_p^2 \propto N(E) \cdot v^2(E)$ is roughly constant is central to the arguments of Wiesman, et. al., (1978) and Ghosh, et. al., (1978) in their discussions of the variations in the superconducting properties of radiation-damaged A15 materials. However, we should emphasize that this approximation deteriorates as one expands the energy range under consideration, whence the precise computed values should be used.

From Table I we see that $\Omega_p(E_F)$ shows relatively small variations amongst the A15 materials, and that generally speaking there is no apparent correlation between T_c and the magnitude of $\Omega_p(E_F)$. However, both *l* and $\rho_1 T$ show very interesting correlations with T_c as shown in Figs. 6 and 7. In these figures the squares represent calculated values of *l* and $\rho_1 T$ plotted against experimental values of T_c. The triangles, for Nb₃Si and Nb₃Ge, are our calculated values of *l* and $\rho_1 T$ and also our *calculated* values of T_c (see Klein, et. al., 1979). As noted previously, using our band structure results and canonical electron-phonon theory (along with the phonon spectrum of Nb₃Sn), both Nb₃Si and Nb₃Ge would not seem to be high-T_c materials in their stoichiometric forms, while at least for Nb₃Ge high T_c's

Table I. Calculated Densities of States and Transport Parameters (at 300K) for A15 Compounds

	T_c^{expt} (K)	$N(E_F)$ (per Ry-cell)	λ_{calc}	$\hbar\Omega_p$ (eV)	$<v^2(E_F)>^{1/2}$ (10^8cm/sec)	$<v(E_F)>$ (10^8cm/sec)	$\hbar\tau_{ep}^{-1}$ (eV)	l (Å)	$\rho_1 T$ (μΩcm)
V_3Al	9.6	188.9	1.09	4.46	0.248	0.233	0.18	9.2	66.2
V_3Ga	16.5	295.6	1.48	4.63	0.205	0.159	0.24	5.6	83.4
V_3Si	17.1	200.2	1.18	4.02	0.210	0.172	0.19	7.2	88.2
V_3Ge	6.1	114.7	0.67	4.14	0.290	0.278	0.11	17.5	47.2
V_3Sn	3.8	123.0	0.52	3.73	0.270	0.260	0.08	21.0	45.1
Nb_3Al	18.6	199.2	2.14	4.58	0.277	0.249	0.35	5.3	123.2
Nb_3Ga	20.3	191.8	1.92	4.71	0.289	0.253	0.31	6.1	104.5
Nb_3Si (a = 5.03Å)		61.42	0.65	4.43	0.460	0.428	0.11	28.7	40.0
Nb_3Si (a = 5.10Å)	>10	69.20	0.64	4.45	0.444	0.411	0.10	28.1	39.0
Nb_3Si (a = 5.20Å)		83.61	0.64	4.34	0.406	0.370	0.10	25.7	41.0
Nb_3Ge	23.2	106.7	0.89	4.35	0.356	0.315	0.14	16.2	56.8
Nb_3Sn	18.0	158.6	1.12	4.00	0.278	0.233	0.18	10.1	84.5
Nb_3Sb	0.2	51.95	0.22	3.74	0.451	0.420	0.04	83.1	19.0

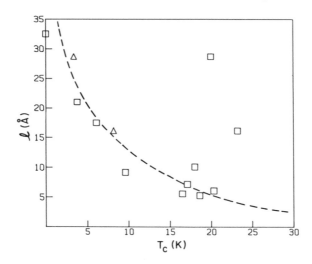

FIGURE 6. Calculated room temperature mean free path, l, versus measured values of T_C (squares) for 11 A15 compounds; in this figure we assume that $T_C = 20$ K for Nb₃Si. Additionally, the triangles denote calculated l and calculated T_C values for Nb₃Ge and Nb₃Si. The dashed line is a guide for the eye. The calculated l for Nb₃Sb is off-scale, having a value of 83 Å with $T_C = 0.2$ K.

have been observed (although the details of the stoichiometry are still being debated). In all other cases, our calculated T_c's are in reasonable agreement with experiment leading us to the conclusion that Nb₃Ge and Nb₃Si have soft phonons, defects, or some other high-T_c producing mechanism that is operative (Klein, et. al., 1979). This point is again emphasized in Figs. 6 and 7 where, using standard electron-phonon theory, Nb₃Ge and Nb₃Si correlate with the other A15 materials only if we use our calculated values of T_c for the pure materials. Leaving aside Nb₃Si and Nb₃Ge, Fig. 6 emphasizes the intuitively satisfying point that those materials with the highest room temperature electron-phonon resistivity also have the highest T_c. In fact $\rho_1 T$ vs. T_c is approximately linear.

From Fig. 7 we see that there is an inverse relationship between the room temperature mean free path and T_c, with the high-T_c materials V₃Si, V₃Ga, Nb₃Al, Nb₃Ga, and Nb₃Sn having values of $l \lesssim 2a_0$, where a_0 is the cubic lattice constant. Another way of putting this is that l is several times the nearest-neighbor separation of transition-metal atoms (along a chain) for the high-T_c compounds, but l is considerably larger (factor 10 for Nb₃Sb) for the low-T_c materials.

Fisk and Webb (1976) first discussed the resistivity saturation observed in A15 materials in terms of the mean free path for electron-phonon scattering becoming comparable to the lattice spacing at approximately room temperature or above. Our results show that such saturation would occur at higher temperatures for the low-T_c materials than for the high-T_c materials (keep in mind that Figs. 6 and 7 are for $T = 300$ K, and $l \propto 1/T$). Indeed it is found experimentally that the

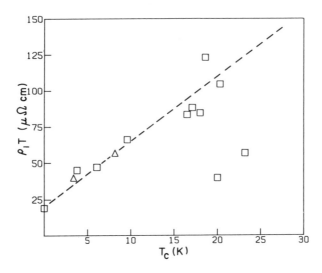

FIGURE 7.   Calculated room temperature electron-phonon resistivity, $\rho_1 T$, versus measured values of T_c (squares) for 11 A15 compounds; in this figure we assume that T_c = 20 K for Nb_3Si. Additionally, the triangles denote calculated $\rho_1 T$ and calculated T_c values for Nb_3Ge and Nb_3Si. The dashed line is a guide for the eye.

saturation for Nb_3Sb compared to Nb_3Sn does occur at somewhat higher temperatures (Fisk and Webb, 1976), but in the absence of a detailed quantitative saturation theory it is not clear whether our numerical results for l give a complete explanation. We should also mention that perhaps an extension of the Mott (1935, 1976) and Wilson (1938) theory of resistivities along the lines of that proposed by Morton, et. al., (1978) may be appropriate.

IV. PSEUDOBINARY A15s — RAISING T_c

A final word regarding predictions of new high-T_c materials which may be culled from our electronic structure results. Recently Papaconstantopoulos, et al., (1979) have shown that the rigid-band model may be applied to our band structure results in the form $(A_{1-x} B_x)_3 C$, where $(x \lesssim 1.0)$. A and B are transition metals from the same row of the periodic table and C is a non-transition-metal. Neglecting small phonon spectra effects, and considering x as small, one may picture atom B as a "Fermi level adjustor" used to shift E_F into a peak in $N(E)$ and hence enhance T_c. For pseudobinaries of the form described this appears to be theoretically justified, as shown by Papaconstantopoulos, et al. (1979). One of the predictions made is that the pseudobinary $(V_{1-x} Ti_x)_3 Ge$, with $x \sim 0.13$ is a good possibility for having $T_c > 15K$. In effect Ti substitutions shift E_F to a lower energy peak in $N(E)$ for V_3Ge and cause a big increase in the theoretical λ and hence T_c. Hopefully, by the time this book is published, such a high-T_c material will be found, for perhaps such a theoretical approach offers an opportunity for predicting even higher T_c materials in the A15 or other structures.

ACKNOWLEDGMENTS

We are grateful to D.U. Gubser and L.F. Mattheiss for their help in various aspects of our A15 work.

REFERENCES

Allen, P.B. (1971). *Phys. Rev. B3*, 305.

Allen, P.B., Pickett, W.E. Ho, K.M., and Cohen, M.L. (1978). *Phys. Rev. Lett. 40*, 1532.

Arbman, G., and Jarlborg, T. (1978). *Solid State Commun. 26*, 857.

Boyer, L.L. (1979). *Phys. Rev. B19*, 2824.

Chakraborty, B., Pickett, W.E., and Allen, P.B. (1976). *Phys. Rev. B14*, 3227.

Fisk, Z., and Webb, G.W. (1976). *Phys. Rev. Lett. 36*, 1084.

Gaspari, G.D., and Gyorffy, B.L. (1972). *Phys. Rev. Lett. 28*, 801.

Ghosh, A.K. Gurvitch, M., Wiesmann, H., and Strongin, M. (1978). *Phys. Rev. B18*, 6116.

Ho, K.M., Pickett, W.E., and Cohen, M.L., (1978). *Phys. Rev. Lett. 41*, 580, 815.

Klein, B.M., Boyer, L.L., Papaconstantopoulos, D.A., and L.F. Mattheiss (1978). *Phys. Rev. B18*, 6411.

Klein, B.M., Boyer, L.L. and Papaconstantopoulos, D.A. (1979). *Phys. Rev. Lett. 42*, 530.

Lehmann, G., and Taut, M., (1972). *Phys. Status Solidi 54*, 469.

Mattheiss, L.F. (1965). *Phys. Rev. 138*, A112.

Mattheiss, L.F. (1975). *Phys. Rev. B12*, 2161.

Mattheiss, L.F., Testardi, L.R., and Yao, W.W. (1978). *Phys. Rev. B17*, 4640.

Morton, N., James B.W., and Wostenholm, G.H. (1978). *Cryogenics 18*, 131.

Mott, N.F. (1935). *Proc. Phys. Soc. 47*, 571.

Mott, N.F. (1976). *Proc. Roy. Soc. A153*, 699.

Papaconstantopoulos, D.A., Gubser, D.V. Klein, B.M., and Boyer, L.L. (1979). Unpublished.

Schweiss, B.P., Renker, B., Schneider, E., and Reichardt, W. (1976). *in* "Superconductivity in d- and f-Band Metals" (D.H. Douglass, ed.), p. 189. Plenum Press, New York.

van Kessel, A.T., Myron, H.W. , and Mueller, F.M. (1978). *Phys. Rev. Lett. 41*, 181, 520(E).

Wiesman, H., Gurvitch, M., Ghosh, A.K., Lutz, H., Kammerer, O.F., and Strongin, M. (1978). *Phys. Rev. B 17*, 122.

Wilson, A.H. (1938). *Proc. Roy. Soc. A167*, 580.

THE ELIASHBERG FUNCTION AND THE SUPERCONDUCTING T_c
OF TRANSITION METAL HEXABORIDES

German Schell
Hermann Winter
Hermann Rietschel

Institut für Angewandte Kernphysik
Kernforschungszentrum Karlsruhe
Karlsruhe, FRG

I. GENERAL

The metal hexaborides of composition MeB_6 consist of two simple cubic sublattices formed by the metal atoms and boron octahedra (fig. 1), their space group being Pm3m. Me stands for rare earth-, alkaline earth metals, or actinides. These substances can be considered as a model for more complicated

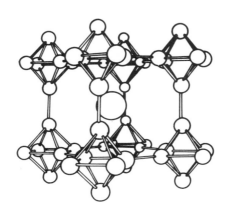

FIGURE 1. *Crystal structure of the hexaborides.*

structures consisting of clusters of atoms. Two compounds of this class are known to be superconducting: LaB_6 with $T_c = .45K$ and YB_6 with $T_c = 7.1K$. We investigate the superconducting properties of these two substances on a microscopic level by calculating their Eliashberg functions $\alpha^2 F(\omega)$.

In order to evaluate the T_c's of these compounds their $\alpha^2 F$'s have to be known in detail, because the individual phonon modes existing in these systems have very different couplings to the electrons. As a starting point for our analysis we use the Gyorffy formula:

$$\alpha^2 F(\omega) = \sum_{\vec{q},\kappa,s} \frac{\eta_s}{6M_s N} \left| \vec{e}^{\,s}_{\vec{q},\kappa} \right|^2 \frac{\delta(\omega - \omega_{\vec{q},\kappa})}{\omega} \tag{1}$$

$$\eta_s = \frac{2m\varepsilon_F}{\pi^2 n(\varepsilon_F)\hbar^2} \sum_{1} 2(1+1)\sin^2(\delta_1^s - \delta_{1+1}^s) \frac{n_1^s n_{1+1}^s}{n_1^{(o)s} n_{1+1}^{(o)s}} \tag{2}$$

The electronic quantities occuring in (1) and (2) are: the Fermi energy ε_F, the total density of states at ε_F ($n(\varepsilon_F)$), the scattering phaseshifts of the different atoms s (δ_1^s), and the partial densities of states (n_1^s, $n_1^{(o)s}$); the phononic quantities: the phonon frequencies for wavenumber $\vec{q}$ and branch κ ($\omega_{\vec{q},\kappa}$), and the polarization vectors ($\vec{e}^{\,s}_{\vec{q},\kappa}$).

In (1) and (2) we give the local version of Gyorffy's formula for brevity where both potentials are at the same atom, whereas in this work $\alpha^2 F$ has been calculated using a nonlocal extension (see appendix).

II. PHONON MODES

The phonon frequencies and eigenvectors have been calculated using a Born von Karman model with nearest neighbour boron-boron and metal-boron force constants. The parameters for LaB_6 have been determined by fitting the position of the energetically highest peak in the phonon density of states curve to Raman experiments (Scholz et al., 1976) and that of the energetically lowest one to neutron scattering experiments (Gompf, 1978). This procedure brings the remaining peaks approximately to the right positions (fig. 2).

The peaks of fig. 2 are connected with certain types of vibrations (at the Γ-point) which we visualize in fig. 3.

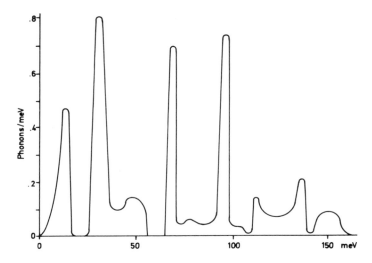

FIGURE 2.   Phononic density of states in LaB$_6$.

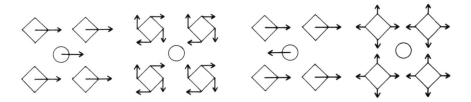

FIGURE 3.   Some phonon modes: a) translational,
b) torsional, c) optical, and d) one of the
octahedron deformation modes.
◇ = one boron octahedron, ○ = metal atom.

III. ELECTRONIC STRUCTURE

In order to get the one particle Green's functions
$g(\vec{r},\vec{r}',\varepsilon)$ we applied the selfconsistent symmetrized cluster
approach described by Ries and Winter (1979) to systems con-
sisting of 68 atoms with a boron octahedron in the centre.
Embedding the clusters into a flat Watson sphere with negative
potential allowed for the calculation of states at negative
energies. A selfconsistency treatment proved necessary because
charge is transferred from the metal- to the boron atoms and

the electronic quantities entering (2) change appreciably in
the course of the about eight iterations needed to attain
selfconsistency when starting with the Matheiss potential
construction.

In fig. 4 we exhibit the total electronic density of
states (DOS) for LaB_6 in the energy range of the valence electrons. The DOS curve of YB_6 which has been calculated independently shows roughly the same structure except that the 5p
electrons of La which give rise to the narrow peak at -.3Ryd
below muffin tin zero in fig. 4 are missing in YB_6. In analogous calculations for the semiconducting compound CaB_6 the
Fermi level turned out to fall in a region of an extremely
low DOS which even diminished with increasing cluster size.

The peaks of fig. 4 can be classified in accord with the
early tight binding considerations of Longuet-Higgins and de
Roberts (1954). Up to $\varepsilon=.4Ryd$ the underlying combinations of
local s-p boron states have intra octahedron binding character,
for ε between .4Ryd and ε_F inter cluster binding dominates,
and only for energies above ε_F the metal d states are admixed
appreciably. There is a good overall agreement with boron K-
spectroscopy (Lyakhovskaya et al., 1970). The values needed
for the evaluation of η_s in (2) are summarized in table I.

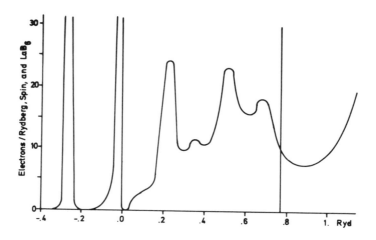

FIGURE 4. Total electronic density of states in LaB_6.

IV. CONCLUSIONS

The results of II and III enable us to calculate $\alpha^2F(\omega)$
according to (3) which is drawn in fig. 5.
The significance of the different phonon modes for superconductivity may be visualized by displaying their contributions to the Mc Millan parameter λ (table II).
As follows from table I the electrons have a considerable
coupling to the boron sublattice and the electronic parameters
η_s are comparable in YB_6 and LaB_6, the main contributions
coming from the low frequency translational and torsional
modes while the coupling to the octahedron deformation modes
is weak due to their high frequencies, invalidating previous
speculations about high T_c's in these substances based upon
their high Debye frequencies. The evaluation of the nonlocal
formula (3) turned out to be of great importance because λ is
reduced by about 25% as compared to the local result. The difference in the T_c's is due to the fact that the low frequency
modes in YB_6 are shifted by a factor of $\sim.75$ with respect to
those in LaB_6 (Gompf, 1978).
T_c has been calculated solving the Eliashberg equation
with the formalism developed by Rainer and Bergmann (1973)
and Allan and Dynes (1975). The value for LaB_6 is .9K in
reasonable agreement with experiment, while for YB_6 we did
not calculate T_c in lack of detailed experimental phonon data
but on account of the findings reported above it should come
out considerably higher.

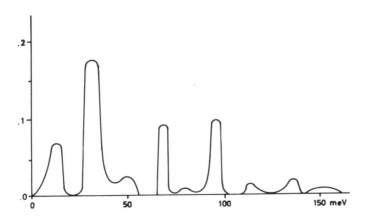

FIGURE 5. Eliashberg function of LaB_6.

TABLE I.   Electronic Quantities Entering Equation (2)

Compound	LaB_6	$\varepsilon_F=.77Ryd$	$n(\varepsilon_F)=9.33Ryd^{-1}$				
Component		La			B		
$\eta_S\ ev/Å^2$		.633			.390		
δ_l	1.46	2.15	.34	.03	.51	.65	.01
$n_l/n_l^{(o)}$	.47	.55	.50	1.20	.21	.51	1.30
Compound	YB_6	$\varepsilon_F=.86Ryd$	$n(\varepsilon_F)=8.14Ryd^{-1}$				
Component		Y			B		
$\eta_S\ ev/Å^2$		.998			.435		
δ_l	1.79	2.45	.55	.01	.46	.72	.01
$n_l/n_l^{(o)}$	.66	.42	.55	1.03	.32	.37	1.32

TABLE II.   Contributions of Different Phonon
Modes to λ

Phonon Energy meV	Type of Vibration	Contribution to λ
13	Translational modes of the metal- and boron sublattices	.11
36	Torsional modes of the octahedra	.12
45	Optical modes of the metal- and boron sublattices	.09
65 90 . .	Vibrations of the boron sublattice connected with deformations of the octahedra	.07

V. APPENDIX

The general formula for $\alpha^2F(\omega)$ can be written in a rather compact manner using the imaginary part of the one particle Green's function at the Fermi level:

$$\alpha^2F(\omega) = \sum_{\vec{q},\kappa,s,s'} e^{\vec{s}}_{\vec{q},\kappa} \vec{\nabla}v^s(\vec{\rho}) \; (\text{Im } g(\vec{\rho},\vec{\rho}',\varepsilon_F))^2$$

$$e^{\vec{s}'}_{\vec{q},\kappa} \vec{\nabla}v^{s'}(\vec{\rho}') \; \delta(\omega-\omega_{\vec{q},\kappa})/\omega \tag{3}$$

where $\vec{\rho},\vec{\rho}'$ are the space coordinates counted from centres s, s' and in the rigid muffin tin approximation $v^s(\vec{\rho})$ is the muffin tin potential at site s. We obtain the g's directly in $\vec{r}$-space by our cluster method.

The local form (1) results from the terms s=s'. Intra-octahedron off diagonal contributions (with respect to s, s') predominantly reduce $\alpha^2F(\omega)$. These findings are in accord with additional calculations we did applying the nonlocal extension of the Gyorffy formula derived by Rietschel (1978), where a spherical approximation is made for the off diagonal elements of the electron scattering matrix.

REFERENCES

Allen, P.B., and Dynes, R.C. (1975). *Phys. Rev.* B12, 905.
Gompf, F. (1978). *In* "Progress Report Teilinstitut Nukleare Festkörperphysik" (K. Käfer, ed.), p. 17. Ges. f. Kernforschung, Karlsruhe (FRG), KfK 2670.
Longuet-Higgins, H.C., and Roberts, M. de V. (1954). *Proc. Roy. Soc. (London)* A224, 336.
Lyakhovskaya, I.I., Zimkina, T.M., and Fomichev, V.A. (1970). *Sov. Phys. - Solid State 12*, 128.
Rainer, D., and Bergmann, G. (1973). *Z. Physik 263*, 59.
Ries, G., and Winter, H. (1979). To appear in *J. Phys. F.*
Rietschel, H. (1978). *Z. Physik B30*, 271.
Scholz, H., Bauhofer, W., and Ploog, K. (1976). *Solid State Commun. 18*, 1539.

SUPERCONDUCTING HYDRIDES
OF TRANSITION METAL ALLOYS[1]

C. B. Friedberg[2]
A. F. Rex
J. Ruvalds

Physics Department
University of Virginia
Charlottesville, Virginia

I.   INTRODUCTION

Soon after the discovery of superconductivity in the
hydrides of thorium (Satterthwaite and Toepke, 1970) and
palladium (Skoskiewicz, 1972), the technique of ion implantation was invoked to produce new superconducting materials
of great interest. $T_c \sim 9$ K was achieved in foils of Pd implanted with hydrogen to high concentration (Stritzker and
Buckel, 1972); an inverse isotope effect was discovered by
implantation studies of PdD for which T_c approaches 11 K
(Stritzker and Buckel, 1972); and of particular importance,
hydrogen-implanted Pd-noble metal alloys were found to exhibit transition temperatures reaching 16.6 K (Buckel and
Stritzker, 1973; Stritzker, 1974). These results have
prompted intense theoretical and experimental investigations
of transition metal hydrides (for a review, see Stritzker
and Wühl, 1978). Although many features of the PdH(D)
systems have been clarified, a great deal of uncertainty
surrounds the current description of related alloy hydrides.

[1]*Supported in part by grants from Research Corporation,
the University of Virginia, and (for theory) the NSF.*
[2]*Present address: Texas Instruments, Houston, Texas.*

For example, the observed suppression of T_c upon the addition
of certain simple metals (Friedberg *et al.*, 1979a) to PdH
is anomalous with respect to existing theories. The case of
the alloy $PdIn_xH_y$ (Friedberg *et al.*, 1979b) is particularly
challenging in that the elastic and (properly scaled) electronic properties of this material might be expected to be
quite similar to those of the high T_c alloy system $PdAg_xH_y$
for which calculations of T_c yield an extremely large enhancement (Papaconstantopoulos *et al.*, 1978).
     We have thus undertaken to examine the superconducting
properties of various new transition metal alloy hydrides.
Since the equilibrium solubility of hydrogen is generally
quite small in these alloys, previous studies of such materials by conventional techniques have been restricted to Pd
deficient samples. For example, in the PdNb system
(Oesterreicher and Clinton, 1976), the addition of hydrogen
to $Nb_{.8}Pd_{.2}$ revealed an increase in T_c from 1.87 K to 2.5 K.
The ion implantation technique, however, allows the fabrication of a wide variety of hydride materials, and we thus
report here the results for $PdNb_xH_y$, $PdTa_xH_y$, $PdBi_xH_y$ and
$PdZn_xH_y$ alloy systems.

II.   EXPERIMENTAL TECHNIQUES

     The transition metal alloy hydride samples were prepared
by the ion implantation of hydrogen into thin films of the
appropriate metallic alloys. The binary alloy films
($PdNb_x$, $PdTa_x$, $PdBi_x$ and $PdZn_x$) were first deposited on
sapphire substrates by electron beam co-evaporation from two
independently controlled sources. The films were typically
0.4 μm in thickness, and the compositions were determined
from the relative rates of evaporation of the two metals.
The Pd evaporation charge contained less than a total of 10
ppm iron and nickel impurities. Once cooled to low temperature, the films were implanted with H_2^+ or H_3^+ ions at 65 keV
and beam current densities in the range 5-10 μa/cm^2. Measurements of the temperature dependence of the resistance of
the implanted films were performed by the four probe technique for various doses of the hydrogen ions and for temperatures above 4.2 K. The sample temperature was maintained
below 15 K throughout the implantation and measurement process. Further details of the sample preparation apparatus
and ion accelerator have been given elsewhere (Friedberg
et al., 1979a,b).

III.   RESULTS AND DISCUSSION

A summary of the experimental results is presented in Figures 1 and 2 in the form of the maximum transition temperature achieved for each sample at an optimum dose of implanted hydrogen ions. For the alloys $Pd_{1-x}Nb_xH_y$, $Pd_{1-x}Ta_xH_y$ and $Pd_{1-x}Bi_xH_y$, the addition of either Nb, Ta, or Bi results in a decrease in T_C with respect to pure PdH, for which $T_C \sim 8.1$ K in these films. This behavior is similar to that observed previously in hydrides of $PdTi_x$ (Stritzker and Wühl, 1978), although in the present data the suppression is not as severe. It should be noted that whereas Pd forms solid solutions with Nb and Bi at the concentrations studied, the $Pd_{.9}Ta_{.1}$ alloy forms a two-phase mixture.

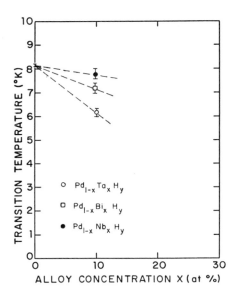

FIGURE 1.   Maximum transition temperatures for $Pd_{1-x}Ta_x$, $Pd_{1-x}Bi_x$, and $Pd_{1-x}Nb_x$ alloy hydrides. The bars denote the width (10% - 90%) of the superconducting transition.

The results for $Pd_{1-x}Zn_xH_y$ samples are shown in Figure 2
along with earlier results (Friedberg *et al.*, 1979b) for
other Pd-simple metal hydrides. Here, the substitution of
Zn for Pd initially causes an *increase* in T_c, reaching an
onset of 9.7 K for x = 0.20. The addition of larger amounts
of Zn then causes T_c to decrease, with no indication of
superconductivity above 4.2 K observed in a sample for which
x = 0.43. These results indicate that the superconducting
properties of $Pd_{1-x}Zn_xH_y$ bear some similarities to the Pd-
noble metal-hydride systems in that there is a significant
enhancement in T_c for reasonably large zinc concentrations.
It is of particular interest that the electronic properties
of $PdZn_xH_y$ and $PdCu_xH_y$ should be quite similar, the main
difference resulting from the additional s-electron which

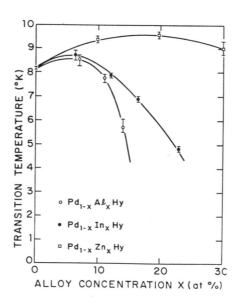

FIGURE 2. *Maximum transition temperatures for several*
Pd-simple metal alloy hydrides. The bars denote the width
(10% - 90%) of the superconducting transition.

is donated to the conduction bands in the case of Zn. Although no detailed calculations have been published for $PdCu_xH_y$, it would seem that a theoretical comparison of these two materials might be in order, especially as a test of the claim (Papaconstantopoulos et al., 1978) that it is the s-like electronic density of states at the Fermi energy which dominates the variation in T_c in similar materials.

ACKNOWLEDGMENTS

We are indebted to C. Traylor for technical assistance.

REFERENCES

Buckel, W., and Stritzker, B. (1973). *Phys. Letters 43A,*403.

Friedberg, C. B., Pickel, J. S., Rex, A. F., and Ruvalds, J. (1979a). *Solid State Commun. 29,* 407.

Friedberg, C. B., Rex, A. F., and Ruvalds, J. (1979b). *Phys. Rev. B19,* in press.

Oesterreicher, H., and Clinton, J. (1976). *J. Solid State Chem. 17,* 443.

Papaconstantopoulos, D. A., Klein, B. M., Economou, E. N., and Boyer, L. L. (1978). *Phys. Rev. B17,* 141.

Satterthwaite, C. B., and Toepke, I. L. (1970). *Phys. Rev. Lett. 25,* 741.

Skoskiewicz, T. (1972). *Phys. Stat. Sol. a 11,* K123.

Stritzker, B. (1974). *Z. Phys. 268,* 261.

Stritzker, B., and Buckel, W. (1972). *Z. Phys. 257,* 1.

Stritzker, B., and Wühl, H. (1978). *In* "Topics in Applied Physics 29" (G. Alefeld and J. Völkl, ed.), "Hydrogen in Metals II", p. 243. Springer-Verlag, New York.

LOW TEMPERATURE NORMAL-STATE
ELECTRICAL RESISTIVITY ANOMALIES
IN SOME DISORDERED d-BAND SUPERCONDUCTORS[1]

R.R. Hake
S. Aryainejad
M.G. Karkut

Department of Physics
Indiana University
Bloomington, Indiana

I. INTRODUCTION

Considerable recent experimental and theoretical effort
has been devoted to the study of the unusual temperature
dependence of the normal-state electrical resistivity of A15
compound superconductors (for a review see Allen, 1979). On
the other hand, relatively little attention has been given to
the related normal-state resistive anomalies in disordered
d-band alloy superconductors. Mooij's (1973) survey of high-
temperature ($300 \leq T \leq 350K$) measurements on about 130 disordered
transition-metal alloys (bulk and thin film, crystalline and
amorphous)showed that the temperature coefficient of resistiv-
ity $\alpha_T \equiv \rho^{-1} \partial \rho / \partial T$ tended to decrease with resistivity ρ,and that
for ρ greater than $150\mu\Omega$cm nearly all alloys displayed anoma-
lous negative α_T. Several years ago we examined the low-
temperature ($1.2 \leq T \leq 4.2K$) normal-state resistive behavior of
one class of negative-α_T, d-band, alloy superconductors (bcc
Ti-base alloys with $106 \leq \rho(4.2K) \leq 195\mu\Omega$cm: $Ti_{84}Mo_{16}$, $Ti_{92}Os_8$,
$Ti_{92}Ru_8$, $Ti_{84}Mn_{16}$, $Ti_{92}Fe_8$) by applying magnetic fields
$H \leq 140kG$ so as to essentially quench all superconductivity. The
measurements (Lue et al., 1975 a,b; Hake et al., 1975; Montgomery, 1976) disclosed (a) linear ρ vs T dependences with α_T
values more negative than those observed at $T \approx 300K$, (b)approx-

[1]*Supported in part by NSF Grant DMR 77-10545-A01.*

values more negative than those observed at T $\approx$ 300K, (b) approximately linear ρ vs H dependences with <u>negative</u> $\alpha_H \equiv \rho^{-1}(\partial \rho / \partial H)_T$ (negative magnetoresistance). In the present paper we report the extension of such high-H helium-temperature measurements to amorphous $Ti_{50}Be_{40}Zr_{10}$ and bcc $V_{80}Al_{20}$.

II. EXPERIMENTAL RESULTS

A. Magnetoresistance

Figures 1 and 2 show longitudinal (H ∥ J) magnetoresistance curves measured for the high-ρ specimens of Table I. The measurements were made using a standard dc 4-lead method, employing techniques and apparatus described by Lue <u>et al</u>. (1975a). Similar results have been obtained for $V_{85}Al_{15}$ and $V_{90}Al_{10}$. The curves are similar to those previously measured[2] at helium temperatures for bcc Ti-base alloys (Lue <u>et al</u>., 1975 a,b): (a) at low H the magnetoresistance is positive and tends to saturate as H increases, (b) the low-H positive magnetoresistance curves tend to sharpen as the temperature is lowered,

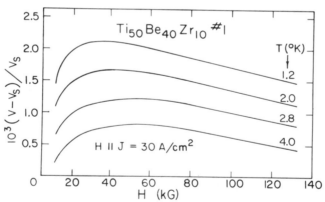

FIGURE 1. Magnetoresistance curves for $Ti_{50}Be_{40}Zr_{10}$. V is the resistive voltage and V_s is a constant nulling voltage applied with a six-dial μV potentiometer. The normalized difference voltage is traced from X-Y recorder plots of $(V-V_s)/V_s$ vs V_{mr}, where V_{mr} is the voltage across a magnetoresistive field sensor.

[2]The previous measurements showed that the magnetoresistance curves were very similar for the H∥J and H⊥J orientations.

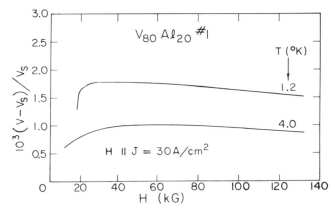

FIGURE 2. Magnetoresistive curves for $V_{80}Al_{20}$, plotted in the same manner as in Fig. 1.

TABLE I. Characteristics of the Alloys Measured

Alloy	Structure	$\rho\ (\mu\Omega cm)$[a] 300K (H=0)	77K (H=0)	4.0K (H=132kG)	T_c[b] (K)
$Ti_{50}Be_{40}Zr_{10}$ #1[c]	amorphous	250.7	262.2	267.9	0.3[d]
$V_{80}Al_{20}$ #1[e]	bcc	143.4	141.3	142.0	< 1.2[b]

[a]Electrical resistivity (absolute values accurate to ± 15%)
[b]Superconducting transition temperature
[c]Allied Chemical Corp. ("Metglas" 2204 ribbon)
[d]Measured by Matey and Anderson (1977) on another specimen
[e]Arc melted (Ti gettered atmosphere, 6 melts)
[b]Present measurements

(c) at high H the magnetoresistance is negative with an approximately linear ρ vs H dependence, (d) as T decreases at constant high H, the resistivity increases [higher $(V-V_s)/V_s$].

For bcc Ti-base alloys with $T_c \approx 3$-4K features (a) and (b) above were interpreted in terms of the H-quenching of superconducting fluctuations at fields well beyond both the upper H_{c2} and sheath H_{c3} critical fields. The magnitude of the deduced fluctuation superconductivity appears to be in reasonable agreement with theory (Ami and Maki, 1978). Similarly,

in the present case, the non-Kohler-rule temperature dependence of the shapes of the positive magnetoresistance curves
suggests H-quenching of remnant superconductivity (not necessarily fluctuation superconductivity) above the bulk T_c of
the specimens, rather than ordinary positive magnetoresistance
due to Lorentz-force bending of electron orbits. The latter
mechanism is invoked by Cochrane and Ström-Olsen (1977) to
explain their observations of positive magnetoresistance up to
45 kG in $Ti_{50}Be_{40}Zr_{10}$, but little information is given on the
details of their measurement.

The high H ($80 \leq H \leq 132$ kG) constant negative slopes of the
1.2K curves of Figs. 1 and 2 yield $\rho^{-1}(\partial\rho/\partial H)_{T=1.2K} = -3.7$,
-7.5 for $V_{80}A\ell_{20}$ and $Ti_{50}Be_{40}Zr_{10}$, respectively, in units of
$10^{-9}G^{-1}$. These values may be compared with the values (in the
same units) $-2.0 \geq \rho^{-1}(\partial\rho/\partial H)_{T=1.2K} \geq -16$ for bcc Ti-base alloys
with $106 \leq \rho(4.2K) \leq 195 \mu\Omega cm$ (Hake et al., 1975).

B. Temperature Dependence of the High-H Electrical Resistivity

Figure 3 shows the normalized change in the electrical
resistivity vs temperature at a constant applied magnetic
field H=132kG for the alloys of Table I. The data are again
rather similar to those obtained previously at helium temperatures for bcc Ti-base alloys (Lue et al., 1975b): roughly
linear ρ vs T with negative slope. The slopes of the curves
at 2K are $\rho^{-1}(\partial\rho/\partial T)_{H=132kG} = -2.5$, -4.3 for $V_{80}A\ell_{20}$ and
$Ti_{50}Be_{40}Zr_{10}$, respectively, in units of $10^{-4}K^{-1}$. These values
may be compared with the 2K values (in the same units) $-2.8 \geq$
$\rho^{-1}(\partial\rho/\partial T)_{H\approx138kG} \geq -5.5$ for bcc Ti-base alloys with $106 \leq \rho$
$(4.2K) \leq 195 \mu\Omega cm$ (Hake et al., 1975).

Comparing temperature and field coefficients, it is note-
worthy that $[\rho^{-1}\partial\rho/\partial(k_BT)]/[\rho^{-1}\partial\rho/\partial(\mu_BH)] = 4.5$, 3.8 for
$V_{80}A\ell_{20}$ and $Ti_{50}Be_{40}Zr_{10}$, consistent with physical expecta-
tions of a value of order one for this ratio.

III. DISCUSSION

The present and past (Lue et al., 1975a,b; Montgomery,
1976) observations of negative α_T coupled with negative α_H in
amorphous $Ti_{50}Be_{40}Zr_{10}$, bcc $V_{80}A\ell_{20}$, and bcc Ti-base alloys
at helium temperatures suggest that such anomalies may be
rather general in high-ρ d-band alloys. Such behavior is
suggestive of a magnetic Kondo effect, but no low-temperature
magnetic or calorimetric manifestation of ordinary localized-
magnetic-moment behavior has ever, to our knowledge, been
observed in bcc Ti-base alloys, even those containing more

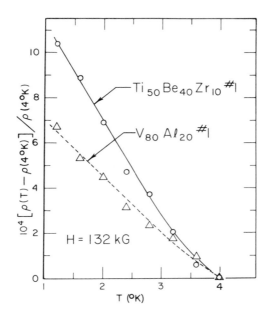

FIGURE 3.   *Normalized resistivity change versus T at a constant applied field H=132kG.*

than 5 at.% Cr, Mn, or Fe (for a bibliography see Lue et al., 1975a).   Lue et al. (1975a,b) and Prekul et al. (1975) suggested that the negative transport coefficients might reflect very high Kondo or localized-spin-fluctuation temperatures such as are thought to occur in some systems such as AℓMn (for a review see Rizzuto, 1974).

Multiple scattering (e.g., Harris et al., 1978), scattering from atomic (Cochrane et al., 1975) or structural (Tseui, 1978) tunneling sites, and modified Ziman liquid-metal theory (Nagel, 1977; Cote and Meisel, 1977) have been advanced to account for negative α_T in highly disordered alloys, but it is not obvious that such approaches can yield a negative α_H.   In addition: (a) the Tseui, Nagel, and Cote-Meisel theories all appear to predict a marked flattening of the ρ vs T curve in the helium temperature range, contrary to the present observations; (b) the structural instability presupposed by Tseui seems unlikely in bcc $V_x Aℓ_{100-x}$ alloys at $x \gtrsim 70$ (Hansen, 1958).   The difficulties in applying the s-d scattering model (e.g., Fradin, 1974) to account for negative α_T in bcc Ti-Mo alloys have been discussed by Hake et al. (1961).

Both prominent negative α_T and α_H are often observed in localized or nearly-localized systems (Mott and Davis, 1971;

McLean et al., 1978; Fukuyama and Yosida, 1979). Jonson and
Girvin (1979) have suggested that the Mooij correlation is a
natural result of the approach to localization. Thus it
seems possible that the presently observed negative coefficients may represent localization precursors. Measurements
over wider temperature ranges are now in progress to clarify
the nature of the transport anomalies.

REFERENCES

Allen, P.B. (1979). This conference.
Ami, S. and Maki, K. (1978). *Phys. Rev. B 18*, 4714.
Cochrane, R.W., Harris, R., Ström-Olsen, J.O., and
    Zuckermann, M.J. (1975). *Phys. Rev. Lett. 35*, 676.
Cochrane, R.W., and Ström-Olsen, J.O. (1977). *J. Phys. F 7*,
    1799.
Fradin, F.Y. (1974). *Phys. Rev. Lett. 33*, 158.
Fukuyama, H., and Yosida, K. (1979). *J. Phys. Soc. Japan 46*,
    102.
Hake, R.R., Leslie, D.H., Berlincourt, T.G. (1961). *J. Phys.
    Chem. Sol. 20*, 177.
Hake, R.R., Montgomery, A.G., Lue, J.W. (1975). "Low Tem-
    perature Physics- LT 14" (M. Krusis and M. Vuorio, eds.),
    vol. 3, p. 122. North Holland, Amsterdam.
Hansen, M. (1958). "Constitution of Binary Alloys." McGraw-
    Hill, New York.
Harris, R., Shalmon, M., Zuckermann, M. (1978). *Phys. Rev. B
    18*, 5906.
Jonson, M., and Girvin, S.M. (1979). Submitted for publica-
    tion.
Lue, J.W., Montgomery, A.G., Hake, R.R. (1975a): *Phys. Rev.
    B 11*, 3393; (1975b): *AIP Conf. Proc. 24*, 432 (1975).
Matey, J.R., and Anderson, A.C. (1977). *Phys. Rev. B 16*,
    3406.
McLean, W.L., Lindenfeld, P., Worthington, T. (1978). *AIP
    Conf. Proc. 40*, 403.
Montgomery, A.G., (1976). Thesis, Indiana Univ.,unpublished.
Mooij, J.H. (1973). *Phys. Stat. Sol (a) 17*, 521.
Mott, N.F., and Davis, E.A. (1971). "Electronic Processes in
    Non-Crystalline Materials." Clarendon Press, Oxford.
Nagel, S.R. (1977). *Phys. Rev. B 16*, 1694.
Prekul, A.F., Rassokhin, V.A., and Volkenshtein, N.V. (1975).
    Sov. Phys. JETP 40, 1134.
Rizzuto, C. (1974). *Rep. Prog. Phys. 37*, 147.
Tsuei, C.C. (1978). *Sol. St. Comm. 27*, 691.

NEW TERNARY MOLYBDENUM CHALCOGENIDES

Ø. Fischer, B. Seeber, M. Decroux

Département de Physique de la Matière Condensée
Université de Genève, 24 Quai E. Ansermet.
CH 1211 Geneva, Switzerland

R. Chevrel, M. Potel, M. Sergent

Laboratoire de Chimie Minérale B, Université de Rennes
Avenue du Général Leclerc, F 35000 Rennes, France

I. INTRODUCTION

Ternary molybdenum chalcogenides of the type $M_xMo_6X_8$ ($0 \leq x < 4$, M = metal, X = S,Se,Te) are known for a series of unusual superconducting and normal state properties, like co-existence of antiferromagnetism and superconductivity and extremely high critical fields.[1-5] The reason why these and other ternary compounds like the $(RE)Rh_4B_4$ compounds[6] show such anomalous behavior is clearly correlated with the increased liberty that nature has in composing structures of ternary com-pounds as compared with binary ones. Thus it is perhaps not surprising that we find qualitatively different properties in such materials as compared with binary compounds and alloys, in the same way as the latter materials show superconducting prop-erties quite different from those of the pure elements.

In this paper we would like to illustrate the point made above by discussing different types of new ternary molybdenum chalcogenides.

II. COMPOUNDS WITH Mo_6-TYPE CLUSTERS

In the study of new materials there are several levels of understanding. In the ternary molybdenum chalcogenides we have probably reached the first level of a rough overall picture. We know the density of states of the phonons,[7] and thanks to tunneling[8] and isotope effect[9] measurement we also have an idea of the electron-phonon coupling. Three band calculations[10-12] have been carried out. These calculations differ in the details but agree in the overall picture: the X-p bands are below the Fermi surface, the Mo-5s band is above and the narrow bands at the Fermi level have mainly 4d-character. The Fermi level is therefore determined by the number of Mo-4d electrons calculated according to the formula

$$M_x^{+\nu}(Mo_6)^{(16 - \nu x)}S_8^{-2} \, ,$$

ν being the valence of the third element M. In this way one finds that the average valence of Mo varies between +2.66 and +2, and correspondingly that the number n of 4d electrons per Mo_6-cluster varies between 20 and 24. We then expect that the properties of these materials are correlated with n. That this is so can be seen from Table 1, where one sees that for n = 22 one finds high-field high-temperature superconductors and for n = 23, both T_c and H_{c2} are generally low. For n = 24 we find semiconductivity, a result which was predicted by the band calculations. This simple classification is to be used with some

TABLE 1. Ternary molybdenum sulfides with 22, 23, and 24 cluster-electrons.

n	Compound	T_C [K]	$H_{c2}(0)$ [kGauss]
22	$PbMo_6S_8$	15.2	~600
22	$SnMo_6S_8$	14.2	~400
~22	$YbMo_6S_8$	9.1	~160
22	$Mo_6S_6Br_2$	13.8	
22	$Mo_6S_6I_2$	14.0	~340
23	YMo_6S_8	2.1	~4.5
23	$LaMo_6S_8$	7.1	~70
23	$GdMo_6S_8$	1.5	~1.3
23	$LuMo_6S_8$	2.1	
24	$Mo_2Re_4S_8$	Semiconducting	

caution since it has been shown that the relative positions of
the Mo-4d subbands are sensitive to the deformation of the Mo_6
cluster (as compared to an ideal octahedron).[10] It is for
instance not clear why $BaMo_6S_8$ and $SrMo_6S_8$ are not superconducting.

In the following we would like to focus our attention on
$Mo_6S_6Br_2$, $Mo_6S_6I_2$ and the compounds with n = 24 electrons. The
binary compound Mo_6S_8 is unstable and the $Mo_6S_6Br_2$ and $Mo_6S_6I_2$
compounds were made in order to stabilize the former and to
demonstrate that the metal M (say Pb or Sn) is not essential
for high temperature and high field superconductivity in these
materials.[13] However, there are two questions one may immediately ask:
   1) How do the two Br-atoms or I-atoms stabilize Mo_6S_8?
   2) Why is T_c so high in spite of the fact that we have
       substituted on the chalcogen site?
Note in comparison that $PbMo_6S_6Se_2$ has only a T_c of 5 K.[14] A
recent structural investigation[15] allows us to answer these
questions. Consider first the system $LaMo_6S_{8-x}Se_x$ shown in
Fig. 1. The rapid decrease of T_c upon substitution at the
chalcogen site and the minimum of T_c close to x = 4 seems to
be a general behavior of all $M_xMo_6X_8$ materials.[5,14] The variation of the lattice parameters upon substitution suggests that

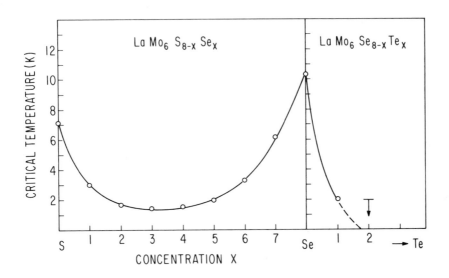

FIGURE 1.   Critical temperature of the series $LaMo_6S_{8-x}Se_x$
$(0 \leq x \leq 8)$ and $LaMo_6Se_{8-x}Te_x$ $(0 \leq x \leq 2)$.

the distribution of the S and Se atoms on the 8 chalcogen
positions is essentially a statistical one. We therefore believe that the lowering of T_c in these materials is a disorder
effect. In the $Mo_6Se_{8-x}Br_x$ system, on the contrary the lattice
parameters show a clear break at x = 2 indicating that the Br
atoms preferentially occupy the 2c sites on the ternary axis.[13]
This hypothesis has now been verified on $Mo_6S_6Br_2$.[15] This
material is an ordered compound with the same symmetry as
Mo_6S_8, and we therefore do not expect any reduction of T_c. As
to the stabilization of Mo_6S_8 by Br, it is brought about by a
strong deformation of the empty S-cube around the origin such
that the two Br ions on the ternary axis practically touch each
other.

As mentioned before, band structure calculations predict a
gap for 24 cluster-electrons. Exactly 24 electrons may be obtained from Mo_6X_8 upon partial substitution of the molybdenum
atoms by either Re or Ru atoms.

And, in fact $Mo_2Re_4S_8$, $Mo_2Re_4Se_8$[16] and $Mo_4Ru_2Se_8$[17] are
semiconducting. However, there is a general trend of the activation energy determined from conductivity to decrease when we
go from sulfur to selenium and tellurium. For the series
$Mo_2Re_4X_{8-x}X_x$ this activation energy extrapolates to zero close
to the compound $Mo_2Re_4Se_6Te_2$ and this compound also constitutes
the phase boundary. The compound $Mo_2Re_4Te_8$ which should be
metallic does not form. It is interesting to notice, however,
that $Mo_4Re_2Te_8$ with n = 22 becomes superconducting at T_c =
3.55 K, whereas Mo_6Te_8 is not superconducting above 1.1 K. In
the case of Ru there exists a complete solid solution
$Mo_4Ru_2Se_{8-x}Te_x$ ($0 \leq x \leq 8$). At about x = 1.2 the compounds
switch from semiconducting to metallic. $Mo_4Ru_2Te_8$ is not
superconducting above 1.1 K but again for n = 22, i.e., in
Mo_5RuTe_8 we find superconductivity ($T_c \simeq 2.0$ K). With Rh substituted for Mo only the telluride forms. It is metallic and
superconductivity has not been observed above 1.1 K.

We would like to point out here that in all materials with
the $PbMo_6S_8$-type structure there is on the average a trend toward lower critical temperatures when we go from the sulfides
to the selenides and to the tellurides. This might be correlated with the disappearance of the band-gap in the tellurides,
discussed above. Both of these trends appear contrary to the
prediction of the band calculations that the Mo bandwidth is
less in the tellurides than in the sulfides.[10] However, a
more detailed experimental study is necessary to clarify the
reasons for the rather different behavior of the sulfides, the
selenides and the tellurides. It is, for instance, unclear if
T_c in $Mo_4Re_2Te_8$ is lowered due to a disorder effect resulting
from a statistical distribution of Mo and Re atoms on the
cluster.

III. COMPOUNDS WITH Mo_9 and Mo_{12} CLUSTERS

One anomaly in the general picture of the ternary molybdenum chalcogenides is that $In_1Mo_6S_8$ exists, but the corresponding selenide which we might expect would be a good superconductor does not form in the well-known $PbMo_6S_8$ type structure. However, when we tried to prepare this compound we found a new superconducting material with $T_c \simeq 4.3$ K. This new compound, which turned out to have the formula $In_3Mo_{15}Se_{19}$ is another extreme high field superconductor.[18,19] The initial slope $(dH_{c2}/dT)T_c$ ranged from 70 kGauss/K to 80 kGauss/K depending on the exact composition. These values are 15-20% higher than the ones observed in $PbMo_6S_8$ and even higher than the ones observed in $LaMo_6Se_8$. We show H_{c2} versus temperature for this compound in Fig. 2. The hexagonal structure of this new compound is similar in many respects to the one of $PbMo_6S_8$, with the difference, however, that it contains a completely new building block, Mo_9Se_{11}, in addition to the usual one, Mo_6Se_8.[20] The formula may therefore be written $In_3(Mo_6Se_8)$-(Mo_9Se_{11}) so that the compound contains an equal number of Mo_6Se_8 and Mo_9Se_{11} units. The new cluster Mo_9 may be considered as the fusion of two Mo_6 clusters along a face (Fig. 3). The overall structure is most easily visualized by looking at the $11\bar{2}0$-projection, shown in Fig. 4. For a detailed structural description we refer to ref. 20. Here we just retain the important feature that we again get a cluster-structure and that the intercluster bonding is realized in exactly the same way as in $PbMo_6S_8$. Thus, we expect qualitatively similar electronic properties in this new compound as for the $M_xMo_6S_8$-type compounds. In particular, the very high critical field found here confirms the importance of the cluster-structure for this property.

Recently, another modification of this structure was found: $In_2Mo_{15}Se_{19}$.[21] This structure is shown in Fig. 5 and as can be seen the stacking of the Mo_6Se_8 and the Mo_9Se_{11} units is different.[22] The lattice symmetry is hexagonal-rhombohedral with rhombohedral lattice parameters $a_R = 20.16$ Å and $\alpha_R = 27.81°$. This compound becomes superconducting at 1.47 K. Contrary to the former compound which only seems to form with In, the latter can be formed with several other metals like K, Tl, Ba. In Table 2 we give the critical temperature of several of these compounds. The initial slope of the critical field was measured in $K_2Mo_{15}S_{19}$ and $Ba_2Mo_{15}Se_{19}$ and we found 34 kGauss/K and 32 kGauss/K, respectively. Although these values are much lower than the ones found in $In_3Mo_{15}Se_{19}$, they are still very high compared with other materials, confirming the general tendency of such cluster compounds to high critical fields. Knowing the existence of the Mo_9Se_{11} building unit it seems

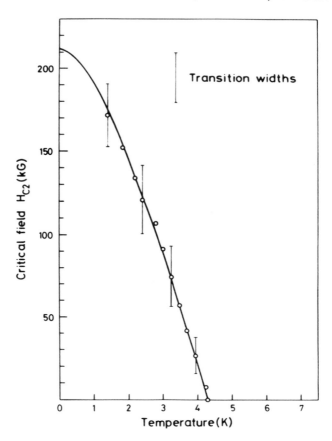

FIGURE 2. Temperature dependence of the critical field
H_{c2} *for* $In_{2.9}Mo_{15}Se_{19}$ *(from ref. 18).*

natural to look for a compound containing only these units. In
this way one falls upon the compounds $K_2Mo_9S_{11}$ and $Tl_2Mo_9S_{11}$.[23]
However, it turns out that these compounds do not contain
Mo_9S_{11} units and that their formulas should rather be written
$M_4(Mo_6S_8)(Mo_{12}S_{14})$. In fact, yet another building unit,
$Mo_{12}S_{14}$ is present.[24] This unit is shown in Fig. 6. The new
cluster Mo_{12} may be considered as a condensation of 3 Mo_6
clusters. The overall structure is hexagonal rhombohedral and
$K_2Mo_9S_{11}$ has the rhombohedral lattice parameters $a_R = 13.13$ Å
and $\alpha_R = 41.3°$. The stacking sequence of the building blocks
is similar to the one in $K_2Mo_{15}S_{19}$ with the main difference
that the Mo_9S_{11} unit has to be replaced by the $Mo_{12}S_{14}$ unit and

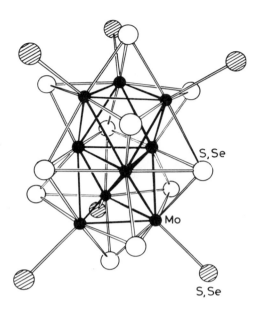

FIGURE 3. The Mo₉Se₁₁ unit. The hatched Se atoms belong to the neighbouring units (from ref. 21).

between the units there are now two K atoms on the ternary axis. We have so far not observed superconductivity in these materials.

IV.   CONCLUDING REMARKS

In the foregoing section we have described a series of new superconducting ternary compounds. The critical temperatures of these compounds are not very high, but the interesting feature of these materials is that they are cluster compounds with very high critical fields in spite of their low critical temperatures. Whether it is possible to find higher critical temperatures is, of course, not clear at the present time, but the results themselves suggest how this search for new compounds can be continued. We expect that in the future further new classes of ternary compounds with interesting properties will be found.

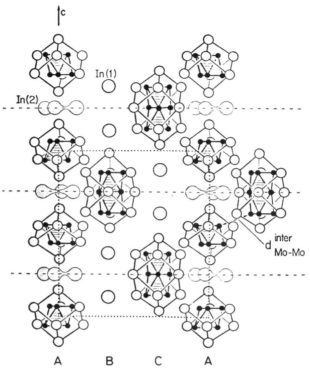

FIGURE 4.   Projection onto a (11$\bar{2}$0) plane of the hexagonal compound $In_3Mo_{15}Se_{19}$ (from ref. 20).

TABLE 2.   Lattice parameters, critical temperature and initial slope of the critical field for compounds containing Mo_9 clusters.

Compound	$A_H[\overset{\circ}{A}]$	$C_H[\overset{\circ}{A}]$	$T_c[K]$	$\frac{dH_{c2}}{dT}$ [kGauss K^{-1}]
$In_3Mo_{15}Se_{19}$	9.80	19.43	4.3	78
$K_2Mo_{15}S_{19}$	9.36	56.22	3.32	34
$K_2Mo_{15}Se_{19}$	9.74	58.16	2.45	—
$Ba_2Mo_{15}Se_{19}$	9.88	57.60	2.75	32
$In_2Mo_{15}Se_{19}$	9.69	58.10	1.50	—
$Tl_2Mo_{15}Se_{19}$	9.80	58.23	1.65	—

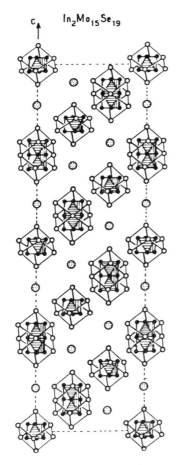

FIGURE 5. Projection onto the (11$\bar{2}$0) plane of the rhombohedral-hexagonal compound $In_2Mo_{15}Se_{19}$ (from ref. 21).

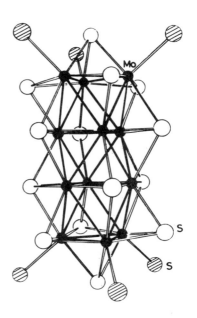

FIGURE 6. The $Mo_{12}S_{14}$ unit. The hatched S-atoms belong to the neighbouring units (from ref. 23).

REFERENCES

1) Chevrel R., Sergent M., and Prigent V. *J. Sol. State Chem.*
   3, 515 (1971)
2) Matthias B.T. Marezio M., Corenzwit E., Cooper A.S. and
   Barz H.E. *Science, 175, 1465 (1972)*
3) Odermatt R., Fischer Ø, Jones H. and Bongi G. *J. Phys. C7,*
   L13 (1974)
4) Ishikawa M.and Fischer Ø. *Sol. State Comm. 23, 37 (1977)*
   and 24, 747 (1977)
5) For a review see Fischer Ø. *Appl. Phys. 16, 1 (1978)*
6) See for instance Maple M.B. *J. Phys. 39, C6-1374 (1978)*
7) Schweiss P., Renker B. and Suck J.B. *J. Phys. 39, C6-356*
   (1978)
8) Poppe U. and Wühl H. *J. Phys. 39, C6-361 (1978)*
9) Culetto F.J. and Pobell F. *J. Phys. 39, C6-354 (1978)*
10) Andersen O.K., Klose W., and Nohl H. *Phys. Rev. B17, 1209*
    (1978)
11) Mattheiss L.F., Fong C.Y. *Phys. Rev. B15, 1760 (1977)*
12) Bullett D.W. *Phys. Rev. Letters 39, 664 (1977)*
13) Sergent M., Fischer Ø., Decroux M., Perrin C. and Chevrel R.
    J. Sol. State Chem 22, 87 (1977)
14) Chevrel R., Sergent M. and Fischer Ø. *Mat. Res. Bull. 10,*
    1169 (1975)
15) Perrin C., Chevrel R., Sergent M. and Fischer Ø. *to be*
    published.
16) Perrin A., Sergent M., and Fischer Ø. *Mat Res. Bull.13,*
    259 (1978)
17) Perrin A., Chevrel R., Sergent M. and Fischer Ø. *to be*
    published.
18) Seeber B., Decroux M. Fischer Ø., Chevrel R. Sergent M.
    and Grüttner A. *Sol. State Comm. 29, 419 (1979)*
19) Chevrel R., Sergent, M., Seeber B., Fischer Ø. Grüttner A.
    and Yvon K. *Mat. Res. Bull. 14, 567 (1979)*
20) Grüttner A., Yvon K., Chevrel R., Potel M., Sergent M. and
    Seeber B. *Acta Cryst. B35, 285 (1979)*
21) Potel M., Chevrel R., Sergent M., Decroux M. and Fischer Ø.
    C.R. Acad. Sciences, Paris. To be published.

    Chevrel R., Potel M., Sergent M., Decroux M. and Fischer Ø.
    To be published.
22) Potel M., Chevrel R. and Sergent M. *To be published.*
23) Chevrel R., Potel M., Sergent M. Decroux M., and Fischer Ø.
    To be published.
24) Potel M., Chevrel R., Sergent M. *To be published.*

MAGNETIC INTERACTIONS IN TERNARY SUPERCONDUCTORS

G. K. Shenoy, B. D. Dunlap, F. Y. Fradin, S. K. Sinha

Argonne National Laboratory[*]
Argonne, Illinois 60439

C. W. Kimball

Northern Illinois University[**]
DeKalb, Illinois

I. INTRODUCTION

There has been a continued effort at Argonne to investigate the properties of ternary superconductors using microscopic tools such as NMR, Mössbauer spectroscopy and neutron
scattering. In the present review we will be dealing with the
magnetic interactions as obtained from Mössbauer spectroscopy
and comparing them with results obtained from other tools.
Our concern will be in understanding (1) why the superconductivity in some of the Chevrel phase ternary compounds is
rather weakly perturbed by the presence of magnetic spins,
and (2) what influence the superconducting forces have on
the magnetic ordering in $ErRh_4B_4$.

II. PAIR-BREAKING INTERACTION

Early experiments by Suhl and Matthias (1959) have
demonstrated the destruction of superconductivity by the
presence of a small quantity of magnetic impurities. For
binary compounds, these effects have been considered in
detail (Maple, 1973). The reason for such an effect is that

[*]*Work supported by the U. S. Department of Energy.*
[**]*Work sponsored by the National Science Foundation.*

the exchange interaction between the s or d type conduction electron spin, σ, and the magnetic impurity with an angular momentum, S, lift the degeneracy of the two states which form a Cooper pair. Such a magnetic interaction is referred to as pair-breaking and is described by

$$\mathscr{H}_m = \frac{1}{N} \sum \Gamma \vec{S} \cdot \vec{\sigma} \qquad (1)$$

where Γ is the exchange coupling constant and N is the total number of magnetic atoms.

The knowledge of Γ is hence crucial in ternary super-conductors in order to understand how superconductivity can exist in the presence of a large concentration of magnetic atoms in the lattice (Fischer, 1978). Within the frame-work of the Abrikosov-Gor'kov theory (1961) of pair-breaking, Γ can be deduced from a variety of experiments. Thus, for example, the suppression of the critical temperature T_c is given by

$$\Delta T_c = -\frac{\pi^2}{2k_B} \rho(E_F) \Gamma^2 S(S+1) \Delta x \qquad (2)$$

Here, Δx is the magnetic impurity concentration and $\rho(E_F)$ is the density of states at the Fermi level.

The energy transferred to the conduction electron bath in a spin-flip process for a paramagnetic impurity is also proportional to the exchange coupling Γ. Thus the spin-flip rate

$$W = \frac{2\pi}{\hbar} (g_J - 1)^2 [\Gamma \rho(E_F)]^2 k_B T \qquad (3)$$

when measured as a function of temperature can yield the value of $|\Gamma \rho(E_F)|$. For the case of $Eu_{0.25}Sn_{0.75}Mo_6S_8$ we have measured the value of W for Eu as a function of T, exhibiting the linear relationship of Eq. (3) (Dunlap et al. 1979). From the slope of W vs T we obtain

$$|\Gamma \rho(E_F)| = 0.0033/\text{atom spin}$$

Here $\rho(E_F)$ does not represent the total electron density of states at the Fermi level but only the projected density of the states with proper component of angular momentum at the impurity atom. This can be deduced, for example, from a detailed band calculation (Jarlborg and Freeman, 1979). It is instructive to compare the above value of $\Gamma \rho(E_F)$ with

that for a Eu atom in superconducting $LaAl_2$, obtained
from ESR experiments (Koopman et al., 1972), viz. 0.05/atom
spin. The two values differ by an order of magnitude.

It is now possible to use the above estimate for the
exchange coupling to calculate the concentration (Δx)
dependence of T_c using Eq. (2). It turns out to be far
from what is experimentally measured (Fischer et al. 1975).
This casts certan doubts about the applicability of the
Abrikosov-Gor'kov theory to the Chevrel phase compounds.
Indeed, the presence of spin-spin interaction could be one
important factor contributing to such a difference.

III. NATURE OF ORDERED MAGNETISM IN $ErRh_4B_4$

The ternary compound $ErRh_4B_4$ is the first material to show
re-entrant superconductivity due to the occurrence of long-
range magnetic order (Fertig et al.,1977). It superconducts
below 8.5 K, and becomes normal below about 0.9 K commensurate
with an onset of magnetic ordering below this temperature. Two
techniques have been used to probe the magnetism of the ordered
state: neutron scattering (Moncton et al., 1977) and the
Mössbauer effect of [166]Er (Shenoy et al., 1979a).

In a neutron diffraction experiment one observes either
new reflections or changes in the intensities of the nuclear
peaks, depending on the magnetic structure. In either case
it is possible to deduce the ordered magnetic moment on the
atom from the diffraction pattern. At 50 mK in $ErRh_4B_4$ this
ordering was found to be ferromagnetic with 5.6 μ_B on Er
atoms and no moment on Rh atoms.

In the Mössbauer studies one measures the nuclear
Zeeman effect. The magnetic field at the nucleus in the
case of a non-S state rare-earth atom is primarily produced
by the orbital current set up by the ordered spin. This
amounts to a few megagauss and is easily measured from a
Mössbauer spectrum. In turn this can be used to deduce
the magnetic moment on the Er atom (Shenoy et al., 1979a).
The hyperfine field measurement in $ErRh_4B_4$ at 100 mK yields
a magnetic moment of 8.3 μ_B on the Er atoms. The difference
from the value of 5.6 μ_B obtained by neutron diffraction is
well outside the experimental error bars of both the studies.

In order to appreciate the implications of these results
it is necessary to know the inherent difference between the
two techniques employed. In neutron diffraction one measures
the long-range correlation between magnetic moments, while
in the Mössbauer studies one measures the single-ion auto-
correlation of the spin. Hence from the Mössbauer effect
one measures the total moment on the Er atom. The neutron
experiment then gives a lower value for one of the following
reasons:

1. Only a component of the magnetic moment shows a long-range order with the remaining 30% component of the moment showing disorder.

2. There is long-range order in some parts of the material forming magnetic domains, while in the remaining parts of the material there is spin disorder which does not contribute to the neutron diffraction.

One can indeed debate the pros and cons of these two explanations, but the important fact is that the nature of magnetic ordering in $ErRh_4B_4$, even well below the magnetic ordering temperature, is far more complex than hitherto suggested.

The fundamental reason for such a behavior is perhaps rooted in the possibility of non-vanishing superconducting forces even well-below the magnetic ordering temperature. Additionally, the anisotropy of dipole forces might produce a desirable behavior for $\chi(q)$ (Anderson and Suhl, 1959). Whether a calculation including various effects would prefer one possibility against the other suggested above remains to be seen (Shenoy et al., 1979b).

REFERENCES

A. A. Abrikosov and L. P. Gor'kov, 1961, *Sov. Phys. JETP* <u>12</u>, 1243.

P. W. Anderson and H. Suhl, 1959, *Phys. Rev.* <u>116</u>, 898.

B. D. Dunlap, G. K. Shenoy, F. Y. Fradin C. D. Barnet, and C. W. Kimball, *J. Magnetism and Mag. Materials,* 1979, (in press).

W. A. Fertig, D. C. Johnston, L. E. DeLong, R. W. McCallum, M. B. Maple, and B. T. Matthias, 1977, *Phys. Rev. Letters* <u>38</u>, 987.

Ø. Fischer, 1978, *Appl. Phys.* <u>16</u>, 1.

Ø. Fischer, M. Decroux, S. Roth, R. Chevrel and M. Sergent, 1975, *J. Phys.* <u>C8</u>, L474.

T. Jarlborg and A. J. Freeman, 1979, (this volume).

G. Koopman, U. Engel, K. Baberschke, and S. Hüfner, 1972, *Sol. State Commun.* <u>11</u>, 1197.

M. B. Maple, 1966, in Magnetism, Vol. V, Ed. H. Suhl
(Academic Press, N.Y.), p. 289.

D. E. Moncton, D. B. McWhan, J. Eckert, G. Shirane and
W. Thomlinson, 1977, *Phys. Rev. Letters* 39, 1164.

G. K. Shenoy, B. D. Dunlap, F. Y. Fradin, C. W. Kimball,
W. Potzel, F. Pröbst, and G. M. Kalvius, 1979a, *J. Appl.
Phys.* 50, 1872.

G. K. Shenoy, B. D. Dunlap, F. Y. Fradin, S. K. Sinha, C. W.
Kimball, W. Potzel, F. Pröbst, and G. M. Kalvius, 1979b
(to be published).

H. Suhl and B. T. Matthias, 1959, *Phys. Rev.* 114, 977.

HEAT CAPACITY AND ELECTRICAL RESISTIVITY OF CHEVREL
PHASE SUPERCONDUCTORS

K. P. Nerz, U. Poppe, F. Pobell, M. Weger,[1] *H. Wühl*[2]

Institut für Festkörperforschung
Kernforschungsanlage Jülich
Jülich, W. Germany

*We report on measurements of the specific heat of
$Mo_{6.3}Se_8$ at $1.5 K \leq T \leq 20 K$ which show substantial deviations
to former data. Furthermore, results for the pressure and
temperature dependence of the electrical resistivity of single
crystal $Pb_{0.9}Mo_{6.4}S_8$ at $12 K \leq T \leq 300 K$ and up to 14 kbar
will be reported. There is a marked change in the pressure
dependence of the residual resistivity and of the temperature
dependent part of the resistivity at around 4 kbar.*

I. SPECIFIC HEAT OF $Mo_{6.3}Se_8$

Some of the earliest indications of the unusual lattice,
electronic, and magnetic properties of Chevrel phase superconductors came from specific heat measurements. But there
are hints that some of these results may have been influenced
by impurity phases in the investigated samples. As a check,
we have measured the specific heat of $Mo_{6.3}Se_8$ by the heat
pulse method at $1.5 K \leq T \leq 20 K$ (Nerz, 1979). Sintered
samples of about 100 mg were glued with 1 mg grease to a sapphire substrate.
    The investigated samples are the ones used earlier for
isotope effect experiments; their preparation has been reported elsewhere (Culetto and Pobell, 1978). X-ray analysis

[1]*Permanent address: Hebrew University, Jerusalem.*
[2]*New address: Universität Karlsruhe.*

501

showed no impurity phases, and we believe that the samples contain at most 3% MoSe$_2$. The measured specific heat is shown in
Fig. 1, and compared there to earlier results of others. The
full line corresponds to the specific heat of pure Mo$_6$Se$_8$ if
we assume that our sample contains 3% MoSe$_2$. The results reported by Fradin et al. (1976) and McCallum et al. (1978) can
be quantitatively explained assuming that their samples contained substantial amounts of MoSe$_2$ and Mo (Nerz, 1979).

From the analysis of our experimental data we obtain
T$_c$ = 6.3 K, γ = 47.5 mJoule/mole K^2, N(0) = 0.72 states/eV spin,
θ_D = 177 K, and (C$_{e,s}$-C$_{e,n}$)/γT$_c$ = 2.25. The value for the
jump of the electronic specific heat at T$_c$ is substantially
larger than the BCS-value of 1.43. Using C$_{e,s}$(T) $\propto$ exp(-Δ/kT),
we find 2Δ = 4.2 kT$_c$ which also exceeds the corresponding BCS-
value. In Fig. 2 it is shown that the specific heat of Mo$_6$Se$_8$

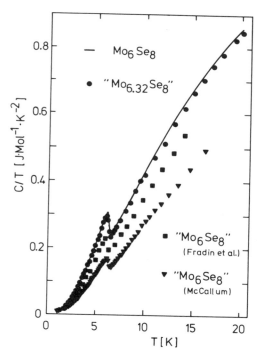

FIGURE 1. *C/T as a function of T for our Mo$_{6.3}$Se$_8$ sample*
(•), and as measured by Fradin et al. (■) and McCallum et al.
(▼). The full line is the result for pure Mo$_6$Se$_8$ if we assume
that our sample contains 3% MoSe$_2$.

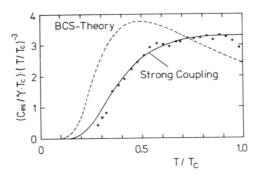

FIGURE 2.   Comparison of $(C_{es}/\gamma T_c) \cdot (T/T_c)^{-3}$ of Mo_6Se_8 to the BCS-theory prediction and to the prediction of Padamsee et al. for a strong coupling superconductor.

follows rather well the equation derived by Padamsee et al. (1973) for a strong coupling superconductor. From a modified McMillan equation (Allen and Dynes, 1975),

$$T_c = \frac{\langle \omega_{log} \rangle}{1.2} \exp\left(-\frac{1.04 \ (1 + \lambda)}{\lambda - \mu^* - 0.62 \ \lambda\mu^*}\right) \quad ,$$

we find $\lambda \simeq 0.8$ using $\langle \omega_{log} \rangle = 12$ meV and $\mu^* = 0.1$. All results confirm the strong coupling character of Mo_6Se_8. Comparing to results given in the literature, the importance of phase pure samples for the analysis of specific heat data is obvious.

II.   ELECTRICAL RESISTIVITY OF $Pb_{0.9}Mo_{6.4}S_8$

Another property of Chevrel phases which we have investigated recently is the pressure and temperature dependence of their electrical resistivity. The measurements were performed on small single crystals (0.01 mm^3) of $PbMo_6S_8$ at $12 \text{ K} \leq T \leq 300$ K and up to 14 kbar. The crystals were prepared by solid state diffusion (Poppe, 1979). Density measurements and chemical analysis gave a formula $Pb_{0.9}Mo_{6.4}S_8$. The crystals show a $T_c = 12$ K and a RRR of about 2. This low RRR might be related to their non-stoichiometry. The absolute resistivity is some hundred $\mu\Omega$ cm. Hydrostatic pressurization was performed in a teflon cell with gasoline as transmitting medium. Pressure was determined from the resistivity of a manganin coil. Because this pressure measurement was not very accurate, we refer

only to the pressure applied at room temperature. During cool-
down the pressure decreases by about 30%.

Figure 3 shows the pressure dependence of the resistance
at 300 K; it decreases linearly at about -1%/kbar. Between
each measurement, the sample was cooled down very slowly to
take low temperature data. This resulted in a small crack
when the pressure was released; this crack may be responsible
for the larger resistance values of points 6-9. For samples
investigated at room temperature only, data were reproducible,
and always confirmed the -1%/kbar dependence.

In Fig. 4, we show the temperature dependence of the resistance ρ at pressures applied at room temperature. There is
a marked difference in the behavior at P < 4 kbar and at
P > 4 kbar. At low pressure, ρ decreases linearly with temperature; only at about 20 K do we see a stronger negative
curvature. A linear temperature dependence of the resistance
at ambient pressure has been observed for single crystal and
sintered $PbMo_6S_8$, but only up to 50 K (Flükiger et al. 1978;
Woollam and Alterovitz, 1978). At higher temperatures and for
other Chevrel compounds, deviations from linearity have been
reported. The resistance at about 12 K, just before the superconducting transition - which we define as the "residual resistivity" - is unchanged by pressure. A significant change of
slope with temperature occurs at pressures above about 4 kbar.
This change results in a drastic decrease of the "residual
resistivity" with increasing pressure. The pressure dependence

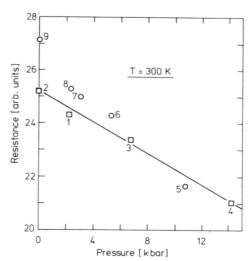

FIGURE 3. *Pressure dependence of the resistivity of a*
$Pb_{0.9}Mo_{6.4}S_8$ *single crystal at room temperature.*

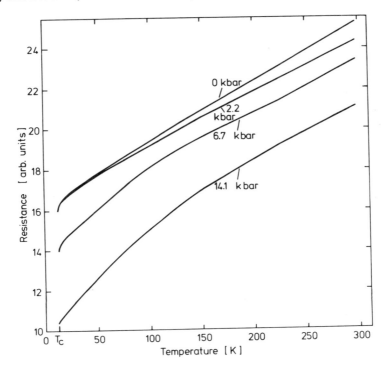

FIGURE 4.   Temperature dependence of the resistivity of a $Pb_{0.9}Mo_{6.4}S_8$ *single crystal at the indicated pressures which were applied at room temperature.*

of the "residual resistivity" is shown in Fig. 5.   The temperature where deviation from linearity occurs shifts to higher
values when pressure is increased.
     We observe also quite a different behavior of the tempera-
ture dependent part of ρ for P < 4 kbar and for P > 4 kbar
(Fig. 6).   The relative change $\{[\rho(300\ K,P)-\rho(12\ K,P)]$ –
$[\rho(300\ K,0)-\rho(12\ K,0)]\}/[\rho(300\ K,0)-\rho(12\ K,0)]$, drops strongly
by about -3%/kbar at small pressures but increases by about
+3%/kbar at higher pressures (see Fig. 6).
     The observed pressure dependence of the transition temper-
ature agrees with Shelton's data (1976) for powdered sample,
showing a maximum at about 4 kbar.
     For the low-pressure range at P < 4 kbar, our results may
be explained in terms of a molecular crystal model where the
clusters are coupled by van der Waals forces (Gutfreund et al.,
1978).   Within this simple model, the residual resistivity, of
course, should be pressure independent, whereas the temperature
dependent part of ρ should show a strong pressure or volume

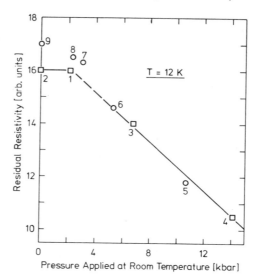

FIGURE 5. Dependence of the resistivity of a $Pb_{0.9}Mo_{6.4}S_8$ single crystal at 12 K as a function of the pressure applied at room temperature.

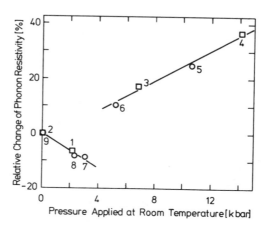

FIGURE 6. Relative change of the temperature dependent part (see text) of the resistivity of a $Pb_{0.9}Mo_{6.4}S_8$ single crystal as a function of the pressure applied at room temperature.

dependence with $\delta \ln \rho_{thermal}/\delta \ln V \sim 14$. The observed decrease of -3%/kbar combined with the compressibility of about -0.25%/kbar (Webb and Shelton, 1978), corresponds to $\delta \ln \rho_{thermal}/\delta \ln V \sim 12$.

At higher pressures, the pressure dependence of the resistance has to be described by a more complicated model. To explain the pressure dependence of T_c Webb and Shelton (1978) have suggested a change in the density of states at the Fermi energy with a maximum at around 4 kbar. Band structure calculations have shown that there are rather narrow d-bands near E_f (Anderson et al., 1978). Applying pressure may shift E_f through the maximum of the d-band, or may broaden this band. Because of the strong deviation from stoichiometry of our samples, the high residual resistivity may result from scattering at ions being charged impurities. This scattering and therefore the residual resistivity decrease if the Fermi velocity increases. An increase in Fermi velocity usually is connected with a decrease of the density of states. This means that the observed pressure dependence of T_c and of $\rho_{residual}$ may have the same origin. For a more detailed discussion one would need volume dependent band structure calculations.

REFERENCES

Allen, P.B., and Dynes, R.C. (1975). *Phys. Rev. B12,* 905.
Anderson, K.O., Klose, W., and Nohl, H. (1978). *Phys. Rev. B17,* 1209.
Culetto, F.J., and Pobell, F. (1978). *Phys. Rev. Lett.40,* 1104.
Flükiger, R., Baillif, R., and Walker, E. (1978). *Mat. Res. Bull. 13,* 743.
Fradin, F.Y., Knapp, G.S., Bader, S.D., Cinader, G., and Kimball, C.W. (1976). In "Superconductivity of d- and f-band Metals," ed. D. H.Douglass, Plenum Press. p. 297.
Gutfreund, H., Weger, H., and Kaveh, M. (1978). *Solid State Commun. 27,* 53.
McCallum, R.W., Woolf, L.D., Shelton, R.N., and Maple, M.B. (1978). *J. de Physique 39,* C6-359.
Nerz, K.P. (1979). Ph.D. thesis, JÜL-Spez-30, Kernforschungsanlage Jülich.
Padamsee, H., Neighbor, J.E., and Shifferman, C.A. (1973). *J. Low Temp. Phys. 12,* 387.
Poppe, U. (1979). Ph.D. thesis, JÜL-Report, Kernforschungsanlage Jülich.
Shelton, R.N. (1976). In "Superconductivity in d- and f-Band Metals," ed. D. H. Douglass, Plenum Press. p. 137.

Webb, G.W., and Shelton, R.N. (1978). *J. Phys. F8*, 261.
Woolam, J.A., and Alterovitz, S.A. (1978). *Solid State Commun. 27*, 571.

MÖSSBAUER STUDY OF METAL CATIONS
IN THE CHEVREL PHASES

Byron Stafford
C.D. Barnet
C. W. Kimball

Northern Illinois University[1]
DeKalb, Illinois

F. Y. Fradin

Argonne National Laboratory[2]
Argonne, Illinois

Mössbauer effect studies at ^{119}Sn and ^{57}Fe sites have been made in $Sn_{1.0}Fe_{0.1}Mo_6S_8$ and $Sn_{1.0}Fe_{0.4}Mo_6S_8$ to examine the Sn- and Fe- motion and electronic states in the high T_c, high H_{c2} superconductor $SnMo_6S_8$ with Fe additions. Above 50K the ^{57}Fe spectra are asymmetric doublets with isomer shift and quadrupolar splitting similar to that of the intermediate temperature results in $Fe_{1.0}Mo_6S_8$. At 4.4K the Fe atoms have begun to magnetically order. The Sn atoms are found to occupy two electronically different sites in $Sn_{1.0}Fe_{0.4}Mo_6S_8$.

I. INTRODUCTION

High critical temperature T_c and high critical field H_{c2} superconductivity are often found in intermetallic compounds

[1]*Work performed under the auspices of the National Science Foundation.*
[2]*Work performed under the auspices of the U.S. Department of Energy.*

which have a tendency to be metastable and to exhibit anomalous phonon properties at low temperature. The recently discovered ternary Chevrel phases with high T_c and H_{c2} display lattice instabilities, very large pressure dependence of T_c and phonon mode softening at low temperature (Fischer, 1978). The tendency toward lattice instability was generally believed to be related to the size of the metal cation M in $M_x Mo_6 S_8$ phases with transformations occurring only in phases containing small cations (such as Cu, Co, Zn).

Recent x-ray work (Yvon, 1978) indicates that the tendency to be "structurally unstable" is a characteristic of these phases independent of cation size and that the degree of instability is correlated with the delocalization of the cations from the center of inversion symmetry in the rhombohedral cell. A compression along the rhombohedral axis leads from a single-well potential (localized) to a multiple-well potential (delocalized) in which two six ring (adjacent) sites become available to the cation, but only three sites in each ring can be occupied simultaneously. It appears that compounds with strongly localized cations have superior superconducting properties (e.g., $PbMo_6 S_8$, $SnMo_6 S_8$) but the reason for the importance of the localization for superconductivity or metastability is yet an open question. Hence, the basic reasons for the wide range of superconducting properties of a given compound as the metal cation concentration changes may be related to electronic or vibrational aspects of delocalization.

The Mössbauer results yield measures of the phonon distribution important to the electron-phonon coupling (McMillan, 1968) and within the context of current theoretical treatments of strong-coupled superconductors to the determination of T_c (Kitchens, Craig and Taylor, 1969; Kimball et al, 1974). The mean-square displacement $<x^2>$ is related to the Mössbauer absorption and the mean square velocity to the thermal shift of the centroid of the spectrum. These quantities can be related to moments of the phonon distribution by model calculations (Kimball, Weber and Fradin, 1976).

Previous measurements of phonon shifting (softening) with decreasing temperature have been reported for [119]Sn in $SnMo_6 S_8$ (Kimball et al, 1976). The Sn atom was found to have highly anisotropic and non-harmonic lattice vibrations in the Chevrel phase.

II. EXPERIMENTAL

In Fig. 1 we show the results of the [57]Fe Mössbauer effect in $Sn_{1.0}Fe_x Mo_6 S_8$ with x = 0.4 at 297K. At this temperature

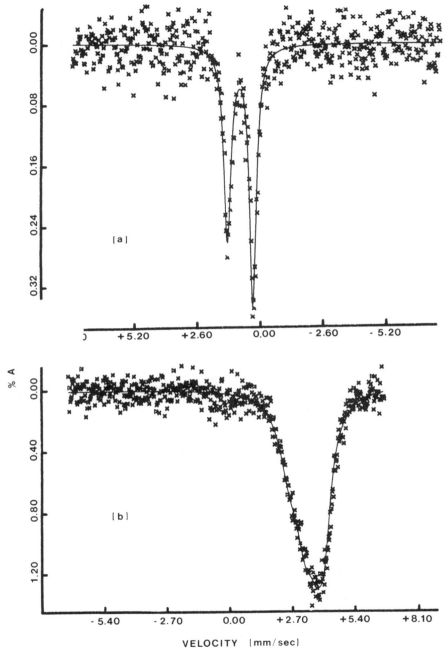

FIGURE 1. (a). Typical Mössbauer absorption spectrum for ^{57}Fe at temperatures above 50K in $Sn_{1.0}Fe_{0.4}Mo_6S_8$. (b). Typical absorption spectrum for ^{119}Sn in $Sn_{1.0}Fe_{0.4}Mo_6S_8$.

the spectrum is characteristic of the quadrupole split pattern
which arises in $Fe_{1.0}Mo_6S_8$ above 100K (Friedt et al, 1979).
This behavior is found for Fe concentrations x of 0.1 at 300K
and 0.4 between 50K and 300K and indicates all Fe atoms are in
nearly equivalent sites that we tentatively assign to an outer
ring, i.e., associated with a channel location. For the x =
0.4 sample we find that some of the iron atoms magnetically
order (see Fig.2) at 4.4°K. The pattern is a mixed quadrupole
and magnetic hyperfine pattern with a magnetic hyperfine field
of ~30kG. Comparing x = 0.4 to x = 1.0, the ordering temperature roughly scales with the square of the iron concentration.
We note that there appears to be a paramagnetic contribution
to the pattern at 4.4°K. This could indicate that there is a
distribution of occupied Fe sites varying perhaps between 1
and 2 per unit cell. The effect of 2 Fe atoms at opposite

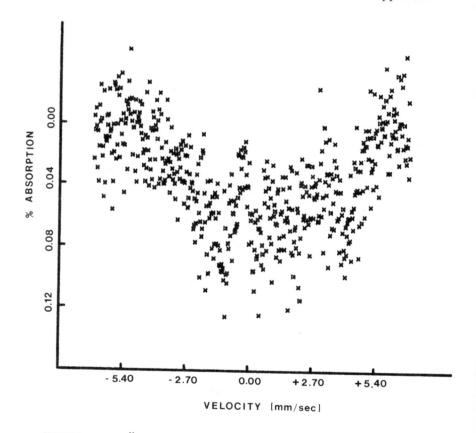

FIGURE 2. Mössbauer absorption spectrum for ^{57}Fe in
$Sn_{1.0}Fe_{0.4}Mo_6S_8$ at 4.4°K which shows the presence of magnetic
hyperfine interaction.

ends of an outer ring gives rise to a small effect in the ^{57}Fe paramagnetic pattern, but a large effect in the magnetic coupling.

The temperature dependence of the ^{119}Sn Mössbauer effect in $Sn_{1.0}Fe_{0.4}Mo_6S_8$ indicates that Sn occupies two electronically different sites. (see Fig.1) One Sn site is quite similar to that of Sn in $SnMo_6S_8$ (Kimball et al, 1976). The other has a stiffer vibrational spectrum at low temperature and is much less anharmonic.

Debye-Scherrer powder patterns indicate the single phase hexagonal (rhombohedral) R3 structure. Lattice parameters were obtained using Cr - K_α radiation and yield $a_0 = 6.523(1)$Å and $\alpha = 89.986°$ for $Sn_{1.0}Fe_{0.4}Mo_6S_8$. Thus the addition of Fe to $Sn_{1.0}Mo_6S_8$ increases the rhombohedral angle by ~0.44° and the cell volume by 0.5%.

We propose a structural model in which the $SnMo_6S_8$ cell-centered-cation structure is modified by the addition of Fe at a channel location. This location for the Fe, which is in intimate contact with the molybdenum d-electrons responsible for superconductivity, would explain why even an addition of 0.1 Fe per formula unit depresses T_c below 1.5K. This is to be contrasted to Eu^{2+} (S = 7/2) which can be substituted for Sn up to 50% in $SnMo_6S_8$ with a ΔT_c of only -0.3K (Fradin et al, 1977).

REFERENCES

Fischer, φ. (1978). *Appl. Phys.* <u>16</u>, 1.
Fradin, F.Y. Shenoy, G.K., Dunlap, B.D., Aldred, A.T., and
    Kimball, C.W. (1977). *Phys. Rev. Letters* <u>38</u>, 719.
Friedt, J.M. et al (1979). (to be published).
Kimball, C.W., Taneja, S.P., Weber, L., and Fradin, F.Y.
    (1974). In "Mössbauer Effect Methodology" (Gruverman, I.,
    Seidel, C.W., and Dieterly, D.K., eds.) Plenum, New York.
Kimball, C.W., Weber, L.W., and Fradin, F.Y. (1976). *Phys.
    Rev.* <u>14</u>, 2769.
Kimball, C.W., Weber, L., Van Landuyt, G., Fradin, F.Y.,
    Dunlap, B.D., and Shenoy, G.K. (1976). *Phys. Rev. Letters*
    <u>36</u>, 412.
Kitchens, T.A., Craig, P.P., and Taylor, R.D. (1969). In
    "Mössbauer Effect Methodology" (Gruverman, I., ed.) vol.5.
    Plenum, New York.
Knapp, G. (1979). (Private Communication).
McMillan, W.C. (1968). *Phys. Rev.* <u>167</u>, 331.
Yvon, K. (1978). *Solid State Comm.* <u>25</u>, 327.

THE INFLUENCE OF THE FORMAL ELECTRIC CHARGE ON THE SIZE OF
THE TRANSITION METAL ATOM CLUSTER IN $Y Rh_4B_4$, $Y Ru_4B_4$ AND
$Pb Mo_6S_8$ RELATED COMPOUNDS

Klaus Yvon
Andreas Grüttner[1]

Laboratoire de Cristallographie aux Rayons X
University of Geneva
Geneva, Switzerland

A single-crystal X-ray study of the superconducting com-
pounds $M Rh_4B_4$ (M = Y, Th), $Y(Ru_{1-x}Rh_x)_4B_4$ and $M Mo_6S_8$
(M = Ag, Cu, Pb, Sn, In, Rare Earths) shows that the size of
the T atom cluster (T = transition metal) changes as a func-
tion of its formal charge. An increase of the formal charge
by one electron leads to an expansion of the T_4 tetrahedron
by about 1% in the borides and to a contraction of the T_6
octahedron by about 2% in the chalcogenides. The relative
displacements of the T atoms due to this electronic effect
are of the same order of magnitude as their r.m.s. amplitudes
of thermal vibrations at low temperatures.

I. INTRODUCTION

Due to their unusual low-temperature properties, the
title compounds are presently the subject of intensive research (Matthias *et al.*,1978; Fischer, 1978). Except for some

[1]*Present address: Institut für Kristallographie*
Universität Freiburg i. Brsg., Freiburg, West-Germany

of the chalcogenides, no single-crystal X-ray study has been
reported so far. The knowledge of accurate atomic coordinates
for these compounds is of interest because their electronic
properties are mainly determined by the 4d electrons whose
energetics at the Fermi level (E_F) depend in a sensitive
manner on the structural arrangement of the transition metal
(T) atoms (Mattheiss & Fong, 1977; Bullett, 1977; Jarlborg *et al.*,
1977; Andersen *et al.*, 1978). In this study we will show that
the T atom arrangement differs slightly from one compound to
another, and that some of these changes are due to differences in the number of valence electrons which are localized
on the metal atom cluster.

II. EXPERIMENTAL AND RESULTS

The following compounds have been examined by single-
crystal X-ray diffraction analysis : $M Rh_4 B_4$ (M = Y, Sm, Th),
$Y (Ru_{1-x} Rh_x)_4 B_4$ (x = 0, 0.3, 0.5, 0.6, 0.85) and $M Mo_6 S_8$
(M = Ag, Cu, Pb, Sn, In, La, Gd, Ho, Er). The structural
parameters obtained for the borides have been summarized in
Table 1, and those of the chalcogenides have been published
elsewhere (Yvon, 1979).

III. DISCUSSION

In view of the important 4d contribution to the electronic density of states at E_F the structural parameters studied
in this work are mainly those which determine the interatomic
distances within and between the T atom clusters. Of particular interest was the question of how these distances are influenced by the variations of the formal charge on the cluster. Changes of the formal charge have been estimated from
crystal chemical arguments as follows. In the borides, the
formal charge of the Ru_4 tetrahedron was said to increase by
4 electrons as Ru (8 valence electrons) was replaced by
Rh (9 valence electrons). In the chalcogenides the formal
charge of the Mo octahedron was said to increase by 1, 2, 3
and 4 electrons as M was replaced by mono-, di-, tri- and
tetravalent cations. The fact that the charge differences defined in this way are always bigger than those obtained by a
charge integration within atomic spheres of given radii shall
not be of major importance on the conclusions to be drawn
later in the text.

TABLE I. Structural data for MT_4B_4 compounds (M = Y, Sm, Th; T = Ru, Rh) at 293 K. Standard deviations are in parentheses, point positions are according to International Tables (1969), centre at origin of unit cell. The Y, Sm and Th atoms occupy the equipoints 2b ($P4_2/nmc$) or 8b ($I4_1/acd$) having no positional freedom.

Compound	space group	a c (Å)	x y(Rh,Ru) z	x y(B) z
$Y\,Rh_4B_4$	$P4_2/nmc$ (No 137)	5.310(2) 7.402(3)	0.25(-) 0.9988(1) 0.3948(1)	0.25(-) 0.080(1) 0.099(1)
$Sm\,Rh_4B_4$	"	5.324(2) 7.449(3)	0.25(-) 0.9989(1) 0.3935(1)	0.25(-) 0.076(1) 0.101(1)
$Th\,Rh_4B_4$	"	5.356(2) 7.538(3)	0.25(-) 0.0006(4) 0.3934(3)	0.25(-) 0.071(5) 0.099(4)
$Y\,Ru_4B_4$	$I4_1/acd$ (No 142)	7.443(3) 14.990(7)	0.1159(2) 0.1023(2) 0.9369(1)	0.832(3) 0.106(2) 0.957(1)
$Y\,(Ru_{.7}Rh_{.3})_4B_4$	"	7.452(3) 14.961(7)	0.1145(1) 0.1019(1) 0.9370(1)	0.834(1) 0.107(1) 0.961(1)
$Y\,(Ru_{.5}Rh_{.5})_4B_4$	"	7.459(3) 14.941(7)	0.1152(2) 0.1016(2) 0.9375(1)	0.828(5) 0.111(5) 0.971(2)
$Y\,(Ru_{.4}Rh_{.6})_4B_4$	"	7.471(5) 14.903(9)	0.1155(3) 0.1002(3) 0.9377(2)	0.850(5) 0.109(5) 0.962(3)
$Y\,(Ru_{.15}Rh_{.85})_4B_4$	"	7.478(3) 14.887(7)	0.1165(3) 0.0999(3) 0.9383(2)	0.827(3) 0.106(3) 0.950(2)

    The dependences of the T-T intracluster distances on the
changes of the formal charge on the T atom clusters have been
illustrated in Fig. 1. One can see that an increase of the
formal charge tends to expand the cluster in the borides and
tends to contract the cluster in the chalcogenides. For the
latter compounds the contraction has already been described
in detail in a previous study (Yvon & Paoli, 1977). It has
been interpreted in terms of a valence-bond model, assuming
that the valence electrons of the M atoms are transferred to
the Mo atoms where they fill molecular orbitals which are
bonding with respect to other Mo atoms of the Mo_6 octahedron.
For the borides no such model has been found as yet. However,
it is clear that strong T-T atom bonds also exist in these
compounds, and that an increase of the number of valence electrons at the T_4 tetrahedron leads to a filling of molecular
orbitals on the T atoms. Contrary to the chalcogenides these
orbitals are probably antibonding with respect to the other
T atoms in the cluster, which explains why the cluster expands

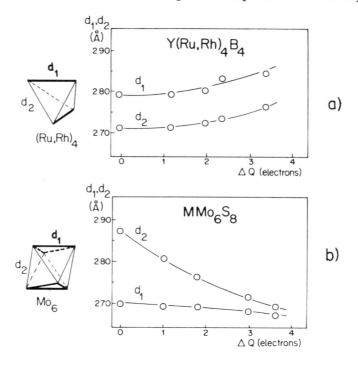

*FIGURE 1. The T-T intra cluster distances d_1 and d_2 as a
function of the increase of the formal charge, ΔQ, on the
$(Rh, Ru)_4$ cluster (a) and Mo_6 cluster (b).*

as one replaces Ru by Rh. An influence of a geometrical factor on this expansion can be neglected to a first approximation because the Ru and Rh atoms have practically the same size [r(Ru) = 1.339, r(Rh) = 1.345 Å (Zachariasen, 1973)].

From these findings it becomes clear that changes of the charge on the T atom cluster can induce quite strong atomic displacements. An increase of the charge by 1 electron, for instance, leads to atomic shifts which are of the order of 0.02 Å in the borides and 0.03 Å in the chalcogenides. It is interesting to compare these values with the r.m.s. amplitudes of thermal vibrations of the T atoms at the superconducting transition temperature. In fact, a rough estimate shows that they have both the same order of magnitude. This indicates that vibrational modes which change the size of the T atom cluster ("breathing modes") can lead to cooperative movements of the conduction electrons and vice-versa.

ACKNOWLEDGMENTS

We thank Drs. R. Flükiger and D.C. Johnston for having put at our disposal some of the samples.

REFERENCES

Andersen, Ø.K, Klose, W., and Nohl, H. (1978). *Phys. Rev.* *B17*, 1209.
Bullett, D.W. (1977). *Phys. Rev. Lett.* *39*, 664.
Fischer, Ø. (1978). *Appl. Phys.* *16*, 1.
International Tables for X-ray Crystallography, Vol. I. Birmingham Press (1969).
Jarlborg, T., Freeman, A.J., and Watson-Yang, T.J. (1977). *Phys. Rev. Lett.* *39*, 1032.
Mattheiss, C.F., and Fong, C.Y. (1977). *Phys. Rev. B15*, 1760.
Matthias, B.T., Patel, C.K.N., Barz, H., Corenzwit, E., and Vandenberg, J.M. (1978). *Physics Lett.*, *68A*, 119.
Yvon, K. (1979). Current Topics in Materials Science, *3*, 53.
Yvon, K., and Paoli, A. (1977). *Solid State Commun.* *24*, 41.
Zachariasen, W.H. (1973). *J. Inorg. Nucl. Chem.*, *35*, 3487.

SELF-CONSISTENT ELECTRONIC BAND STRUCTURE OF CHEVREL PHASE COMPOUNDS[1]

T. Jarlborg

Department of Physics and Astronomy
Northwestern University
Evanston, Illinois

A. J. Freeman

Department of Physics and Astronomy
Northwestern University
Evanston, Illinois
and
Argonne National Laboratory
Argonne, Illinois

As described elsewhere in these proceedings, the Chevrel phase molybdenum chalcogenides, MMo_6S_8 (or Se_8), exhibit a number of unusual properties including high T_c superconductivity and the highest upper critical fields, H_{c2}, known. They also exhibit magnetism and its coexistence with superconductivity and the phenomenon of re-entrant magnetism at low temperatures (see, for example, Fischer, 1978, and references cited therein).

From the theoretical side, the complexity of the crystal structures has made difficult even non-self-consistent energy band studies of the Chevrel phase ternaries. Instead, early studies were carried out with approximate molecular cluster-tight binding (Mattheiss and Fong, 1977; Andersen et al.,1978)

[1]*Supported by the AFOSR (grant No.76-2948), the NSF (grant Nos. DMR 77-23776 and DMR 77-22646 and through the Northwestern University Materials Research Center, NSF grant No. 76-80847), and the DOE.*

and (limited basis set) localized orbital methods (Bullett, 1977). In the case of the ternary rhodium borides, our self-consistent Linear Muffin Tin Orbital (LMTO) energy band studies (Jarlborg et al., 1977; Freeman et al., 1978; Freeman and Jarlborg, 1979) have provided a qualitative understanding of the underlying electronic structure and some of the basic phenomena observed for these materials. All 18 atoms/unit cell were included and total and partial (by atom type and by orbital angular momentum) densities of states (DOS) were used to discuss qualitatively the origin of magnetism in the first part of the series (M = Gd,Tb,Dy,Ho), superconductivity in the second part (M = Er,Tm,Lu) and the re-entrant state of magnetism at low temperature in $ErRh_4B_4$. These results were also used to predict a possible mixed state in $ErRh_4B_4$ at temperatures above the re-entrant magnetism temperature – a state which may have been seen in the magnetization vs magnetic field measurements of Ott et al. (1979). We here report results of the first __ab initio__ self-consistent LMTO energy band studies of several of the MMo_6X_8 compounds which included all electrons and all 15 atoms in the unit cell. These results, mostly in the form of total and partial (by atom type and orbital angular momentum) DOS, are used to interpret a number of experiments.

The band calculations have been performed by use of the LMTO method which combines high accuracy with moderate computing costs. The scheme of calculation is essentially the same as has been used for calculations on A15 compounds (Jarlborg, 1979; Arbman and Jarlborg, 1978) and ternary borides (Jarlborg et al., 1977). The full LMTO formalism was used to obtain self-consistent band structures for the first compound ($SnMo_6S_8$), while for the others, canonical band calculations were used in the initial stage of self-consistency to reduce the number of required LMTO iterations. In the iterations, 10 k-points were used, while the final converged bands were determined at 35 k-points of the irreducible Brillouin-zone. The degree of convergence is about 5 mRy for states below the Fermi level.

The crystal structure data of Marezio et al. (1973) on $PbMo_6S_8$ has been used in all our Chevrel phase calculations, but without distortion of the Mo-octahedron and with a rhombohedral angle of 90°. With these approximations, the structure coefficients to the LMTO calculations need only be determined once for this undistorted but characteristic Chevrel structure and can be used for all compounds independent of a uniform scaling of the lattice. Further, if one considers spherically symmetric potentials around each site, this leaves only 4 types of atoms in the structure, whereas including the distortion would make most of the 15 atoms per cell inequivalent. In this simplified structure there is one Sn site, six equivalent Mo

sites, but two types of S-sites. The S atoms along the diagonal (around which the smaller Mo_6S_8 cube is tilted in the unit
cell cube) have different surroundings (and potential) than the
other six S atoms. The Mo_6S_8 cluster itself is close packed,
but around the Sn sites and along the unit cell cube edges
there is considerable open space, so touching spheres fill only
about 42 percent of the total volume. This problem is partly
reduced by using many plane waves in the construction of the
correction matrix to the overlapping spheres.

The potential used the Hedin-Lundqvist treatment of exchange and correlation. The charge densities from the core
states as well as from partly occupied f-states were recalculated in each iteration by using the actual MT-potential. The
basis set included s, p and d states for all atoms, while in
the three-center terms in the LMTO matrices $\ell_{max} \leq 3$. This
results in an eigenvalue problem of dimension 135×135.

As an example of our results, to be described in detail
elsewhere, we show in the Figure the total DOS for $SnMo_6S_8$.

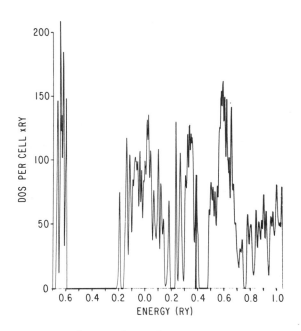

FIGURE 1. Total DOS in units of states per cell, Ry and
spin, for $SnMo_6S_8$ obtained from a k-space integration routine
using 35 first principle points of the irreducible Brillouin
zone. The DOS includes a $\sim$ 7 mRy Gaussian broadening function.
The Fermi energy falls on the peak of 0.395 Ry.

We see that there is considerable structure in the total
DOS, particularly around E_F, which arises from the Mo-4d
electrons. From a partial DOS calculation, we find that
there is a high 4d DOS at E_F which is favorable for superconductivity. There is a distinct gap in the DOS just above
E_F which falls in the middle of the Mo-d bands and a smaller
band gap between the Mo-d and S-p states as was also inferred
or seen in the earlier calculations. As for most other high
T_C compounds, the Fermi level falls just below the middle of
the bonding-antibonding "gap" of the metal d-states. Usually
for Mo compounds the d bands are occupied up to the "gap"
region where the DOS is low, but in the Chevrel compounds a
large charge transfer from Mo to S (about 1 electron per Mo
atom) was found to occur. The DOS per atom at the Fermi
level is 50-60 percent of that for the best superconducting
A15 compounds. Thus the conditions for fairly high T_C values
are present. This is the case for both compounds studied
here, so an explanation of the fact that stoichiometric
$EuMo_6S_8$ is not superconducting cannot be found directly from
the DOS, but has to take the magnetic moments of the Eu atoms
into account. The coupling of the Eu 4f electrons to the
conduction electrons is found to be very weak. There is a
large charge transfer to the cluster and Eu has essentially
no occupied conduction bands with the Eu 6s band falling
high above E_F. These results agree with the Mössbauer
measurements of Dunlap et al. (1979) which show that the
product of the exchange coupling and density of states,
$|J N(E_F)|$, is roughly one order of magnitude smaller than
that measured in binary superconductors like Eu in $LaAl_2$ and
an isomer shift result which is typical for Eu^{2+} in an ionic
compound without conduction electron contributions.

Detailed results and analysis, including rigid muffin-tin
model calculations of λ, will be given elsewhere.

ACKNOWLEDGMENTS

We are grateful to B. Dunlap for helpful discussions,
encouragement and support and to F. Y. Fradin, D. D. Koelling,
W. Pickett and G. Shenoy for helpful discussions.

REFERENCES

Andersen, O. K., Klose, W. and Nohl, H., *Phys. Rev. B17*, 1209
    (1978).
Arbman, G. and Jarlborg, T., *Solid State Commun. 26*, 857
    (1978).

Bullett, D. W., *Phys. Rev. Lett. 39*, 664 (1977).
Dunlap, B. D., Shenoy, G. K., Fradin, F. Y., Kimball, C. W.
    and Barnett, C., *Bull. Am. Phys. Soc. 24*, 389 (1979);
    J. Mag. and Magn. Matls. (to appear).
Fischer, Ø., *Appl. Phys. 16*, 1 (1978).
Freeman, A. J., Jarlborg, T. and Watson-Yang, T. J., *J. Mag.
    and Magn. Matls. 7*, 296 (1978).
Freeman, A. J. and Jarlborg, T., *J. Appl. Phys. 3*, 1876 (1979).
Jarlborg, T., *J. Phys. F: Met. Phys. 9*, 289 (1979).
Jarlborg, T., Freeman, A. J. and Watson-Yang, T., *Phys. Rev.
    Lett. 39*, 1032 (1977).
Marezio, M., Dernier, P. D., Remeika, J. P., Corenzwit, E. and
    Matthias, B. T., *Mat. Res. Bull. 8*, 657 (1973).
Mattheiss, L. F. and Fong, C. Y., *Phys. Rev. B15*, 1760 (1977).
Ott, H. R., Fertig, W. A., Johnston, D. C., Maple, M. B. and
    Matthias, B. T., *J. Low Temp. Phys. 33*, 159 (1978).